D1239076

OFFSHORE STRUCTURAL ENGINEERING

Silhouette of an offshore platform in Alaska's Cook Inlet.
(Courtesy Atlantic Richfield Company and American Petroleum Institute.)

OFFSHORE STRUCTURAL ENGINEERING

THOMAS H. DAWSON

Associate Professor of Ocean Engineering
Department of Naval Systems Engineering
United States Naval Academy

PRENTICE-HALL, INC. / *Englewood Cliffs, New Jersey 07632*

Library of Congress Cataloging in Publication Data

Dawson, Thomas H.
Offshore structural engineering.

Bibliography: p.
Includes index.
1. Offshore structures—Design and construction.
I. Title.
TC1665.D38 1983 627'.98 82-16157
ISBN 0-13-633206-4

Printed in the United States of America

10 9 8 7 6 5 4 3 2 1

ISBN 0-13-633206-4

Prentice-Hall International, Inc., *London*
Prentice-Hall of Australia Pty. Limited, *Sydney*
Editora Prentice-Hall do Brasil, Ltda., *Rio de Janeiro*
Prentice-Hall Canada Inc., *Toronto*
Prentice-Hall of India Private Limited, *New Delhi*
Prentice-Hall of Japan, Inc., *Tokyo*
Prentice-Hall of Southeast Asia Pte. Ltd., *Singapore*
Whitehall Books Limited, *Wellington, New Zealand*

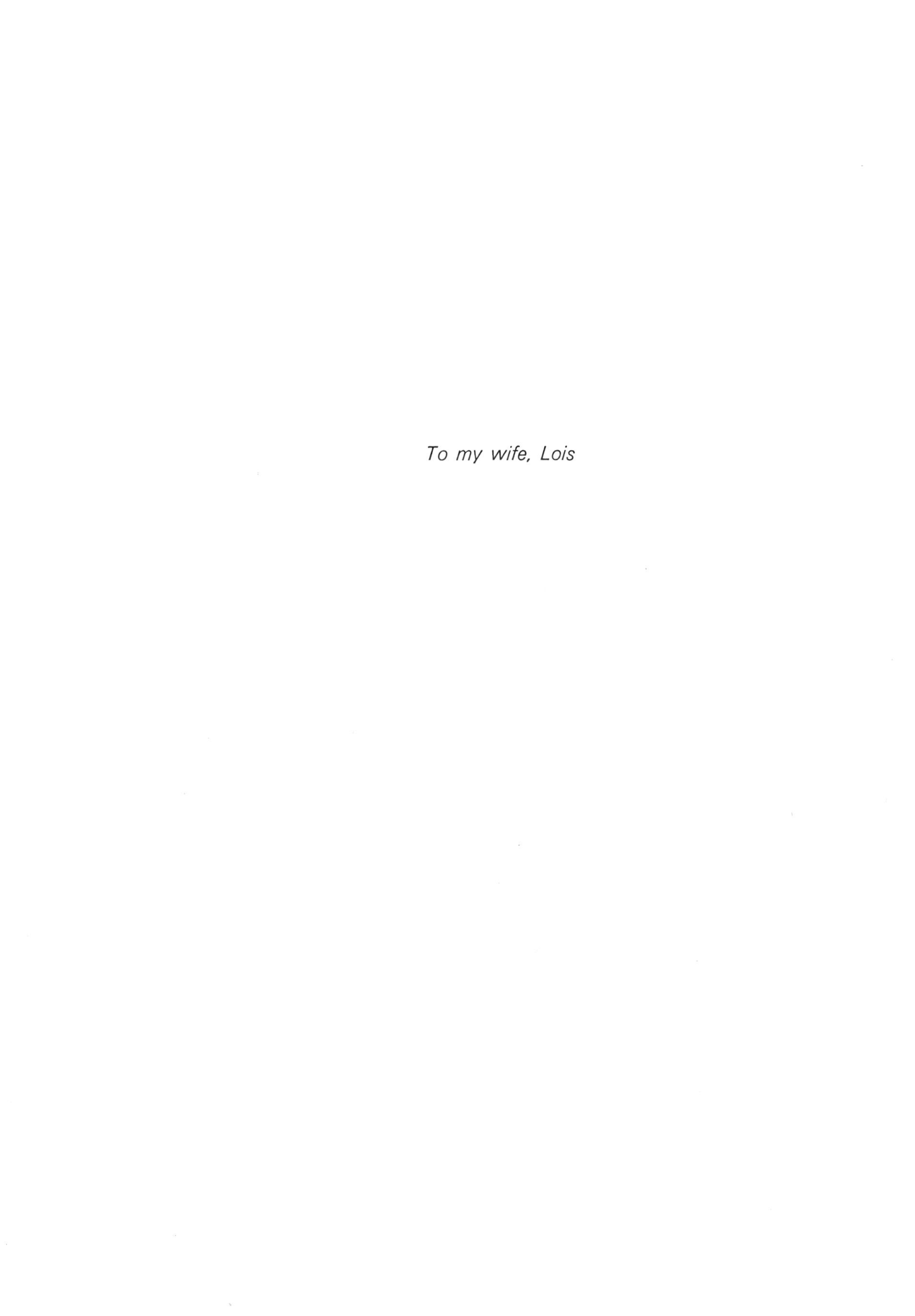

To my wife, Lois

CONTENTS

Chapter 5 FOUNDATION ANALYSIS 229

Chapter 6 DYNAMICS OF STRUCTURES 281

Appendix MATRIX ALGEBRA 331

INDEX 339

PREFACE

THIS BOOK is intended primarily as an introduction to the structural engineering of fixed offshore platforms. It deals with basic structural theory as well as with those aspects, such as wave loading, pile support, and dynamic response, that that are of special importance to the subject.

The presentation is developed entirely from fundamentals acquired from basic undergraduate courses in engineering mechanics. The mathematical level is similarly limited to that obtained in conventional calculus and differential-equations courses. Throughout the book, discussions of theory are followed by numerical examples designed to clarify details and illustrate proper applications.

A general introduction is provided in Chapter 1, followed by treatment of the structural theory of trusses and frames in Chapter 2. In presenting this theory, I have found it convenient to use modern matrix methods and to emphasize computer solutions for complex problems. Such an approach has the distinct advantage of allowing the concise development of theory and the direct application of such theory to practical problems, without the need for further study of approximate solution techniques. For the reader who is not familiar with matrix algebra, a brief but adequate discussion of this topic has been included as an appendix.

The structural theory explained in Chapter 2 is followed in Chapter 3 by consideration of the environmental loadings that can exist on offshore structures. Topics include wind and wave forces and their engineering description, along with other loadings from current, pressure, floating ice, and mudslides.

Subjects covered in these two chapters are brought together in Chapter 4, where static analysis of fixed offshore structures is considered. In addition to steel frame structures, some attention is also given to concrete structures, prestressed or reinforced to withstand environmental loadings. Another aspect of this chapter is consideration of some basic structural dynamics that can be used for assessing the applicability of static analysis to a particular design.

Chapter 5 deals with the analysis of foundations for offshore structures. A brief introduction to soil mechanics is given, and engineering formulas for the design and analysis of piles and footings are described. Approximate methods for incorporating soil-structure response into the analysis of offshore structures are also discussed.

Finally, Chapter 6 extends the brief treatment of structural dynamics of Chapter 4 to include more detailed topics, such as time-dependent joint loading from wave forces, multiple-degree-of-freedom dynamic analysis, dynamic response in irregular random seas, and response of offshore structures to earthquake loadings.

With such converage, this book can be used for an undergraduate course for ocean engineering majors or for an elective course for civil and mechanical engineering students. The book should also be of interest to practicing engineers now involved in ocean engineering, since it provides treatment of topics not previously available under a single cover.

Annapolis T.H.D.

ACKNOWLEDGMENTS

I AM GRATEFUL to a number of people for their assistance in preparing this book Mr. Frank Domingues of McDermott Incorporated gave generously of his time in several conversations regarding current engineering practice of offshore structures. Mr. Thomas Langan of McDermott Incorporated read an early draft of the manuscript and made valuable suggestions for its improvement. Commander Ronald Erchul, CEC, USN, also reviewed portions of the manuscript and offered many helpful comments.

For their assistance in finding photographs and illustrations, I thank Mr. Kenneth Salzer of McDermott Incorporated, Mr. Richard Drew of the American Petroleum Institute, Mr. Jack Holleran of Exxon Company USA, Mr. Joseph Maranto of Mobil Oil Corporation, and Dr. William Cox of McClelland Engineers, Incorporated.

At Prentice-Hall, I am grateful to my editor, Mr. Bernard Goodwin, for his confidence in me and his patience in waiting for a long overdue manuscript.

I also thank Mr. Nicholas Romanelli and the staff of the Editorial-Production Department for their excellent contributions to the book.

To my former student, Ensign Marc Rolfes, USN, I express my appreciation for his comments on an early draft of the manuscript and for his assistance with some of the examples.

I also thank my colleague, Professor Michael McCormick, for his invaluable advice and assistance.

Finally, I want to record my gratitude to Dr. Bruce Davidson, Academic Dean at the Naval Academy, for creating the environment here that makes educational ventures like this one possible.

T.H.D.

OFFSHORE STRUCTURAL ENGINEERING

Offshore platform development in the Gulf of Mexico. Upper: first offshore steel platform, installed in 20 ft of water in 1947. Lower: Shell's COGNAC platform installed in 1000 ft of water in 1978. (Courtesy of McDermott Incorporated.)

1

INTRODUCTION

OFFSHORE STRUCTURAL ENGINEERING is a relatively new field of engineering concerned with the design and installation of fixed offshore platforms for various purposes. It dates from about 1947, when the first steel structure was installed in the open waters of the Gulf of Mexico and differs from conventional structural engineering mainly in the special problems that have to be considered in the transportation of the structure to the offshore site, its installation there, and its ability to withstand the severe environmental loadings experienced during its intended life. The chief driving force behind this new technology has come from the oil industry and its need for fixed platforms for exploitation of the extensive hydrocarbon deposits existing offshore. Its use is, however, not limited solely to this industry and important applications also exist for military and navigational purposes.

This book is intended to give an introduction to the basic methods of analysis used in offshore structural engineering. The present chapter gives a brief overview of the subject. This is followed by detailed topical coverage in the remaining chapters.

1.1 DESIGN OF FIXED OFFSHORE STRUCTURES

The design of offshore structures parallels in many ways that of land-based structures but with the very special additional requirement that the offshore structure must be constructed in one area and installed in another. Although it is difficult to catalog the many steps taken in the creative design of an offshore structure, the following sequence of events is usually involved to one degree or another:

1. Identify need.
2. Evaluate environmental and local site conditions.
3. Make preliminary design proposals with major emphasis given to the method of installation to be employed.
4. Evaluate these in terms of economics, construction and installation difficulties, foundation requirements, and so on, and select final design form.
5. Size and detail the chosen form to carry required loads and environmental forces.
6. Finally, evaluate the design to ensure that it can withstand the loadings associated with transportation to and installation at the offshore site.

The need for a fixed offshore structure is usually specified simply as a requirement at a designated offshore site for an operational deck having a prescribed minimum working area and carrying a prescribed minimum weight loading. The choice of the structure to support this deck in an economical way then depends primarily on the installation procedure and the environmental and local site conditions. Environmental conditions normally involve the wind, current, and waves likely to be experienced by the structure, together with possible hazzards from floating ice and earthquakes. Local site conditions involve the water depth and seafloor characteristics, the latter being important for foundational requirements of the structure.

With the foregoing information available, preliminary designs can be developed around possible installation procedures using the required operational loads and rough estimates of the major environmental loadings arising at the site location. In most cases, it is, in fact, the wave forces that dominate these designs, although exceptions can exist if floating ice or earthquakes pose severe hazards to the structure. Of these various preliminary designs, some may easily be rejected as economically unfeasible or impractical from construction or installation considerations. Others may require closer evaluation in terms of the many steps involved (including cost) in converting the structure from idea to reality. Ultimately, though, a final design form and installation procedure are chosen and the design detailed and sized to carry the necessary loads. The sizing of the structure to carry the operational loads is a relatively direct matter, but the inclusion of environmental loadings generally requires several trial-and-error iterations because a change in the size of a structural member changes the magnitude of the environmental loading exerted on it and on the structure as a whole. Finally, the structural design must be examined to ensure that it can withstand the forces associated with transportation and placement of the structure at its intended site. These forces can be significant and neglect of them could cause major failure of the structure before it is set in operation.

1.2 EXAMPLES OF FIXED OFFSHORE STRUCTURES

Template Structures

By far the most common type of fixed offshore structure in existence today is the *template*, or *jacket*, *structure* illustrated in Fig. 1.1. This type of structure consists of a prefabricated steel substructure that extends from the seafloor to above the water surface and a prefabricated steel deck located atop the substructure. The deck is supported by pipe piles driven through the legs of the substructure into the seafloor. These piles not only provide support for the deck but also fix the structure in place against lateral loadings from wind, waves, and currents.

The offshore structure shown in Fig. 1.1 is a modern example of template structures designed for waters to a depth of about 350 ft. Template structures have been installed off Louisiana in the Gulf of Mexico since the late 1940s. According to Lee (1968), the first such structure was placed in 20 ft of water in 1947, and this was soon followed by a second structure in 50 ft of water. By the 1950s, offshore structures had reached waters of 100-ft depth. These early template structures, although relying on a substructure and piles to support an operating deck, were not of the compact form illustrated in Fig. 1.1. Instead, they consisted of several small substructures placed side by side, with piles driven through the numerous, closely spaced legs of the substructures.

By the mid-1960s, the template designs had evolved to the general form shown in Fig. 1.1, and structures had been placed in waters more than 300 ft deep. During the 1970s, the water depth for template structures was more than doubled with the installation of a structure in 1000 ft of water. Today, many hundreds of modern template platforms exist in the Gulf of Mexico, and the installation of new ones continues. Although record-breaking structures exist, the vast majority of these structures are in waters of depth less than 300 ft.

The construction and installation of a template structure plays a central role in its design (Lee 1980). The substructure is usually prefabricated on its side at a waterside facility and then placed horizontally on a flat-topped barge

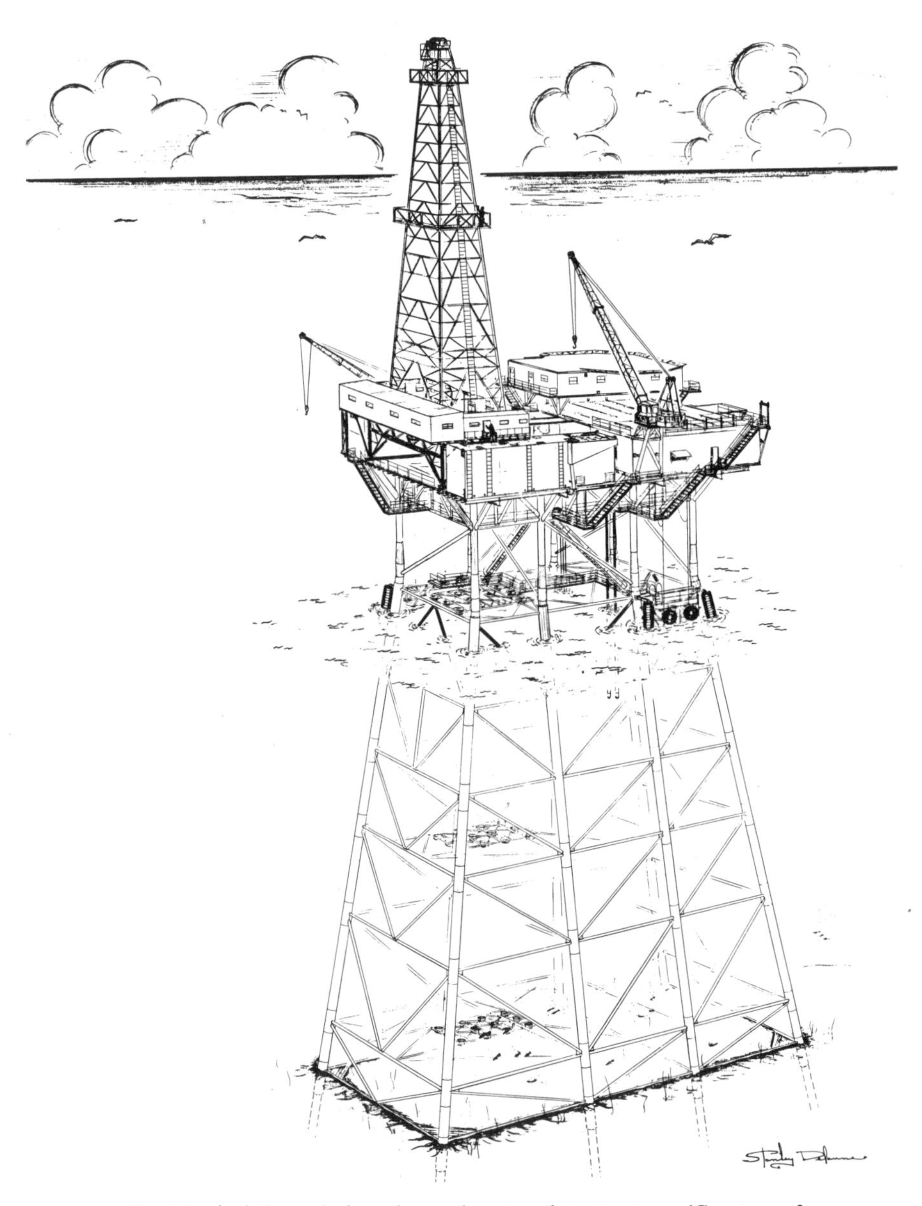

Fig. 1.1. Artist's rendering of a modern template structure. (Courtesy of McDermott Incorporated.)

and towed to its offshore location. At the installation site, the substructure is then slid off the barge and uprighted with the help of a derrick barge and allowed to sink vertically to the seafloor. Once the substructure is in place, pipe piles are inserted through its legs and driven into the seafloor by means of a pile driver supported on a surface vessel. After the piles are driven to predetermined depths, they are cut off at the top of the substructure and the prefabricated deck stabbed into the piles and connected with field welds. In its completed form, the deck weight is carried entirely by the piles, with the substructure providing bracing against their lateral movement. This installation procedure is illustrated in Fig. 1.2.

An inspection of the construction and installation procedures outlined in Fig. 1.2 reveals some of the many factors that must be considered in developing a satisfactory design of a template structure. For instance, the weight of the substructure, if a single unit, cannot exceed the capacity of lifting equipment at the waterside construction facility, nor can its weight and dimensions exceed the capacity of the tow-out barge. The substructure must also be designed to withstand the lifting loads involved in placing it on the barge. At the offshore site, it must be of adequate design to withstand the launching forces. When in water, the structure must also be designed to float unassisted and, with the assistance of a derrick barge, to be uprighted and sunk to the seafloor by selective flooding of its members. The design depths of the piles, driven through the legs of the substructure to fix it in place and, ultimately, to provide support for the deck assembly, must not exceed the capability of pile-driving hammers. Finally, the weight of the prefabricated deck, or major sections of it, must not exceed the capacity of available derrick barges used to place the deck atop the substructure.

In the case of very large substructures designed for deep-water installation, the procedure illustrated in Fig. 1.2 has been modified to facilitate handling by constructing the substructure in two or more sections, towing these separately to the offshore site and assembling them there. The separate parts of the substructure may either be joined at sea before uprighting and sinking to the bottom, or the separate parts may be sunk on top of one another and joined in their final vertical position.

Another variation of the installation procedure illustrated in Fig. 1.2 that is used for large substructures is to design the substructure to be self-floating so as to eliminate the need for launching from a barge. This is usually achieved by making the legs of the substructure sufficiently large to provide buoyancy to the entire assembly. After launching from a waterside facility, the structure is then towed to the offshore site, the legs selectively flooded, and the substructure allowed to settle to the seafloor in its vertical position.

In contrast, for lightweight substructures designed for relatively shallow waters, the launching of the substructure from the barge may be dispensed with and the substructure lifted directly off the tow-out barge with the help of a derrick barge brought alongside.

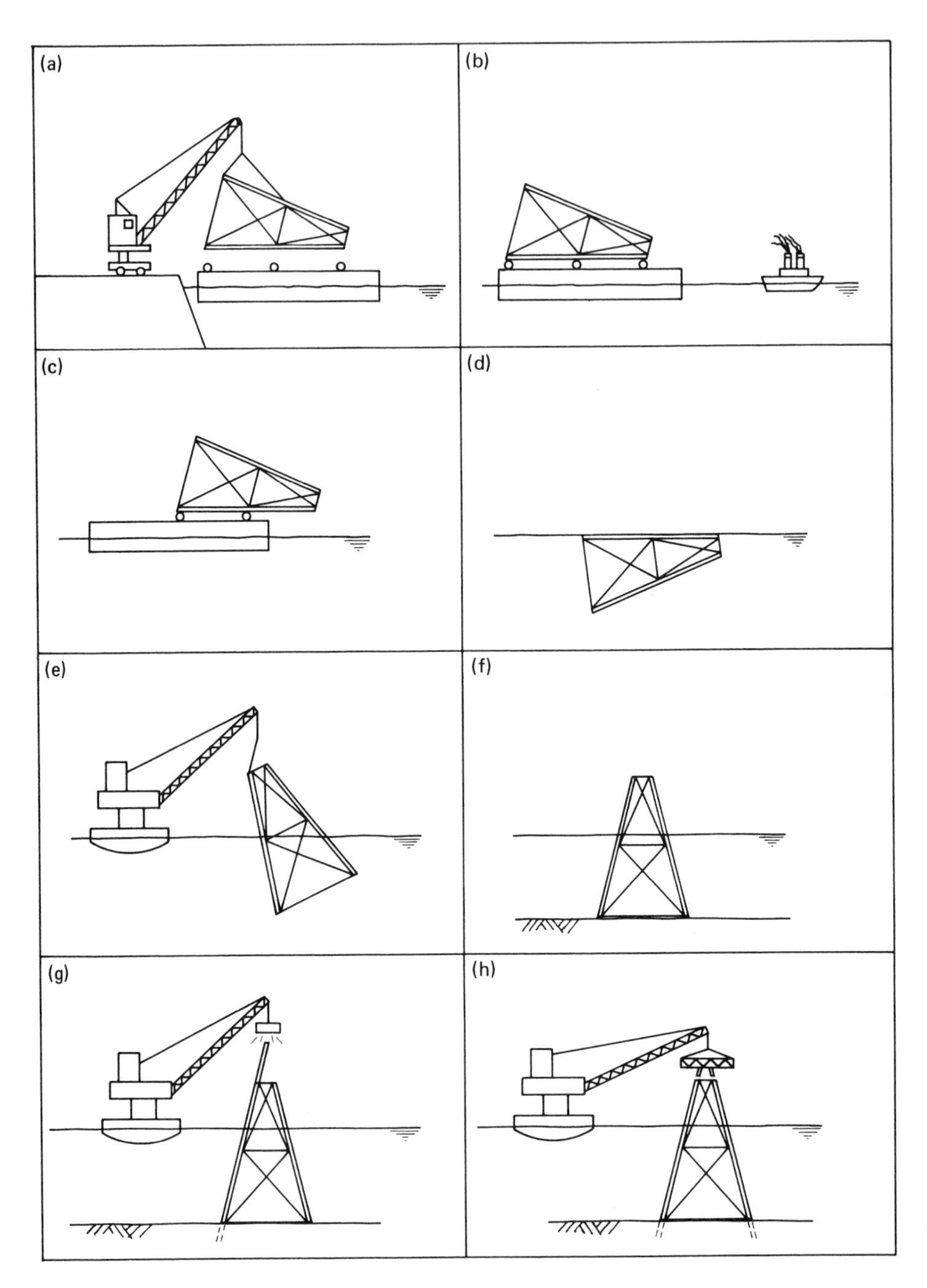

Fig. 1.2. Installation procedure for template structures.

The template structure shown in Fig. 1.1 represents a somewhat standard eight-leg design for structures in waters to about 350 ft depth. A minor modification is frequently employed when the structure is located on very soft sediment. This involves the use of additional *skirt piles* around the base of the structure to provide further support. These are driven through short columns attached to the base of the substructure using a follower (driving rod) from the surface. The skirt piles are then bonded to the columns by pumping concrete (grout) in the annular space between them.

A typical oil drilling and production platform constructed according to the procedures described above is shown in Fig. 1.3. This structure is located off Louisiana in about 300 ft of water in the Gulf of Mexico. The deck measures approximately 60 × 120 ft and, with operating equipment, weighs about 2 million pounds. The weight of the substructure is about 4 million pounds. The eight pipe piles driven through the legs of the substructure have outside diameters of 4 ft and wall thicknesses of about 1 in. In addition to these, four skirt piles are placed around the base of structure. All piles are driven 200 to 300 ft into the seafloor. The structure is designed to withstand a resultant lateral force of about 3 million pounds from wind, waves, and currents during extreme hurricane conditions. Because the wave forces are greatest near the water surface, this

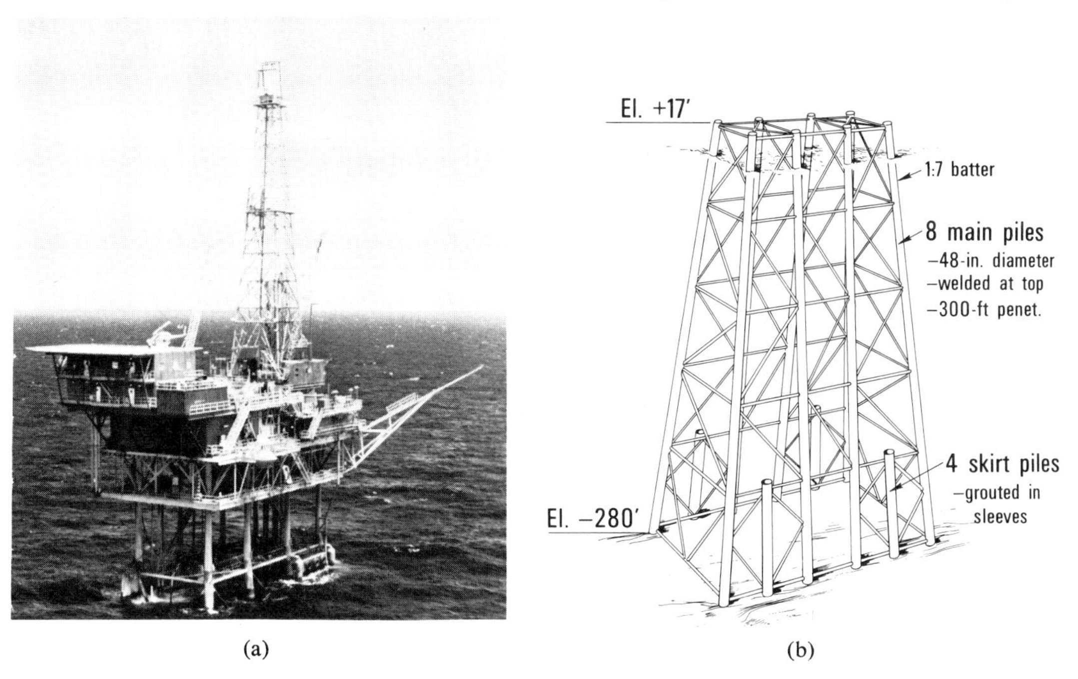

(a) (b)

Fig. 1.3. Typical offshore template structure off Louisiana in the Gulf of Mexico: (a) installed structure; (b) substructure illustrating skirt piles. [(a) Courtesy of the American Petroleum Institute; (b) courtesy of McClelland Engineers, Inc., after McClelland, 1974.]

resultant force acts near the top of the structure. The structure is therefore also designed to withstand a base-overturning moment of the order of 700 million foot-pounds. These loads and moments are five to seven times those caused by extreme winds on a typical 25-story, 300-ft-tall building on land.

For structures designed for waters greater than about 350 ft, two variations of the basic eight-leg template design have been considered. The first has been to increase the number of legs of the structure so that, with skirt piles, the structure can carry additional deck loads and resist the increased lateral loading and overturning moment. A second modification has been based on the observation that, with taller structures and increased base widths, the interior piles become less effective in resisting overturning moments. As an alternative to the eight-pile structure, consideration has thus been given to eliminating interior piles and placing all piles near the four exterior corners of the structure (Lee, 1980).

Offshore template structures for oil production have been installed not only off the Louisiana coast, but also off California. These structures are similar to those installed in the Gulf of Mexico, except that they are designed to withstand possible earthquake loadings in addition to those arising from wind, waves, and currents. An interesting discussion of the installation of one of these structures has been given by Mashburn and Hubbard (1967). Template structures have also been installed at many other locations around the world, including the hostile North Sea, where intense oil drilling and production activities are presently under way.

Apart from oil production activities, offshore template structures have also found military application. Indeed, the first major use of fixed offshore platforms for other than oil drilling and production activities was for defense purposes and involved the Texas Towers located off the northeastern United States. These platforms formed part of the Early Warning Defense System of the United States during the 1950s and allowed radar surveillance and early detection of possible hostile aircraft over a large approach area that could not otherwise be covered. Figure 1.4 shows one of these towers, several of which were located offshore in the Atlantic.

A more recent military application has been the installation of four template-type structures off the coast of North Carolina in 1977 for support of the United States Navy's Air Combat Maneuvering Range. This range allowed a new, safe, economical way to train Navy pilots over a large unpopulated area using radar-tracking equipment located on the decks of the platforms to record the pilot's performance. Figure 1.5 shows one of these towers. They are similar in design to the oil production platforms except that, as in the case of the Texas Towers, only three support legs are needed because of the reduced deck weights involved.

During the 1960s, the United States Coast Guard installed several template-type structures off the Atlantic coast to provide navigational aid to ships. Like the Texas Towers, these structures represented a departure from the conventional use of offshore structures for oil production purposes and illustrate yet another

Fig. 1.4. Texas Tower used for defense purposes in the 1950s. These towers were located off the northeastern United States in the Atlantic Ocean.

application of offshore structural engineering. The design of these towers has been discussed by Fowler (1965) and by Ruffin (1965). The structures were installed to replace aging lightships. Although economics was the principal reason for replacing the lightships with structures, their fixed position and functional reliability in all kinds of weather made them an even more attractive alternative. Figure 1.6 shows one of these light towers, located near the mouth of the Chesapeake Bay off the Virginia coast.

Ice-Resistant Structures

The first major variation in the design of offshore platforms occurred during the 1960s when structures were designed for Cook Inlet, Alaska. Here, large

Fig. 1.5. Illustration of platforms off the coast of North Carolina, used by the U.S. Navy in support of its Air Combat Maneuvering Range. (Courtesy of the U.S. Navy.)

Fig. 1.6. Offshore light tower, installed off the Atlantic coast of the United States for navigational purposes. (Courtesy of the U.S. Coast Guard.)

sheets of floating ice, moving with the tide, can crash into a structure and exert loadings more severe than those from storm winds, waves, and currents. The design of structures for this area accordingly involved eliminating diagonal and horizontal bracings in the tidal range, where they could be struck by the floating ice, and supporting the deck on four large-diameter columns. Each column was supported in turn by several piles in a circular group (McClelland, 1974). These structures have come to be known as *ice-resistant structures.* Figure 1.7 shows one of these structures in operation, together with details of the substructure below the water surface.

Although designed mainly for ice loadings, these structures retain the same basic installation concept as the template structure. The substructure is pre-

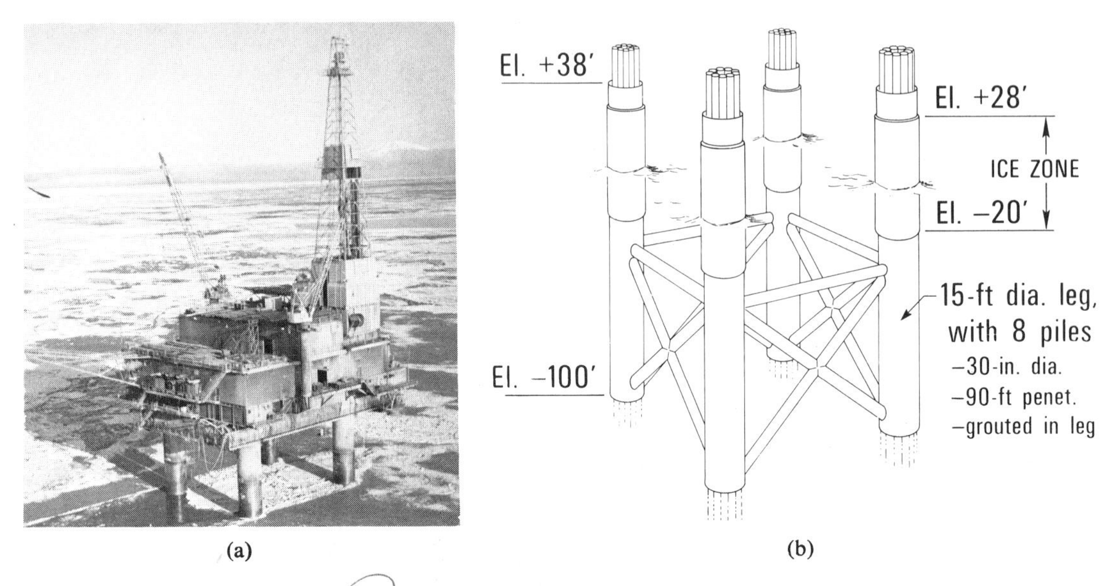

Fig. 1.7. Ice-resistant structure designed for Cook Inlet, Alaska: (a) actual platform; (b) illustration of substructure details. [(a) Courtesy of McDermott Incorporated; (b) courtesy of McClelland Engineers, Inc., after McClelland, 1974.]

fabricated at a waterside facility and floated to its offshore site, uprighted, and fixed with piles driven through its legs. Because of the large-diameter legs of the substructure and the buoyancy they provide when unflooded, no tow-out barge is needed and the substructure can be towed directly in the water.

In addition to the four-legged platforms, a single-leg (Monopod) structure has also been installed in Cook Inlet by floating it to location on horizontal tanks attached at the base of the structure. To fix the structure and support the deck, main piles were driven through the leg of the structure. Skirt piles were also driven around its base (Fig. 1.8).

Gravity Structures

Template structures, as described earlier, are especially suited to soft-soil regions such as the Gulf of Mexico, where deeply driven piles are needed to fix the structure in place and carry the required deck loadings. In regions where hard soil conditions exist and pile driving is more difficult, an alternative structural form has been developed which, in place of piles, relies on its own weight to hold it in place against the large lateral loads from wind, waves, and current. These structures have large foundational elements which, when ballasted, contribute significantly to the required weight and which spread this

Fig. 1.8. Illustration of an ice-resistant Monopod platform installed in Cook Inlet, Alaska in 1966. (Courtesy of the Marathon Oil Company and the American Petroleum Institute.)

weight over a sufficient area of the seafloor to prevent failure. Such structures are generally referred to as *gravity structures.*

In their more popular form, gravity structures are constructed with reinforced concrete and consist of a large cellular base surrounding several unbraced columns which extend upward from the base to support a deck and equipment above the water surface. Structures of this kind were installed in the North Sea

Fig. 1.9. Illustration of a concrete gravity platform used in the North Sea. (Courtesy of Exxon Company USA.)

during the mid-1970s. Figure 1.9 illustrates the main features of these structures. This particular structure is referred to as a CONDEEP (concerete deep-water) structure and was designed and constructed in Norway.

The construction of concrete gravity platforms is altogether different from that employed for template-type structures. Figure 1.10 illustrates the typical construction sequence employed for the North Sea structures. As illustrated, the base is constructed in a drydock, after which it is floated out and moored in a deep-water harbor. The construction is then completed by slip-forming the large towers in a continuous operation until they are topped off. The structure is next ballasted down and a steel prefabricated deck floated over the structure and attached to the top of the towers. Additional deck modules are then set in place and the entire structure refloated for towing to its offshore site, where it is again ballasted down to its final operating position.

Figure 1.11 shows a typical concrete gravity platform under tow and at its final offshore site in the North Sea in approximately 390 ft of water. The base structure consists of 16 hollow concrete cells, 66 ft in diameter and 164 ft high, with wall thicknesses of about 2 ft. The three main columns extend upward a distance of 327 ft above the top of the base cells, with outside diameters tappered from 66 to 39 ft. The deck weight, with equipment, is approximately 50 million pounds and the weight of the support structure is approximately 600 million pounds.

One advantage of the gravity structure over the template type is the reduced time needed for on-site installation. This is especially important in hostile areas such as the North Sea, where unpredictable weather conditions make it highly desirable to limit the construction time needed to fix the structure in place. Another advantage is the very large deck weights that can be carried by the massive concrete columns. With equipment, the deck of the structure shown in Fig. 1.11 weighs more than 20 times that of typical steel template structures.

The construction of large gravity structures of the type described above obviously requires deep harbors and deep tow-out channels. The Norwegian fjords have provided such conditions for the North Sea platforms.

Not all gravity structures need, of course, to be made of concrete or to be of the mammoth scale of those discussed above. Steel gravity platforms have, for example, been designed and installed off Nigeria, where the presence of rock close to the seafloor ruled out the possibility of using piles to fix the structure in place against lateral loadings (Watt, 1978).

A very simple illustration of a steel gravity platform is also provided by the towers installed by the U.S. Air Force off Florida in the Gulf of Mexico for purposes of monitoring pilot training and performance. These towers consist of a steel tubular column supported by a boxlike steel base containing rock ballast. Figure 1.12(a) shows one of these towers being towed to its offshore site, and Fig. 1.12(b) shows the structure before being sunk to the seafloor by flooding the base. Figure 1.12(c) shows the structure in its final operating

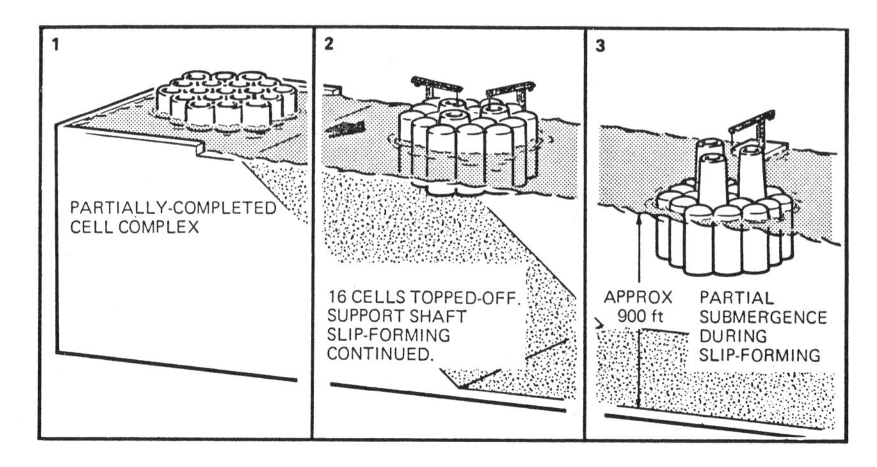

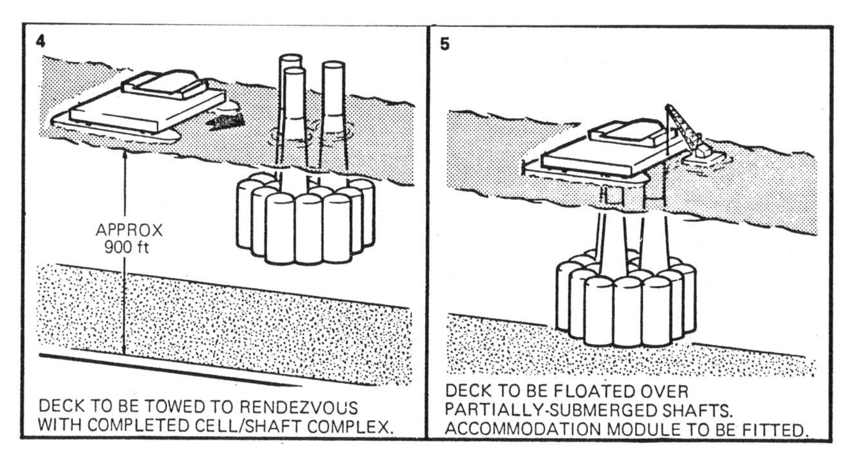

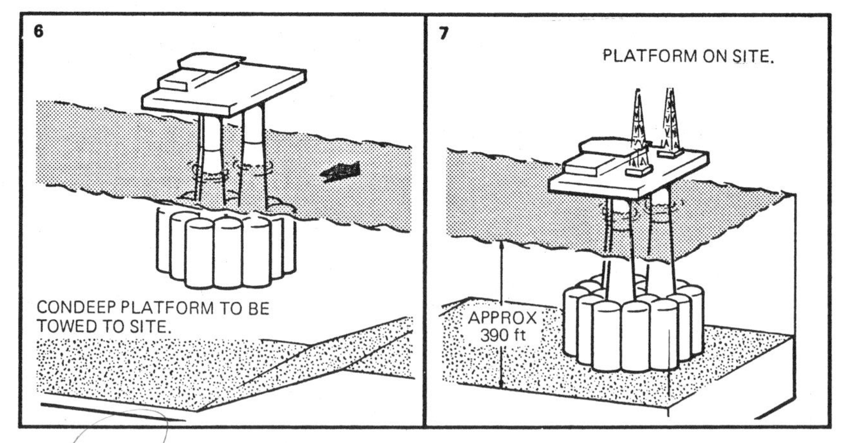

Fig. 1.10. Illustration of construction and installation procedures for concrete gravity platforms. (Courtesy of the Mobil Oil Corporation.)

(a)

(b)

Fig. 1.11. Offshore gravity structure (a) being towed to site and (b) in place and operating in the North Sea. (Courtesy of the Mobil Oil Corporation.)

(a)

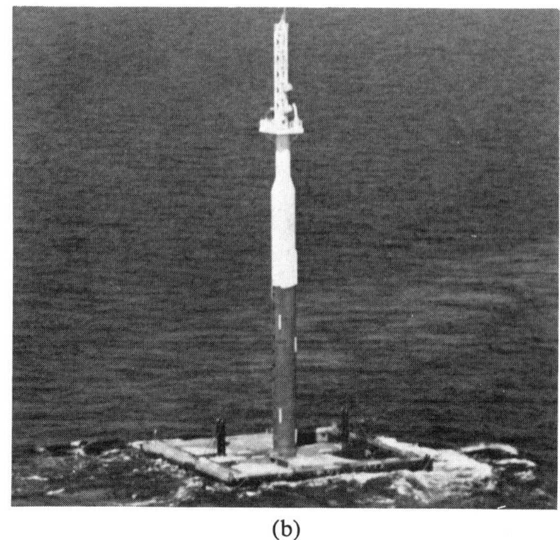

(b)

(c)

Fig. 1.12. Steel gravity platform off Florida, used by the U.S. Air Force in monitoring pilot training: (a) structure being towed to the offshore site; (b) structure before sinking to the sandy seafloor; (c) structure in place and operating. (Courtesy of the U.S. Air Force.)

position. The total height of the structure is approximately 200 ft, with about 100 ft beneath the water surface. The seafloor in this region of the Gulf consists of a thick layer of sand which provides firm support for the structure.

Deep-Water Design Forms

For water depths greater than about 1000 ft, the weight and foundation requirements of traditional offshore structures make them less attractive than other design forms. Two such forms are the guyed tower and tension-leg platform.

The *guyed-tower* concept is illustrated in Fig. 1.13. It consists of a uniform cross-sectional support structure held upright by several guy lines that run to clump weights on the ocean floor. From the clump weights, the lines then run to conventional anchors to form a dual stiffness mooring system. Under normal operating loads, the clump weights remain on the seafloor and lateral motion of the structure is restrained. However, during a severe storm, the clump weights are lifted off the seafloor by loads transferred from the structure to the clump weights through the guy lines. This action permits the tower to absorb the environmental loadings on it by swaying back and forth without overloading the guy lines. The guyed-tower concept is presently considered to be applicable to water depths of about 2000 ft.

Figure 1.14 illustrates the *tension-leg* concept. In this design, vertical members are used to anchor the platform to the seafloor. This upper part of the structure is designed with a large amount of excessive buoyancy so as to keep the vertical members in tension. Because of this tension, the platform remains virtually horizontal under wave action. Lateral excursions are also limited by the vertical members, since such movements necessarily cause them to develop a restoring force. A major advantage of the tension-leg concept is its relative cost insensitivity to increased water depths. At the present time, it appears that the main limitation on the tension-leg platform arises from dynamic inertia forces associated with the lateral oscillations of the platform in waves. These become significant at water depths of about 3000 ft (Lee, 1981).

1.3 ANALYSIS OF FIXED OFFSHORE STRUCTURES

We noted earlier that the design of an offshore structure progresses through various stages from preliminary designs based on rough estimates of environmental loadings to a final design form which must be sized to carry the loadings exerted during installation and when operating at its intended offshore site. This sizing is carried out using a detailed analysis of the structure. The magnitude

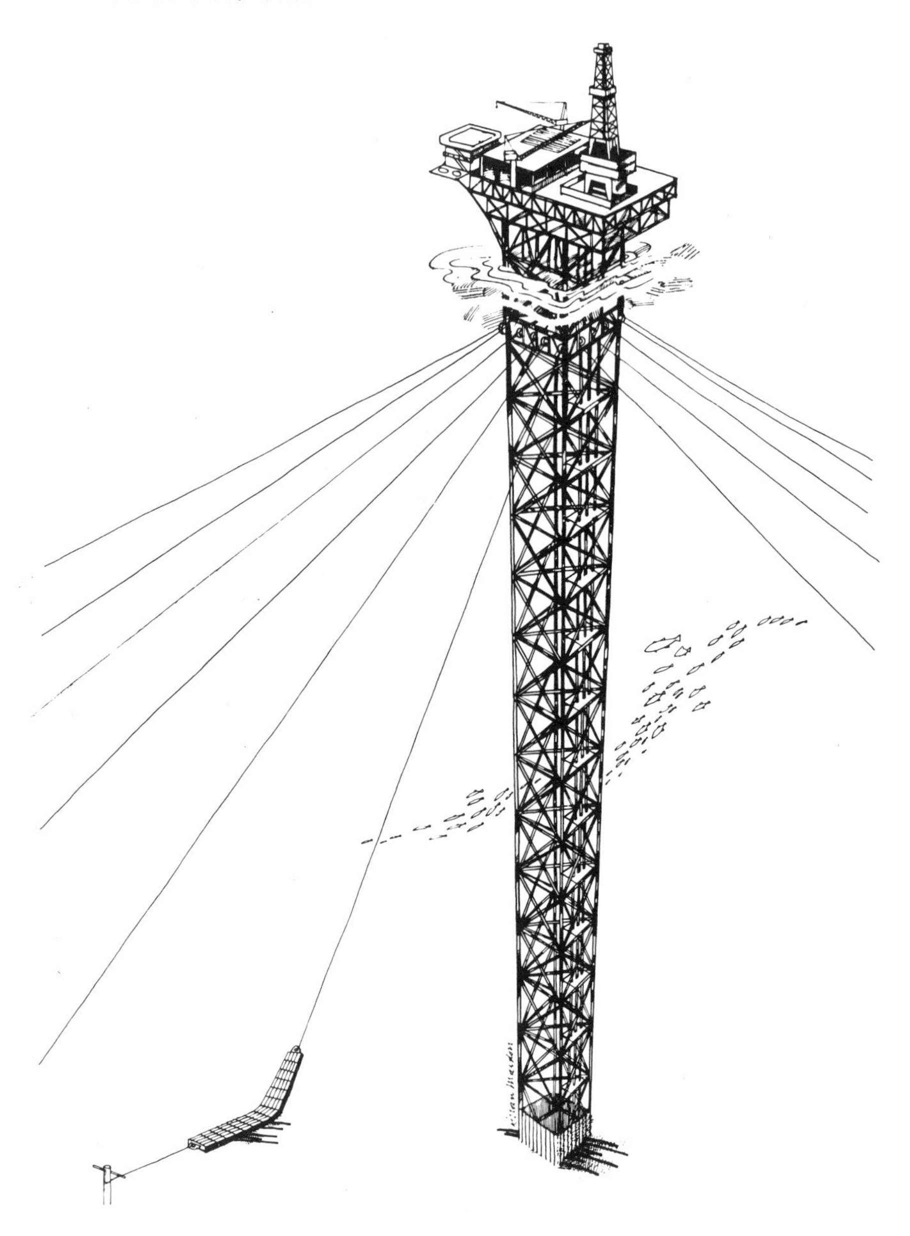

Fig. 1.13. Guyed-tower concept for deep water. (Courtesy of Exxon Company USA.)

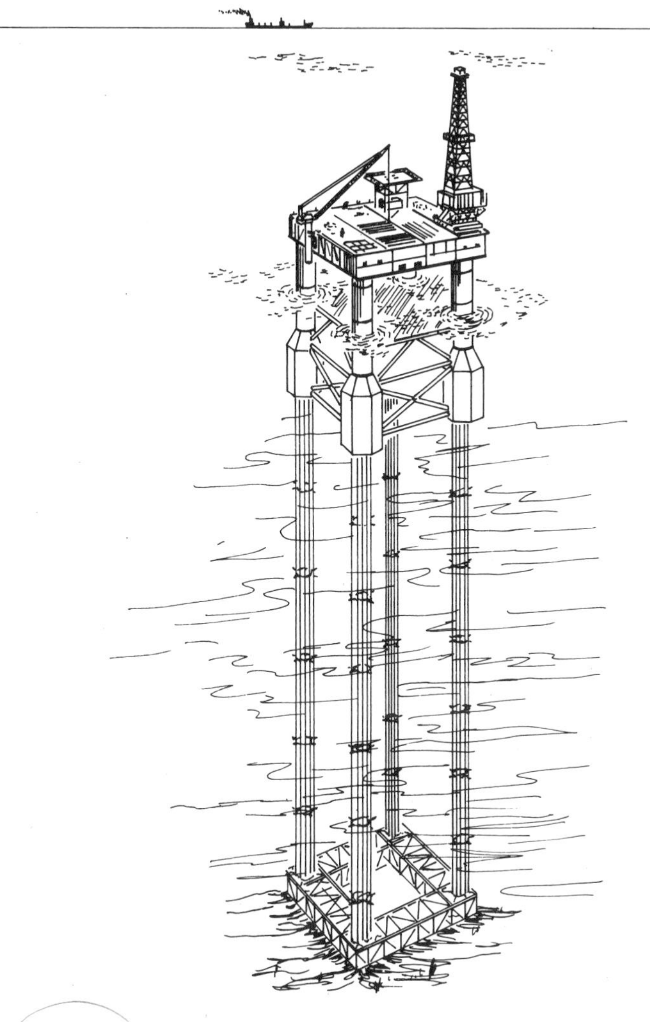

Fig. 1.14. Tension-leg concept for deep water. (Courtesy of McDermott Incorporated.)

of the loadings, as well as the structural response, depends on the size of the individual members, so that the analysis of the structure must first be carried out using initial estimates of them. Member sizes that are found inadequate in this first analysis are then changed and the analysis repeated. To keep the eventual fabrication of the structure as simple as possible, individual member sizes are usually not optimized for efficiency in carrying the loads. Rather, only the size of classes of members are chosen, based on the critically loaded member of the class. For example, with a template substructure, the bracing members between the main legs will normally all have the same dimensions even though some will carry less stress than others.

In the process described above, attention is first given to the severe environmental and operational loads likely to be experienced at the offshore site and classes of members of the structure sized to carry these loads. The proposed structure is then analyzed for loadings experienced during its installation. These loadings may require additional changes in the size of classes of members, or even additional members. If so, the design must be analyzed again for environmental and operating loads to ensure that the changes still allow the members to carry these loads. Such a check is necessary because changing the size, say, of all bracing members will change the loads not only on these members but also on the legs of the structure. If further changes are needed, the process is repeated.

For steel-framed structures, much of the foregoing analysis is conveniently carried out using modern methods of matrix computer analysis, as developed for steel-framed buildings. Stresses within the members are calculated and compared with acceptable levels. With reinforced-concrete gravity structures, the same methods may also be used for determining the structural loads transmitted to the individual members and these used, in turn, in a further analysis of the adequacy of reinforcement.

To perform such an analysis it is, of course, necessary to have a reasonable representation of the forces acting on the structure. In certain cases, such as the lifting and launching of a template substructure, these are defined simply by the weights of the various members of the structure. Operational loads and in-place structural weight are similarly well defined. The conversion of given environmental site conditions to forces on the structure presents, however, more of a problem, and these must be determined by detailed calculations using appropriate formulas.

The most common analysis of fixed offshore structures under combined operational and severe environmental loadings is based on equilibrium considerations of the structure and the maximum loadings that can be exerted on it at any instant. This static analysis is normally satisfactory for structures in waters less than about 300 ft deep since they are generally rigid enough to allow neglect of inertia forces associated with the back-and-forth acceleration of the structure under the variable wave loadings. In the case of structures supported by piles on soft sediments, the analysis must take into account the interaction of the structure and piles at the seafloor. The vertical loads from the analysis are also used to determine the depth that the piles should be driven to carry them.

Finally, for structures designed for water depths greater than about 300 ft, or for structures with considerable flexibility because of their particular form, it can happen that appreciable inertia forces will exist. In this case, the design must then be checked for possible overstress from dynamic loadings. The same is true when earthquake loadings pose a hazzard to the proposed structure, as these can cause a shaking of the base of the structure accompanied by appreciable inertia forces.

The emphasis of this book is on the various steps of analysis outlined above. In Chapter 2 we discuss matrix methods of analysis, as now widely employed in structural engineering. This is followed in Chapter 3 by a discussion of environmental loadings on offshore structures and the engineering methods used to estimate them. In Chapter 4 we discuss static methods of analysis for offshore structures, with particular attention directed toward steel template-type structures and concrete gravity structures. The details of the soil response to the overhead structure are considered in Chapter 5, and the analysis of structures having appreciable inertia effects is considered in Chapter 6.

REFERENCES

Fowler, J. W. (1965). Construction of the Chesapeake Light Station, *Civil Engineering*, Vol. 35 (November), p.76.

Lee, G. C. (1968). Offshore Structures—Past, Present, Future and Design Considerations, *Offshore*, Vol. 28, No. 6, pp. 45–55.

Lee, G. C. (1980). Recent Advances in Design and Construction of Deepwater Platforms—Part I, *Ocean Industry*, November, pp. 71–80.

Lee, G. C. (1981). Recent Advances in Design and Construction of Deepwater Platforms—Part 2, *Ocean Industry*, February, pp. 78–82.

Mashburn, M. K., and J. L. Hubbard (1967). An Ocean Structure, *Proceedings, Conference Civil Engineering in the Oceans*, ASCE, San Francisco, pp. 183–202.

McClelland, B. (1974). Design of Deep Penetration Piles for Ocean Structures, *Journal of the Geotechnical Engineering Division*, ASCE, Vol. 100, pp. 709–747.

Ruffin, J. V. (1965). Steel Offshore Towers Replace Lightships, *Civil Engineering*, Vol. 35 (November), p. 72.

Watt, B. J. (1978). Basic Structural Systems—A Review of Their Design and Analysis Requirements, *Numerical Methods in Offshore Engineering* (ed. O. C. Zienkiewicz et al.), John Wiley & Sons, Inc., New York, pp. 1–42.

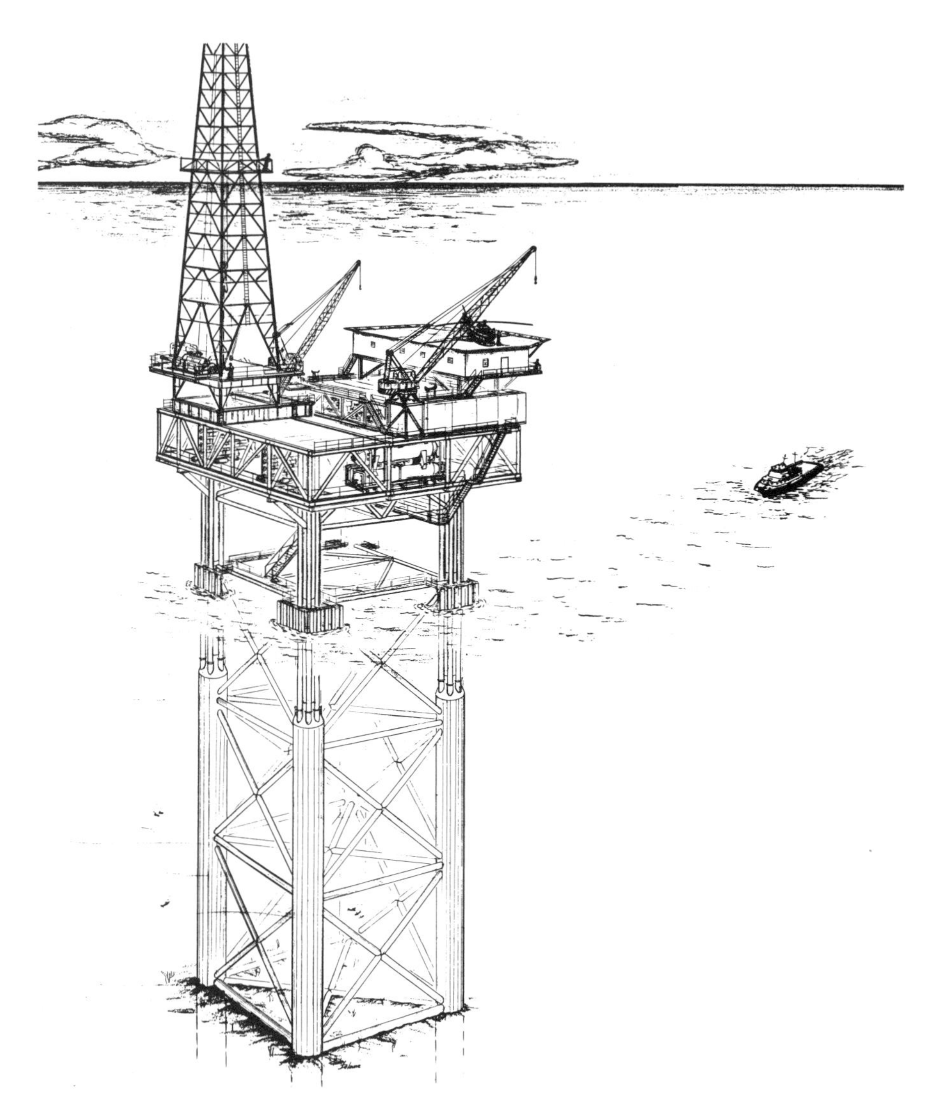

Four-legged offshore platform designed for approximately 400 ft of water. (Courtesy of McDermott Incorporated.)

2

MATRIX STRUCTURAL ANALYSIS

MODERN OCEAN STRUCTURES such as those illustrated in Chapter 1 generally consist of a collection of members assembled together in such a way as to provide an efficient means for resisting applied forces and carrying the required loads. The force–deflection response of the individual members making up the structure is well known from solid mechanics. A knowledge of how the individual members respond is, however, clearly insufficient by itself to allow discussion of the response of the structure as a whole. For this purpose it is necessary to appeal to methods of structural analysis.

Before the advent of high-speed computers in the 1950s, method of analysis of a complex structure generally involved lengthy, time-consuming hand calculations in order to determine the response of the overall structure and the stresses and deflections of the individual members. In recent times, however, all of this has changed and complex structures can now be analyzed rapidly using digital computers.

The use of computers in structural analysis is most conveniently carried out when the structural theory is expressed in matrix formulation. In the present chapter, we accordingly present this formulation using the *direct stiffness method* of structural analysis as described earlier by Martin (1966) and Willems and Lucas (1968), among others. For those not already familar with matrix algebra, a brief but adequate discussion of this topic is given in the Appendix.

2.1 ELEMENTS OF MATRIX FORMULATION

We begin our discussion of matrix structural analysis by considering the elastic tensile or compressive response of a single structural member such as illustrated in Fig. 2.1. The ends of the bar are denoted by coordinates 1 and 2 having positive directions as indicated and the forces and displacements at these ends are denoted by f_1, f_2, and U_1, U_2, respectively. The length of the bar is denoted by l and its cross-sectional area by A. The extension Δl of the bar is equal to $U_2 - U_1$ and the associated tensile force f_2 is, from solid mechanics, equal to $EA\ \Delta l/l$, where E denotes the elastic (Young's) modulus of the material. From statics, we also have that $f_1 = -f_2$, so that the force-displacement relations for the bar are expressible as

$$\begin{aligned} f_1 &= -k(U_2 - U_1) \\ f_2 &= k(U_2 - U_1) \end{aligned} \tag{1}$$

where $k = EA/l$ denotes the *stiffness* of the member.

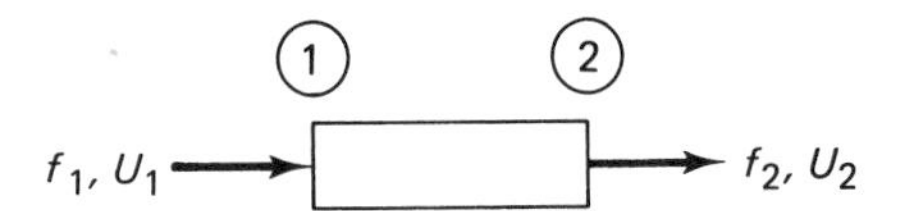

Fig. 2.1. Structural member subjected to axial loads.

These equations may be written in matrix notation as

$$\begin{Bmatrix} f_1 \\ f_2 \end{Bmatrix} = \begin{bmatrix} k & -k \\ -k & k \end{bmatrix} \begin{Bmatrix} U_1 \\ U_2 \end{Bmatrix} \tag{2}$$

or, more compactly, as

$$\{f\} = [K]\{U\} \tag{3}$$

where the elements of $\{f\}$, $[K]$ and $\{U\}$ are given by the corresponding elements of equation (2) and the matrix $[K]$ is referred to as the *stiffness matrix of the member*.

When the displacements U_1 and U_2 are known, or specified, equation (2) can be used directly to determine the corresponding forces f_1 and f_2. On the other hand, however, when the forces f_1 and f_2 are specified (with, of course, $f_1 = -f_2$ as required by equilibrium), it is easily seen that the equation cannot be solved for the displacements U_1 and U_2 but, rather, only for their difference. The physical reason for this is simply that forces applied to the ends of the bar do not fix absolutely the displacements, but only the extension of the bar, that is, the displacement of one end relative to the other. Thus, in order to solve equation (2) for displacement, we must specify at least one boundary condition on the displacement in order to preclude the possibility of rigid-body movement. At the other end, we may of course specify either force or displacement. Notice, though, that we cannot specify both a force *and* a displacement at either end, since if we specify the displacement, we must take whatever value of force is required to maintain it; or if we specify a force, we must take whatever value of displacement is associated with it.

If, for example, we use the condition $U_1 = 0$, together with the force boundary condition $f_2 = F$, then equation (2) yields immediately

$$U_2 = \frac{F}{k}, \qquad f_1 = -F \tag{4}$$

Next consider the more complex case involving two bars, as shown in Fig. 2.2. Our objective here is the same as in the previous discussion, namely the determination of equations connecting the forces and displacements at the indicated coordinates and the solution of these when appropriate boundary conditions are specified.

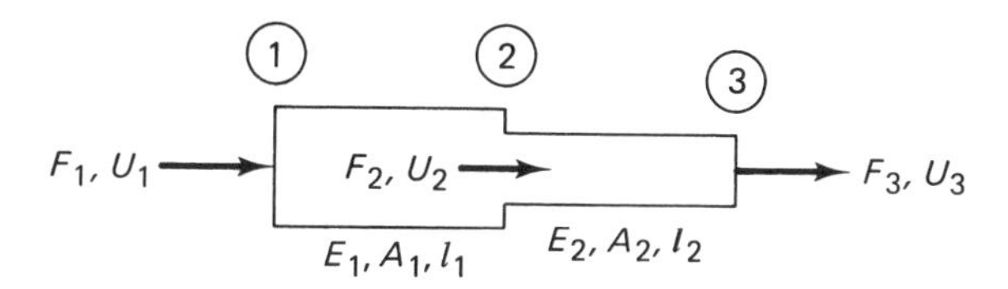

Fig. 2.2. Two-bar structure subjected to axial loads.

To solve this problem, let us first construct free-body diagrams of bars 1 and 2 in the manner indicated in Fig. 2.3. The notation is as follows: Internal forces are denoted by f_1^1, f_1^2, etc., with the superscript denoting the bar and the subscript denoting the coordinate shown previously in Fig. 2.2. For example, f_2^1 denotes the internal force acting on bar 1 at coordinate 2. External forces are denoted by F_1, F_2, F_3, the subscripts denoting the coordinates location.

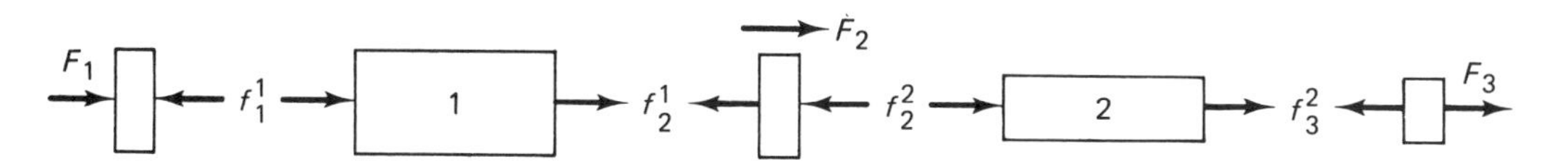

Fig. 2.3. Free-body diagrams.

These forces may either be applied forces or reaction forces arising from a support where displacements are restrained.

From equation (2), we may write the following force–displacement relations for the individual members:

$$\begin{aligned} f_1^1 &= -k_1(U_2 - U_1) \\ f_2^1 &= k_1(U_2 - U_1) \\ f_2^2 &= -k_2(U_3 - U_2) \\ f_3^2 &= k_2(U_3 - U_2) \end{aligned} \tag{5}$$

where $k_1 = E_1A_1/l_1$ and $k_2 = E_2A_2/l_2$.

In addition, from equilibrium calculations applied to the free bodies of the bar faces and interface, we have

$$\begin{aligned} F_1 &= f_1^1 \\ F_2 &= f_2^1 + f_2^2 \\ F_3 &= f_3^2 \end{aligned} \tag{6}$$

Hence, on combining equations (5) and (6), we have

$$\begin{aligned} F_1 &= k_1U_1 - k_1U_2 \\ F_2 &= -k_1U_1 + (k_1 + k_2)U_2 - k_2U_3 \\ F_3 &= -k_2U_2 + k_3U_3 \end{aligned} \tag{7}$$

These equations may be expressed in matrix notation as

$$\begin{Bmatrix} F_1 \\ F_2 \\ F_3 \end{Bmatrix} = \begin{bmatrix} k_1 & -k_1 & 0 \\ -k_1 & k_1 + k_2 & -k_2 \\ 0 & -k_2 & k_2 \end{bmatrix} \begin{Bmatrix} U_1 \\ U_2 \\ U_3 \end{Bmatrix} \tag{8}$$

or in compact notation as

$$\{F\} = [K]\{U\} \tag{9}$$

where the elements of $\{F\}$, $[K]$, and $\{U\}$ are given by the corresponding elements in equation (8), and the matrix $[K]$ is referred to as the *stiffness matrix of the structure*.

Inspection of the structural stiffness matrix of equation (8) reveals that this matrix is *symmetric*; that is, the first, second, and third rows are equal, respectively, to the first, second, and third columns. A similar statement can also be seen to be true for the individual member stiffness matrices and illustrates a general property of elastic systems, namely that a given displacement applied at, say, coordinate 2 of a structure causes a force at coordinate 1 equal to that which would have been caused at coordinate 2 were the same displacement to be applied at coordinate 1 rather than 2.

Consider now the use of equation (8) for determining the response of the structure. For this purpose we must, of course, specify boundary conditions on the structure; that is, we must specify either a force *or* a displacement at each coordinate location. As in the previous discussion of a single bar, we must, however, specify at least one boundary condition on the displacement if we are to solve equation (8) for any displacements.

If we specify the boundary conditions $U_1 = 0$, $F_2 = F_2$, $F_3 = F_3$, we see immediately that equation (8) reduces to

$$F_1 = -k_1 U_2 \tag{10}$$

and

$$\begin{Bmatrix} F_2 \\ F_3 \end{Bmatrix} = \begin{bmatrix} k_1 + k_2 & -k_2 \\ -k_2 & k_2 \end{bmatrix} \begin{Bmatrix} U_2 \\ U_3 \end{Bmatrix} \tag{11}$$

Thus, equation (11) may be used to solve for U_2 and U_3 in terms of F_2 and F_3 and equation (10) then used to solve for the unknown force F_1 required to maintain the boundary condition.

In extracting equations (10) and (11) from (8), a convenient procedure is to delete the first column of the stiffness matrix of equation (8), since each element of this column multiplies the zero displacement U_1, and then separate the equation into two reduced equations, the first having the unknown reaction force and the other having the known applied forces. From the second, we may find the unknown displacements and from the first, we may, in turn, determine the unknown reaction force.

Symbolically, we may write equation (11) as

$$\{F\} = [K]\{U\} \tag{12}$$

with $\{F\}$, $[K]$, and $\{U\}$ now having respective elements given by equation (11). The solution for the displacements U_2, U_3 is then expressible symbolically as

$$\{U\} = [K]^{-1}\{F\} \tag{13}$$

where $[K]^{-1}$ denotes the inverse of $[K]$ and is expressible for this case as

$$[K]^{-1} = \begin{bmatrix} \dfrac{1}{k_1} & \dfrac{1}{k_1} \\ \dfrac{1}{k_1} & \dfrac{1}{k_1} + \dfrac{1}{k_2} \end{bmatrix} \tag{14}$$

Expanding equation (13), we thus have

$$\begin{aligned} U_2 &= \frac{F_2}{k_1} + \frac{F_3}{k_1} \\ U_3 &= \frac{F_2}{k_1} + \left(\frac{1}{k_1} + \frac{1}{k_2}\right)F_3 \end{aligned} \tag{15}$$

and, from equation (10), that

$$F_1 = -F_2 - F_3 \tag{16}$$

The internal forces are given by equation (5) as

$$\begin{aligned} f_2^1 &= -f_1^1 = F_2 + F_3 \\ f_3^2 &= -f_2^2 = F_3 \end{aligned} \tag{17}$$

The corresponding stresses σ_1 and σ_2 in members 1 and 2 can be obtained by dividing the internal forces by the respective sectional areas A_1 and A_2. With positive stress denoting tension, we have $\sigma_1 = f_2^1/A_1 = (F_2 + F_3)/A_1$ and $\sigma_2 = f_3^2/A_2 = F_3/A_2$.

Notice that because of the choice of the single displacement boundary condition $U_1 = 0$ in this example, the structure is *statically determinate*, that is, the reaction force F_1 and the internal forces f_1^1, f_2^1, etc., could all be determined using the equations of statics rather than the matrix formulation. With these, the displacements could then be determined separately using the member stiffness equations.

If, however, we consider displacement boundary conditions $U_1 = U_3 = 0$ together with the force condition $F_2 = F$, the structure then becomes *statically indeterminate* and the equations of statics must be combined with the stiffness equations of the members, as in our matrix formulation, before the reaction and internal forces can be determined. In this case, the matrix formulation of equation (8) gives

$$\begin{aligned} U_2 &= \frac{F}{k_1 + k_2} \\ F_1 &= \frac{-k_1}{k_1 + k_2}F \\ F_3 &= \frac{-k_2}{k_1 + k_2}F \end{aligned} \tag{18}$$

and the stiffness equations of the individual members give

$$\begin{aligned} f_1^1 &= -f_2^1 = \frac{-k_1}{k_1 + k_2}F \\ f_2^2 &= -f_3^2 = \frac{k_2}{k_1 + k_2}F \end{aligned} \tag{19}$$

Thus, we see that the matrix formulation makes no distinction between statically determinate and indeterminate structures and handles both in the same manner. A distinct advantage of the matrix formulation is that it allows systematic determination of unknown displacements and forces. A disadvantage for statically determinate structures is the need for simultaneous solution of the reduced matrix equation for the deflections by means of matrix inversion rather than by direct algebraic determination. The general availability of digital computers and their facility for calculating matrix inverses minimizes this problem, however, even for structures requiring the inverse of very large stiffness matrices.

Direct Stiffness Method

It is important to observe that the general stiffness matrix of the structure above, as given in equation (8), is expressible as the sum of two separate matrices in the form

$$\begin{bmatrix} k_1 & -k_1 & 0 \\ -k_1 & k_1 + k_2 & -k_2 \\ 0 & -k_2 & k_2 \end{bmatrix} = \begin{bmatrix} k_1 & -k_1 & 0 \\ -k_1 & k_1 & 0 \\ 0 & 0 & 0 \end{bmatrix} + \begin{bmatrix} 0 & 0 & 0 \\ 0 & k_2 & -k_2 \\ 0 & -k_2 & k_2 \end{bmatrix} \tag{20}$$

The first matrix on the right-hand side of this equation is clearly the stiffness matrix of member 1, with a zero third row and third column added to indicate its nondependence on coordinate 3 of the system. Similarly, the second matrix on the right-hand side of the equation is the stiffness matrix of member 2 with a zero first row and first column added to indicate its nondependence on coordinate 1 of the entire system. This observation is easily seen to be true for any number of members and, accordingly, suggests a simple way to obtain the stiffness matrix of a structure from the individual member stiffness matrices, namely the direct addition of these, with zero rows and columns added for system coordinates not involved with the individual member. This method is known in matrix structural analysis as the *direct stiffness method.*

With practice, the construction of the stiffness matrix of a structure can, in fact, be achieved by this method without formally expanding the individual member stiffness matrices with rows and columns of zeros; one can simply determine by inspection the proper location of the nonzero elements of the member stiffness matrices and add them to obtain the structural stiffness matrix. The basis for the direct stiffness method is, of course, simply the equilibrium requirement that the internal forces balance the applied forces at the joints of the structure.

EXAMPLE 2.1-1. A winch–cable–payload system is shown in Fig. 2.4. The payload has an effective weight of 2000 lb. The cable weighs 0.66 lb/ft. Determine the displacements and internal forces along the cable using the foregoing direct stiffness method of analysis.

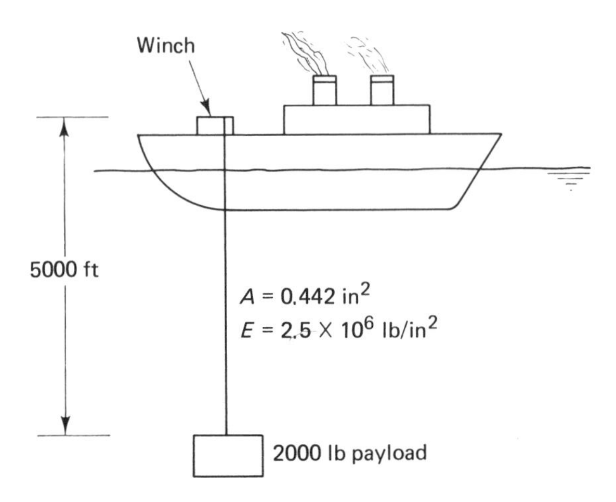

Fig. 2.4

To solve this problem, we first note that the previous discussion did not consider forces acting between the ends of the bar such as exists here with the nonnegligible cable-weight loading. To apply the foregoing procedures, it is thus necessary to divide the cable into a number of segments and assume the weight loading of each segment to be concentrated equally at its ends; that is, half the segment weight applied at one end and half at the other. Obviously, this approach will give an approximate description which will become better and better as the number of segments (and the computational difficulty) is increased. The analysis of a continuous member by artificial division into a number of segments is known as a *finite element analysis*.

In the particular problem at hand, suppose that we divide the cable into four segments, each of length $l = 1250$ ft. Labeling the coordinates as shown in Fig. 2.5, we may establish the matrix stiffness equation for this structure using the direct stiffness method. We find that

$$\begin{Bmatrix} F_1 \\ F_2 \\ F_3 \\ F_4 \\ F_5 \end{Bmatrix} = \begin{bmatrix} k & -k & 0 & 0 & 0 \\ -k & 2k & -k & 0 & 0 \\ 0 & -k & 2k & -k & 0 \\ 0 & 0 & -k & 2k & -k \\ 0 & 0 & 0 & -k & k \end{bmatrix} \begin{Bmatrix} U_1 \\ U_2 \\ U_3 \\ U_4 \\ U_5 \end{Bmatrix}$$

Fig. 2.5

4 @ 1250 ft

1
2
3
4
5

Fig. 2.5

where $k = EA/l = 884$ lb/ft. The boundary conditions, with the weight of each cable segment concentrated at its ends, are expressible as

$$U_1 = 0, \qquad F_2 = F_3 = F_4 = 825 \text{ lb}, \qquad F_5 = 2412.5 \text{ lb}$$

Applying these to the matrix equations above, we find (by deleting the first column of the stiffness matrix and separating the equation into reduced equations containing the unknown force F_1 and the known forces F_2, F_3, F_4) the following reduced equations:

$$F_1 = -kU_2$$

and

$$\begin{Bmatrix} F_2 \\ F_3 \\ F_4 \\ F_5 \end{Bmatrix} = \begin{bmatrix} 2k & -k & 0 & 0 \\ -k & 2k & -k & 0 \\ 0 & -k & 2k & -k \\ 0 & 0 & -k & k \end{bmatrix} \begin{Bmatrix} U_2 \\ U_3 \\ U_4 \\ U_5 \end{Bmatrix}$$

Using the boundary conditions on the forces, the last equation above may be inverted (solved) by hand, or more conveniently with a digital computer, to find the following deflection values:

$$U_2 = 5.53 \text{ ft}, \qquad U_3 = 10.13 \text{ ft}, \qquad U_4 = 13.79 \text{ ft}, \qquad U_5 = 16.52 \text{ ft}$$

From the first of the reduced equations above, we then find that

$$F_1 = -4889 \text{ lb}$$

Remembering that this force represents the reaction force R at coordinate 1 plus one-half the weight of element 1–2 (i.e., $F_1 = R + 412.5$), we thus find the reaction force exerted by the winch to be

$$R = -5300 \text{ lb}$$

which, of course, is consistent with the equilibrium requirement that the reaction force at the winch balance the weight of the cable and payload.

The internal forces acting at the ends of each segment may finally be determined using the stiffness matrices of the individual segments. For example, for segment 1–2, we have

$$\begin{Bmatrix} f_1^1 \\ f_2^1 \end{Bmatrix} = \begin{bmatrix} k & -k \\ -k & k \end{bmatrix} \begin{Bmatrix} U_1 \\ U_2 \end{Bmatrix}$$

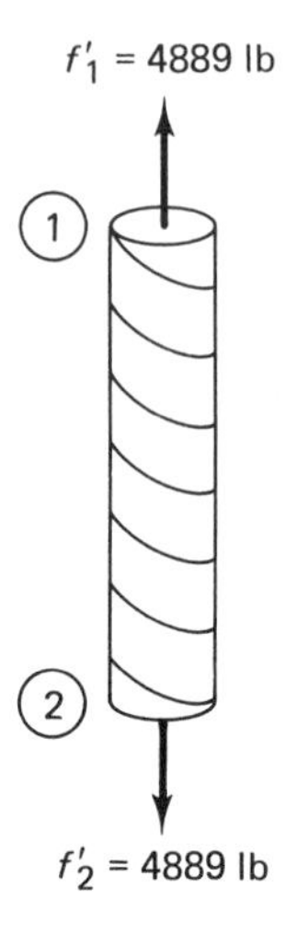

Fig. 2.6

Remembering that $U_1 = 0$, we have (Fig. 2.6):

$$f_1^1 = -f_2^1 = -4889 \text{ lb}$$

Similarly, for the other segments, we find

$$f_2^2 = -f_3^2 = -4067 \text{ lb}$$
$$f_3^3 = -f_4^3 = -3235 \text{ lb}$$
$$f_4^4 = -f_5^4 = -2413 \text{ lb}$$

The stresses σ_1, σ_2, σ_3, and σ_4 within the four segments may, of course, be determined from these internal forces by dividing by the cross-sectional area of the cable. We find

$$\sigma_1 = 11{,}060 \text{ lb/in}^2, \qquad \sigma_2 = 9200 \text{ lb/in}^2$$
$$\sigma_3 = 7320 \text{ lb/in}^2, \qquad \sigma_4 = 5460 \text{ lb/in}^2 \quad \#$$

2.2 PLANE TRUSSES

Plane structures having axially loaded members are generally referred to as *plane trusses* and consist of straight members which are inclined to one another in a single plane and which carry only tension or compression loading. From statics, it can easily be seen that the conditions for this last requirement to be met are that applied forces exist only at the joints connecting the members, that no applied moments exist, and that the members are free to rotate at the joints so as to prevent restraint moments.

To analyze such structures, we consider now the case of an axially loaded bar inclined at an angle α with the horizontal as shown in Fig. 2.7. We choose horizontal and vertical coordinate axes x and y and inclined axes $\bar{x}$ and $\bar{y}$, as shown. The first are referred to as the *system axes* and the second as the *member axes*.

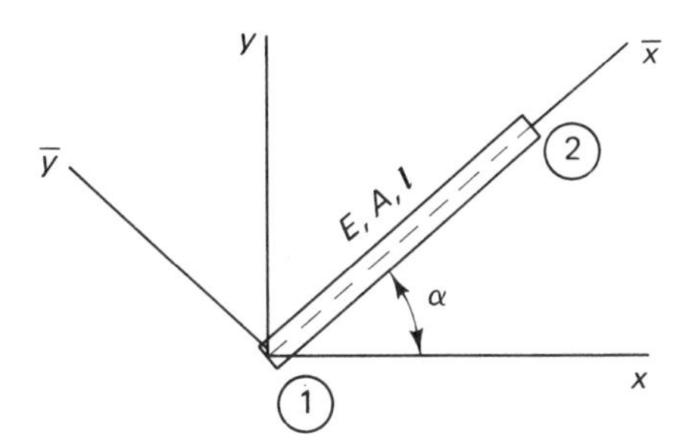

Fig. 2.7. Inclined plane-truss structural member.

Let f_{1x}, f_{1y} and f_{2x}, f_{2y} denote x- and y-force components at positions 1 and 2 of the bar, respectively, and let $\bar{f}_{1x}, \bar{f}_{1y}$ and $\bar{f}_{2x}, \bar{f}_{2y}$ denote corresponding $\bar{x}$- and $\bar{y}$-force components at these positions. From geometry, we may easily establish the following transformation equations between the force components at position 1.

$$\begin{aligned} \bar{f}_{1x} &= f_{1x} \cos \alpha + f_{1y} \sin \alpha \\ \bar{f}_{1y} &= -f_{1x} \sin \alpha + f_{1y} \cos \alpha \end{aligned} \tag{21}$$

and

$$\begin{aligned} f_{1x} &= \bar{f}_{1x} \cos \alpha - \bar{f}_{1y} \sin \alpha \\ f_{1y} &= \bar{f}_{1x} \sin \alpha + \bar{f}_{1y} \cos \alpha \end{aligned} \tag{22}$$

with similar equations applying at position 2.

In matrix notation, these become

$$\begin{Bmatrix} \bar{f}_{1x} \\ \bar{f}_{1y} \\ \bar{f}_{2x} \\ \bar{f}_{2y} \end{Bmatrix} = \begin{bmatrix} \cos \alpha & \sin \alpha & 0 & 0 \\ -\sin \alpha & \cos \alpha & 0 & 0 \\ 0 & 0 & \cos \alpha & \sin \alpha \\ 0 & 0 & -\sin \alpha & \cos \alpha \end{bmatrix} \begin{Bmatrix} f_{1x} \\ f_{1y} \\ f_{2x} \\ f_{2y} \end{Bmatrix} \tag{23}$$

and

$$\begin{Bmatrix} f_{1x} \\ f_{1y} \\ f_{2x} \\ f_{2y} \end{Bmatrix} = \begin{bmatrix} \cos \alpha & -\sin \alpha & 0 & 0 \\ \sin \alpha & \cos \alpha & 0 & 0 \\ 0 & 0 & \cos \alpha & -\sin \alpha \\ 0 & 0 & \sin \alpha & \cos \alpha \end{bmatrix} \begin{Bmatrix} \bar{f}_{1x} \\ \bar{f}_{1y} \\ \bar{f}_{2x} \\ \bar{f}_{2y} \end{Bmatrix} \tag{24}$$

Notice that the square matrix of the second of these is just the matrix of the first with rows and columns interchanged; that is, the second is just the *transpose* of the first. Thus, if $[R]$ denotes the first matrix, then $[R]^T$ may be used to denote the second. Hence,

$$\begin{aligned} \{\bar{f}\} &= [R]\{f\} \\ \{f\} &= [R]^T\{\bar{f}\} \end{aligned} \tag{25}$$

The discussion above has been concerned with force components at positions 1 and 2 of the member. Obviously, however, similar results also apply to the respective x- and y-displacement components U_1, V_1 and U_2, V_2 at positions 1 and 2 and the corresponding $\bar{x}$- and $\bar{y}$-displacement components $\bar{U}_1$, $\bar{V}_1$ and $\bar{U}_2$, $\bar{V}_2$. We may thus also write

$$\begin{aligned} \{\bar{U}\} &= [R]\{U\} \\ \{U\} &= [R]^T\{\bar{U}\} \end{aligned} \tag{26}$$

where the displacement matrices are given by

$$\{\bar{U}\} = \begin{Bmatrix} \bar{U}_1 \\ \bar{V}_1 \\ \bar{U}_2 \\ \bar{V}_2 \end{Bmatrix}, \qquad \{U\} = \begin{Bmatrix} U_1 \\ V_1 \\ U_2 \\ V_2 \end{Bmatrix}$$

Now assume member 1–2 to be subjected to a uniaxial loading as in the previous discussions. We have

$$\bar{f}_{1y} = \bar{f}_{2y} = 0 \tag{27}$$

and, from our earlier discussions,

$$\{\bar{f}\} = [\bar{K}]\{\bar{U}\} \tag{28}$$

where $[\bar{K}]$ denotes the stiffness matrix of the member, relative, of course, to member axes.

Using equations (25) and (26), the latter relation may be written, alternatively, as

$$[R]\{f\} = [\bar{K}][R]\{U\} \tag{29}$$

Premultiplying both sides of this equation by $[R]^T$ and noticing that

$$[R]^T[R] = \begin{bmatrix} 1 & 0 & 0 & 0 \\ 0 & 1 & 0 & 0 \\ 0 & 0 & 1 & 0 \\ 0 & 0 & 0 & 1 \end{bmatrix} = [I] \tag{30}$$

we have

$$\{f\} = [R]^T[\bar{K}][R]\{U\} \tag{31}$$

so that by comparison with $\{f\} = [K]\{U\}$, the stiffness matrix $[K]$ of the member associated with the system axes is given by

$$[K] = [R]^T[\bar{K}][R] \tag{32}$$

This equation provides the general transformation relation for the stiffness matrix between member and system axes. Since $[\bar{K}]$ may be written from equation (2) of our earlier considerations simply as

$$[\bar{K}] = \frac{EA}{l}\begin{bmatrix} 1 & 0 & -1 & 0 \\ 0 & 0 & 0 & 0 \\ -1 & 0 & 1 & 0 \\ 0 & 0 & 0 & 0 \end{bmatrix} \tag{33}$$

with zero second and fourth rows and columns added to indicate nondependence on the force and displacement components in the y-direction, we may expand equation (32) to find $[K]$ expressible explicitly as

$$[K] = \frac{EA}{l}\begin{bmatrix} \lambda^2 & \lambda\mu & -\lambda^2 & -\lambda\mu \\ \lambda\mu & \mu^2 & -\lambda\mu & -\mu^2 \\ -\lambda^2 & -\lambda\mu & \lambda^2 & \lambda\mu \\ -\lambda\mu & -\mu^2 & \lambda\mu & \mu^2 \end{bmatrix} \tag{34}$$

where $\lambda = \cos\alpha$ and $\mu = \sin\alpha$.

The expression above gives the stiffness matrix of a member with respect to a set of system axes. The usefulness of this expression becomes apparent when we consider several nonaligned members making up a structural system. In this case, we may then refer the stiffness matrix of each member to common system axes and add like elements using the direct stiffness method discussed earlier to obtain the stiffness matrix of the entire structure. By assigning appropriate boundary conditions involving the force *or* displacement components at each joint (usually conveniently specified relative to the system axes), we can then use the stiffness equation of the structure to determine the remaining unknown force and displacement components. As in the case of uniaxial structures, we must, however, specify displacement components sufficient to allow unique determination of the remaining ones; that is, we must specify sufficient displacements to preclude the possibility of rigid-body movement of the structure.

Once the joint displacements are known, we may also calculate the internal forces acting at the ends of each member. For a given member, we have, in particular, from equation (34) the force–deflection equation relative to system axes expressible as

$$\begin{Bmatrix} f_{1x} \\ f_{1y} \\ f_{2x} \\ f_{2y} \end{Bmatrix} = \frac{EA}{l} \begin{bmatrix} \lambda^2 & \lambda\mu & -\lambda^2 & -\lambda\mu \\ \lambda\mu & \mu^2 & -\lambda\mu & -\mu^2 \\ -\lambda^2 & -\lambda\mu & \lambda^2 & \lambda\mu \\ -\lambda\mu & -\mu^2 & \lambda\mu & \mu^2 \end{bmatrix} \begin{Bmatrix} U_1 \\ V_1 \\ U_2 \\ V_2 \end{Bmatrix} \tag{35}$$

Considering, for example, the components f_{2x} and f_{2y}, we find from this equation that

$$\begin{aligned} f_{2x} &= \frac{EA}{l}[\lambda^2(U_2 - U_1) + \lambda\mu(V_2 - V_1)] \\ f_{2y} &= \frac{EA}{l}[\lambda\mu(U_2 - U_1) + \mu^2(V_2 - V_1)] \end{aligned} \tag{36}$$

From equation (23), we also have the axial member force $S_{1-2} = \bar{f}_{2x}$ given by

$$S_{1-2} = \lambda f_{2x} + \mu f_{2y} \tag{37}$$

so that, on using equation (36) and the identity $\lambda^2 + \mu^2 = 1$, we find

$$S_{1-2} = \frac{EA}{l}[\lambda(U_2 - U_1) + \mu(V_2 - V_1)] \tag{38}$$

which expresses the axial internal force of the member 1–2 in terms of system displacements. More generally, we may write the equation giving the axial internal force in member m–n as

$$S_{m-n} = \frac{EA}{l}[\lambda(U_n - U_m) + \mu(V_n - V_m)] \tag{39}$$

We note that if S_{m-n} is positive, the member will be in tension; if negative, then compression. The corresponding tensile or compressive stress is obtained by dividing S_{m-n} by the sectional area A of the member.

Sign Convention

The sign convention to be used with equation (34) follows from our original consideration of the bar of Fig. 2.7 and takes the following form. Let α denote the angle between the horizontal system axis x and the member axis $\bar{x}$, with $\bar{x}$ positive in the direction from end coordinate m to end coordinate n of a bar m–n. Then equation (34) will give the stiffness matrix $[K]_{mn}$ such that

$$\begin{Bmatrix} f_{mx} \\ f_{my} \\ f_{nx} \\ f_{ny} \end{Bmatrix} = \begin{bmatrix} \\ K_{mn} \\ \\ \end{bmatrix} \begin{Bmatrix} U_m \\ V_m \\ U_n \\ V_n \end{Bmatrix} \tag{40}$$

EXAMPLE 2.2-1. Consider the simple structure shown in Fig. 2.8. The two members consist of hollow aluminum tubes having 3-in outside diameter and $\frac{1}{8}$-in wall thickness. Assuming that $E = 10 \times 10^6$ lb/in², we wish to determine the stresses and deflections resulting from the 5000-lb applied load.

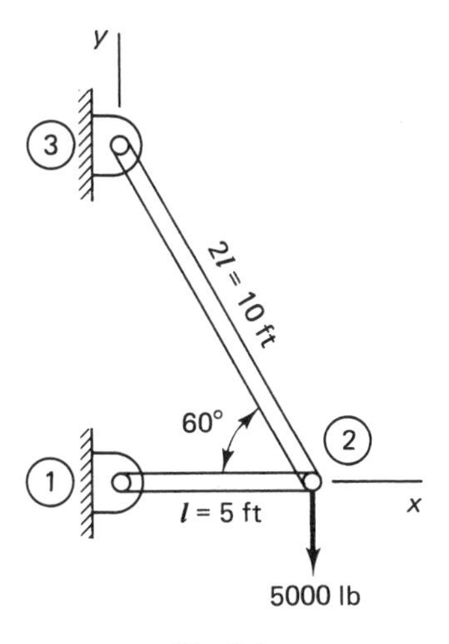

Fig. 2.8

To calculate the stiffness matrix of each member relative to system axes, x, y, we may first construct Table 2.1. From equation (34), we then easily find

Table 2.1

Member	α	λ	μ	λ^2	μ^2	$\lambda\mu$
1–2	0	1	0	1	0	0
2–3	120°	−0.500	0.866	0.250	0.750	−0.433

the following member stiffness matrices:

$$[K]_{1-2} = \frac{EA}{l}\begin{bmatrix} 1 & 0 & -1 & 0 \\ 0 & 0 & 0 & 0 \\ -1 & 0 & 1 & 0 \\ 0 & 0 & 0 & 0 \end{bmatrix}$$

$$[K]_{2-3} = \frac{EA}{2l}\begin{bmatrix} 0.250 & -0.433 & -0.250 & 0.433 \\ -0.433 & 0.750 & 0.433 & -0.750 \\ -0.250 & 0.433 & 0.250 & -0.433 \\ 0.433 & -0.750 & -0.433 & 0.750 \end{bmatrix}$$

The force–deflection response of member 1–2 does not depend on the force or displacement at coordinate 3; hence, the matrix $[K]_{1-2}$ may be expanded to show this nondependence by including fifth and sixth rows and columns of zeros. Similarly, the response of member 2–3 does not depend on the forces or displacements at coordinate 1 and the matrix $[K]_{2-3}$ may be expanded to show this nondependence by including first and second rows and columns of zeros. Adding these (expanded) member stiffness matrices (taking into account the scalar multipliers EA/l and $EA/2l$), we then easily find the structural stiffness matrix expressible, from the direct stiffness method, as

$$[K] = \frac{EA}{2l}\begin{bmatrix} 2 & 0 & -2 & 0 & 0 & 0 \\ 0 & 0 & 0 & 0 & 0 & 0 \\ -2 & 0 & 2.250 & -0.433 & -0.250 & 0.433 \\ 0 & 0 & -0.433 & 0.750 & 0.433 & -0.750 \\ 0 & 0 & -0.250 & 0.433 & 0.250 & -0.433 \\ 0 & 0 & 0.433 & -0.750 & -0.433 & 0.750 \end{bmatrix}$$

With this matrix, the structural response is thus described by

$$\{F\} = [K]\{U\}$$

where $[K]$ is as above and the system forces $\{F\}$ and displacements $\{U\}$ are given by

$$\{F\} = \begin{Bmatrix} F_{1x} \\ F_{1y} \\ F_{2x} \\ F_{2y} \\ F_{3x} \\ F_{3y} \end{Bmatrix}, \qquad \{U\} = \begin{Bmatrix} U_1 \\ V_1 \\ U_2 \\ V_2 \\ U_3 \\ V_3 \end{Bmatrix}$$

To solve this equation for the displacements when forces are specified, we need to consider boundary conditions as in all previous discussions. In the present problem (Fig. 2.8) these are expressible as

$$U_1 = V_1 = 0, \qquad U_3 = V_3 = 0, \qquad F_{2x} = 0, \qquad F_{2y} = -5000 \text{ lb}$$

and we may extract the following reduced equation from the general force–deflection system equation:

$$\begin{Bmatrix} F_{2x} \\ F_{2y} \end{Bmatrix} = \frac{EA}{2l}\begin{bmatrix} 2.250 & -0.433 \\ -0.433 & 0.750 \end{bmatrix}\begin{Bmatrix} U_2 \\ V_2 \end{Bmatrix}$$

With F_{2x} and F_{2y} as specified, we may invert this equation to find U_2 and V_2 and then obtain F_{1x}, F_{1y} and F_{3x}, F_{3y} from the remaining part of the general force–deflection matrix equation, that is, from

$$\begin{Bmatrix} F_{1x} \\ F_{1y} \\ F_{3x} \\ F_{3y} \end{Bmatrix} = \frac{EA}{2l} \begin{bmatrix} -2 & 0 \\ 0 & 0 \\ -0.250 & 0.433 \\ 0.433 & -0.750 \end{bmatrix} \begin{Bmatrix} U_2 \\ V_2 \end{Bmatrix}$$

With $E = 10 \times 10^6$ lb/in², $A = 1.13$ in², and $l = 60$ in, we find

$$U_2 = -0.0153 \text{ in}, \qquad V_2 = -0.0796 \text{ in}$$

and

$$F_{1x} = 2885 \text{ lb}, \qquad F_{1y} = 0$$

$$F_{3x} = -2885 \text{ lb}, \qquad F_{3y} = 5000 \text{ lb}$$

Having the displacements, we may determine the internal member forces using equation (39). We find (Fig. 2.9)

$$S_{1-2} = -2885 \text{ lb}, \qquad S_{2-3} = 5775 \text{ lb}$$

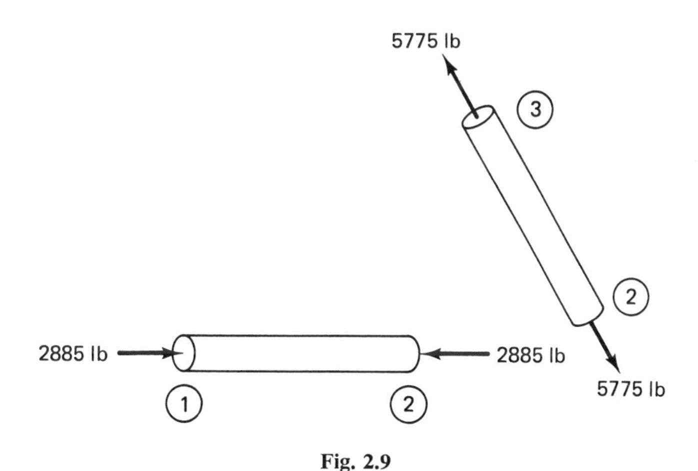

Fig. 2.9

The stress in each member is finally determined by dividing the respective internal force by the member's sectional area. We find

$$\sigma_{1-2} = -2550 \text{ lb/in}^2, \qquad \sigma_{2-3} = 5110 \text{ lb/in}^2 \quad \#$$

EXAMPLE 2.2-2. Show that the equilibrium equations applied to joint 2 of the structure of Example 2.2-1 will yield, with the member stiffness equations, the same reduced equation for determining unknown displacements as found using the direct stiffness method.

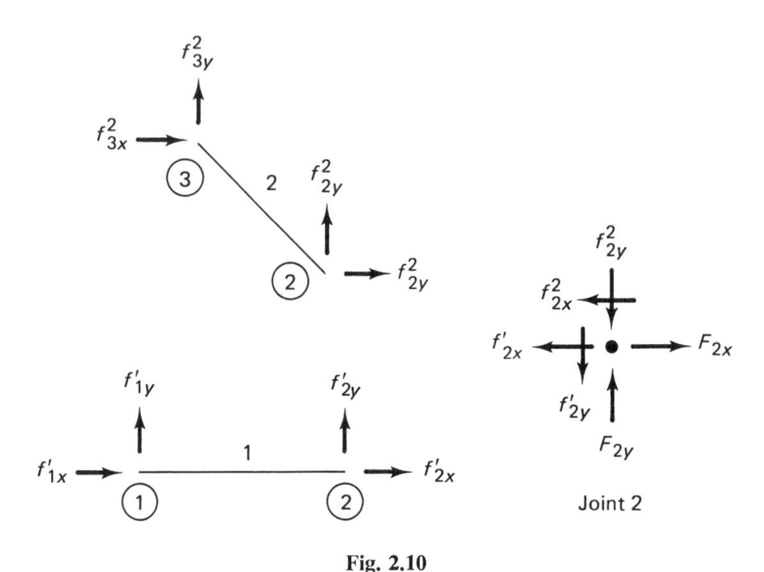

Fig. 2.10

Free-body diagrams of the members and joint 2 are shown in Fig. 2.10. Applying the equilibrium equations

$$\sum F_x = 0, \qquad \sum F_y = 0$$

to the free-body diagram of joint 2 gives

$$F_{2x} = f^1_{2x} + f^2_{2x}$$
$$F_{2y} = f^1_{2y} + f^2_{2y}$$

where superscripts 1 and 2 denote, respectively, bars, 1–2 and 2–3. From the stiffness equations of the individual members, we also have, with $U_1 = V_1 = U_3 = V_3 = 0$,

$$f^1_{2x} = 0, \qquad f^1_{2y} = \frac{EA}{l} U_2$$

$$f^2_{2x} = \frac{EA}{2l}(0.250U_2 - 0.433V_2)$$

$$f^2_{2y} = \frac{EA}{2l}(-0.433U_2 + 0.750V_2)$$

Combining these with the results from the equilibrium analysis, we have, in matrix notation

$$\begin{Bmatrix} F_{2x} \\ F_{2y} \end{Bmatrix} = \frac{EA}{2l} \begin{bmatrix} 2.250 & -0.433 \\ -0.433 & 0.750 \end{bmatrix} \begin{Bmatrix} U_2 \\ V_2 \end{Bmatrix}$$

which is the same reduced equation for determining the unknown displacements as found in Example 2.2-1 using the direct stiffness method. By applying similar procedures to joints 1 and 3, the corresponding reduced equation for determining the unknown reaction forces can be determined. #

EXAMPLE 2.2-3. The deck of an offshore platform together with equipment weighs 375 kips (375,000 lb). It is supported by four corner piles and two side trusses, as indicated in the two-dimensional sketch of Fig. 2.11. The truss members are steel ($E = 30 \times 10^6$ lb/in²) with sectional areas $A = 1.5$ in². As an approximation, each support point (five on each side face) may be assumed to take $W/10 = 37.5$ kips of the load. Under this condition, we wish to determine the deflections, reactions, and internal-member forces of the truss 1–2–3–4 assuming zero displacements at coordinates 1 and 3.

Fig. 2.11

To solve this problem, we proceed, as in the preceding example, and first construct the matrix stiffness equation for the structure using the direct stiffness method. Choosing x and y axes with x horizontal and positive to the right and y vertical and positive upward, we find

$$[K] = \frac{EA}{l_1}\begin{bmatrix} 1.476 & 0.381 & -1 & 0 & 0 & 0 & -0.476 & -0.381 \\ 0.381 & 0.305 & 0 & 0 & 0 & 0 & -0.381 & -0.305 \\ -1 & 0 & 2 & 0 & -1 & 0 & 0 & 0 \\ 0 & 0 & 0 & 1.250 & 0 & 0 & 0 & -1.250 \\ 0 & 0 & -1 & 0 & 1.476 & -0.381 & -0.476 & 0.381 \\ 0 & 0 & 0 & 0 & -0.381 & -0.305 & 0.381 & -0.305 \\ -0.476 & -0.381 & 0 & 0 & -0.476 & 0.381 & 0.952 & 0 \\ -0.381 & -0.305 & 0 & -1.250 & 0.381 & -0.305 & 0 & 1.860 \end{bmatrix}$$

where $l_1 = 25$ ft. This matrix connects the forces and displacements at the indicated coordinates according to the equation

$$\{F\} = [K]\{U\}$$

where

$$\{F\} = \begin{Bmatrix} F_{1x} \\ F_{1y} \\ F_{2x} \\ \cdot \\ \cdot \\ \cdot \\ F_{4y} \end{Bmatrix}, \qquad \{U\} = \begin{Bmatrix} U_1 \\ V_1 \\ U_2 \\ \cdot \\ \cdot \\ \cdot \\ V_4 \end{Bmatrix}$$

The boundary conditions are expressible as

$$U_1 = V_1 = U_3 = V_3 = 0$$
$$F_{2x} = F_{2y} = F_{4x} = 0$$
$$F_{4y} = -37.5 \text{ kips}$$

Using these, we obtain the following reduced equations:

$$\begin{Bmatrix} F_{1x} \\ F_{1y} \\ F_{3x} \\ F_{3y} \end{Bmatrix} = \frac{EA}{l_1}\begin{bmatrix} -1 & 0 & -0.476 & -0.381 \\ 0 & 0 & -0.381 & -0.305 \\ -1 & 0 & -0.476 & 0.381 \\ 0 & 0 & 0.381 & -0.305 \end{bmatrix}\begin{Bmatrix} U_2 \\ V_2 \\ U_4 \\ V_4 \end{Bmatrix}$$

and

$$\begin{Bmatrix} F_{2x} \\ F_{2y} \\ F_{4x} \\ F_{4y} \end{Bmatrix} = \frac{EA}{l_1}\begin{bmatrix} 2 & 0 & 0 & 0 \\ 0 & 1.250 & 0 & -1.250 \\ 0 & 0 & 0.952 & 0 \\ 0 & -1.250 & 0 & 1.860 \end{bmatrix}\begin{Bmatrix} U_2 \\ V_2 \\ U_4 \\ V_4 \end{Bmatrix}$$

Inverting this last equation and substituting for the applied forces, we find

$$U_2 = U_4 = 0$$
$$V_2 = V_4 = -0.0342 \text{ ft}$$

The reaction forces may next be determined from the first of the reduced equations above as

$$F_{1x} = -F_{3x} = 23.4 \text{ kips}$$
$$F_{1y} = F_{3y} = 18.7 \text{ kips}$$

Finally, the internal forces in each member may be determined from equation (39) as

$$S_{1-2} = S_{2-3} = S_{2-4} = 0$$
$$S_{1-4} = S_{3-4} = -30.0 \text{ kips}$$

The stress σ_{m-n} in member *m–n* is obtained from the corresponding internal force by dividing by the load-carrying sectional area of the member. We have

$$\sigma_{1-2} = \sigma_{2-3} = \sigma_{2-4} = 0$$
$$\sigma_{1-4} = \sigma_{3-4} = -20{,}000 \text{ lb/in}^2$$

If the material is ordinary construction steel having a yield stress of, say, 36,000 lb/in^2, we thus see that the truss as designed has a *factor of safety* of $36/20 = 1.8$ with respect to yield failure. #

In this example, as well as the preceding one, the weights of the individual members were neglected. If we wish to include them in the analysis, we may calculate the weight of each member and place half this weight at each end joint as an applied downward force. This will allow inclusion of the weight of the structure in the axial loads carried by the members in a satisfactory manner. It will not, however, account for the (generally small) bending stresses induced by the actual distributed weight loading on a member.

2.3 SPACE TRUSSES

The discussion above was limited to two-dimensional nonalignments of structural members subjected to tension or compression loading and gave rise to plane-truss problems. To extend the work to three dimensions for space-truss problems, we consider the member 1–2 shown in Fig. 2.12.

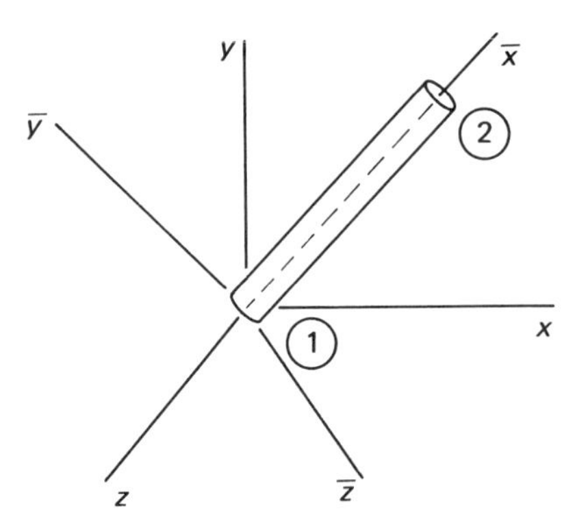

Fig. 2.12. Inclined space-truss member.

The local or member axes are denoted by $\bar{x}$, $\bar{y}$, $\bar{z}$ and the reference or system axes by x, y, z. In addition, we adopt the following notation for the angles between the two sets of axes:

$$\alpha_{x\bar{x}} = \text{angle between } x, \bar{x}$$
$$\alpha_{y\bar{x}} = \text{angle between } y, \bar{x}$$
$$\vdots$$
$$\alpha_{z\bar{z}} = \text{angle between } z, \bar{z}$$

From geometry, the equations expressing the force components at positions 1 and 2 of the member relative to system and member axes are easily seen to be similar to the earlier two-dimensional equations and expressible in matrix notation as

$$\begin{aligned} \{\bar{f}\} &= [R]\{f\} \\ \{f\} &= [R]^T\{\bar{f}\} \end{aligned} \tag{41}$$

where the column matrices $\{\bar{f}\}$ and $\{f\}$ are given by

$$\{\bar{f}\} = \begin{Bmatrix} \bar{f}_{1x} \\ \bar{f}_{1y} \\ \bar{f}_{1z} \\ \bar{f}_{2x} \\ \bar{f}_{2y} \\ \bar{f}_{2z} \end{Bmatrix}, \qquad \{f\} = \begin{Bmatrix} f_{1x} \\ f_{1y} \\ f_{1z} \\ f_{2x} \\ f_{2y} \\ f_{2z} \end{Bmatrix} \tag{42}$$

and where the matrix [R] is given by

$$[R] = \begin{bmatrix} \lambda_{\bar{x}} & \mu_{\bar{x}} & \nu_{\bar{x}} & 0 & 0 & 0 \\ \lambda_{\bar{y}} & \mu_{\bar{y}} & \nu_{\bar{y}} & 0 & 0 & 0 \\ \lambda_{\bar{z}} & \mu_{\bar{z}} & \nu_{\bar{z}} & 0 & 0 & 0 \\ 0 & 0 & 0 & \lambda_{\bar{x}} & \mu_{\bar{x}} & \nu_{\bar{x}} \\ 0 & 0 & 0 & \lambda_{\bar{y}} & \mu_{\bar{y}} & \nu_{\bar{y}} \\ 0 & 0 & 0 & \lambda_{\bar{z}} & \mu_{\bar{z}} & \nu_{\bar{z}} \end{bmatrix} \tag{43}$$

with

$$\begin{aligned} \lambda_{\bar{x}} &= \cos(\alpha_{x\bar{x}}) \\ \mu_{\bar{x}} &= \cos(\alpha_{y\bar{x}}) \\ \nu_{\bar{x}} &= \cos(\alpha_{z\bar{x}}) \\ &\cdot \\ &\cdot \\ &\cdot \\ \nu_{\bar{z}} &= \cos(\alpha_{z\bar{z}}) \end{aligned}$$

Similarly, for the displacements, we have

$$\begin{aligned} \{\bar{U}\} &= [R]\{U\} \\ \{U\} &= [R]^T\{\bar{U}\} \end{aligned} \tag{44}$$

where

$$\{\bar{U}\} = \begin{Bmatrix} \bar{U}_1 \\ \bar{V}_1 \\ \bar{W}_1 \\ \bar{U}_2 \\ \bar{V}_2 \\ \bar{W}_2 \end{Bmatrix}, \qquad \{U\} = \begin{Bmatrix} U_1 \\ V_1 \\ W_1 \\ U_2 \\ V_2 \\ W_2 \end{Bmatrix} \tag{45}$$

As in the earlier two-dimensional treatment, we have the force–deflection equation for the member expressible as

$$\{\bar{f}\} = [\bar{K}]\{\bar{U}\} \tag{46}$$

Using equations (41) and (44), this may be written as

$$[R]\{f\} = [\bar{K}][R]\{U\} \tag{47}$$

On premultiplying both sides of the expression by $[R]^T$ and noticing that

$$[R]^T[R] = \begin{bmatrix} 1 & 0 & 0 & 0 & 0 & 0 \\ 0 & 1 & 0 & 0 & 0 & 0 \\ 0 & 0 & 1 & 0 & 0 & 0 \\ 0 & 0 & 0 & 1 & 0 & 0 \\ 0 & 0 & 0 & 0 & 1 & 0 \\ 0 & 0 & 0 & 0 & 0 & 1 \end{bmatrix} = [I] \tag{48}$$

this equation becomes

$$\{f\} = [R]^T[\bar{K}][R]\{U\} \tag{49}$$

so that by comparison with $\{f\} = [K]\{U\}$ we have the stiffness of the member given in terms of system axes by

$$[K] = [R]^T[\bar{K}][R] \tag{50}$$

Taking $[\bar{K}]$ as

$$[\bar{K}] = \frac{EA}{l}\begin{bmatrix} 1 & 0 & 0 & -1 & 0 & 0 \\ 0 & 0 & 0 & 0 & 0 & 0 \\ 0 & 0 & 0 & 0 & 0 & 0 \\ -1 & 0 & 0 & 1 & 0 & 0 \\ 0 & 0 & 0 & 0 & 0 & 0 \\ 0 & 0 & 0 & 0 & 0 & 0 \end{bmatrix} \tag{51}$$

and using equations (43) and (50), we find

$$[K] = \frac{EA}{l}\begin{bmatrix} \lambda^2 & & & & & \\ \lambda\mu & \mu^2 & & & \text{symmetric} & \\ \lambda\nu & \mu\nu & \nu^2 & & & \\ -\lambda^2 & -\lambda\mu & -\lambda\nu & \lambda^2 & & \\ -\lambda\mu & -\mu^2 & -\mu\nu & \lambda\mu & \mu^2 & \\ -\lambda\nu & -\mu\nu & -\nu^2 & \lambda\nu & \mu\nu & \nu^2 \end{bmatrix} \tag{52}$$

where λ, μ, and ν are written for $\lambda_{\bar{x}}$, $\mu_{\bar{x}}$, $\nu_{\bar{x}}$ and where "symmetric" means that elements k_{ij} are equal to elements k_{ji}.

In solving space-truss problems with the help of this expression, we proceed exactly as in the two-dimensional case and write the stiffness matrix for each member relative to system axes, adding zero rows and columns to indicate non-dependence on system coordinates not involved. We then add these matrices using the direct stiffness method and employ appropriate boundary conditions to solve for displacements and unknown reaction forces. Finally, internal forces may be established using equations of the form

$$S_{m-n} = \bar{f}_{nx} = \lambda f_{2x} + \mu f_{2y} + \nu f_{2z} \tag{53}$$

where f_{2x}, f_{2y}, and f_{2z} are determined from equation (49) when applied to member m–n. The corresponding tensile or compressive stress is then determined by dividing S_{m-n} by the sectional area of the member as in the case of plane-truss members.

2.4 PLANE FRAMES

Section 2.3 was concerned with members subjected to axial loadings only. In certain structures, called *frames*, the members are also subjected to transverse forces and bending moments. To analyze these structures, we thus need to extend our basic matrix formulation for trusses to include the additional loadings.

In contrast with trusses, where members are assumed free to rotate at connecting joints, frames are generally assumed to have *rigid joints*, that is, joints which, on loading, preserve the relative angular spacing of the attached members at the joints. An example is the right-angle joint illustrated in Fig. 2.13. On loading, the 90° angular spacing of the members at the joint is maintained even though the joint rotates. With rigid joints, all members connected at the joint will therefore suffer the same rotation at the joint as the joint itself.

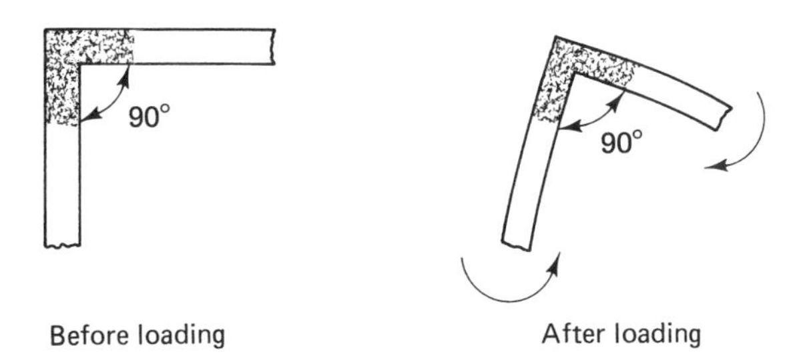

Fig. 2.13. Illustration of rigid joint: (a) before loading; (b) after loading.

In the present section we consider the matrix formulation for the analysis of *plane frames*, that is, frames having all their members lying in a single plane. Joints connecting members will be assumed rigid in the sense described above.

Member Stiffness Matrix

Consider the member 1–2, shown in Fig. 2.14, having moments and vertical forces m_1 and f_{1y} at position 1 and m_2 and f_{2y} at position 2. The vertical forces are assumed positive when acting in the positive y-direction and the moments

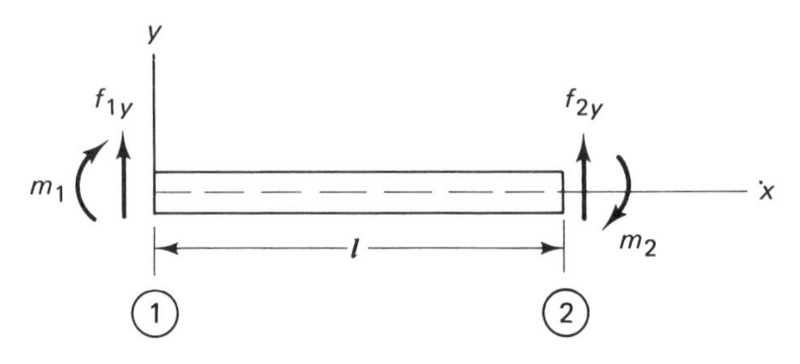

Fig. 2.14. Structural member subjected to bending moments and shear forces.

are assumed positive when acting clockwise. Restricting attention to the case where the cross section of the member is symmetric about a vertical axis through its centroid, the vertical deflection $v(x)$ of the centroidal x-axis of the member is known from solid mechanics to be governed by the equation

$$EI\frac{d^2v}{dx^2} = M(x) \tag{54}$$

where E dènotes Young's modulus, I denotes the moment of inertia of the cross section of the member about its centroid, and $M(x)$ denotes the internal moment, assumed positive counterclockwise when acting on a right-hand face as illustrated in Fig. 2.15. From statics, we have

$$M = m_1 + xf_{1y} \tag{55}$$

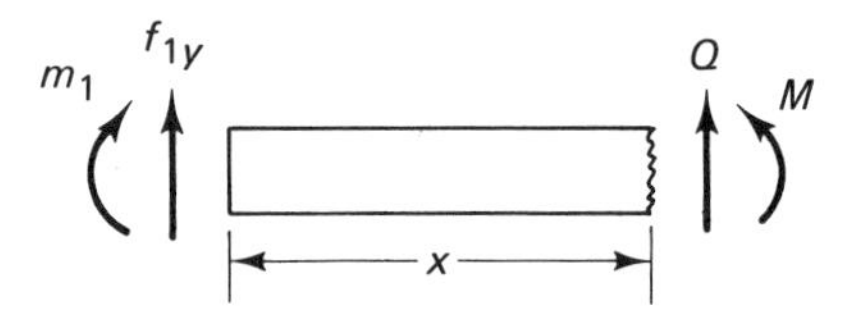

Fig. 2.15

Combining the two equations above and intergrating, we find (for constant EI) that

$$EI\frac{dv}{dx} = m_1x + f_{1y}\frac{x^2}{2} - EI\theta_1 \tag{56}$$

and

$$EIv = \frac{m_1x^2}{2} + f_{1y}\frac{x^3}{6} - EI\theta_1x + EIV_1 \tag{57}$$

where the boundary conditions

$$v = V_1, \qquad \frac{dv}{dx} = -\theta_1 \tag{58}$$

have been used at $x = 0$, that is, at position 1. Note that θ_1 is positive in the same sense that m_1 is.

On substituting the conditions

$$v = V_2, \qquad \frac{dv}{dx} = -\theta_2 \tag{59}$$

at place 2 ($x = l$), we also have from equations (56) and (57) that

$$\begin{aligned} EI\theta_2 &= -m_1 l - f_{1y}\frac{l^2}{2} + EI\theta_1 \\ EIV_2 &= \frac{m_1 l^2}{2} + f_{1y}\frac{l^3}{6} - EI\theta_1 l + EIV_1 \end{aligned} \tag{60}$$

These equations may be solved for m_1 and f_{1y} to obtain

$$\begin{aligned} m_1 &= \frac{6EI}{l^2}(V_2 - V_1) + \frac{2EI}{l}(2\theta_1 + \theta_2) \\ f_{1y} &= -\frac{12EI}{l^3}(V_2 - V_1) - \frac{6EI}{l^2}(\theta_1 + \theta_2) \end{aligned} \tag{61}$$

In addition, from equilibrium of the entire member (Fig. 2.14), we have

$$\begin{aligned} m_2 &= -m_1 - lf_{1y} \\ f_{2y} &= -f_{1y} \end{aligned} \tag{62}$$

so that

$$\begin{aligned} m_2 &= \frac{6EI}{l^2}(V_2 - V_1) + \frac{2EI}{l}(\theta_1 + 2\theta_2) \\ f_{2y} &= \frac{12EI}{l^3}(V_2 - V_1) + \frac{6EI}{l^2}(\theta_1 + \theta_2) \end{aligned} \tag{63}$$

Hence, the matrix stiffness equation for the member may be written as

$$\begin{Bmatrix} f_{1y} \\ m_1 \\ f_{2y} \\ m_2 \end{Bmatrix} = \frac{EI}{l^4}\begin{bmatrix} 12l & -6l^2 & -12l & -6l^2 \\ -6l^2 & 4l^3 & 6l^2 & 2l^3 \\ -12l & 6l^2 & 12l & 6l^2 \\ -6l^2 & 2l^3 & 6l^2 & 4l^3 \end{bmatrix}\begin{Bmatrix} V_1 \\ \theta_1 \\ V_2 \\ \theta_2 \end{Bmatrix} \tag{64}$$

In addition to the transverse forces and moments acting at each end of the member, we may also consider axial forces. From Section 2.1 we know the form of the stiffness equation for these forces, so that we may immediately expand equation (64) to include them, i.e.,

$$\begin{Bmatrix} f_{1x} \\ f_{1y} \\ m_1 \\ f_{2x} \\ f_{2y} \\ m_2 \end{Bmatrix} = \begin{bmatrix} \frac{EA}{l} & & & & & \\ 0 & \frac{12EI}{l^3} & & \text{symmetric} & & \\ 0 & -\frac{6EI}{l^2} & \frac{4EI}{l} & & & \\ -\frac{EA}{l} & 0 & 0 & \frac{EA}{l} & & \\ 0 & -\frac{12EI}{l^3} & \frac{6EI}{l^2} & 0 & \frac{12EI}{l^3} & \\ 0 & -\frac{6EI}{l^2} & \frac{2EI}{l} & 0 & \frac{6EI}{l^2} & \frac{4EI}{l} \end{bmatrix} \begin{Bmatrix} U_1 \\ V_1 \\ \theta_1 \\ U_2 \\ V_2 \\ \theta_2 \end{Bmatrix} \tag{65}$$

Arbitrarily Oriented Member

As in the case of uniaxially loaded members, we may consider the stiffness matrix of a member inclined with respect to system axes. We take $\bar{x}$, $\bar{y}$ to denote member axes and x, y to denote system axes, as shown in Fig. 2.16. The trans-

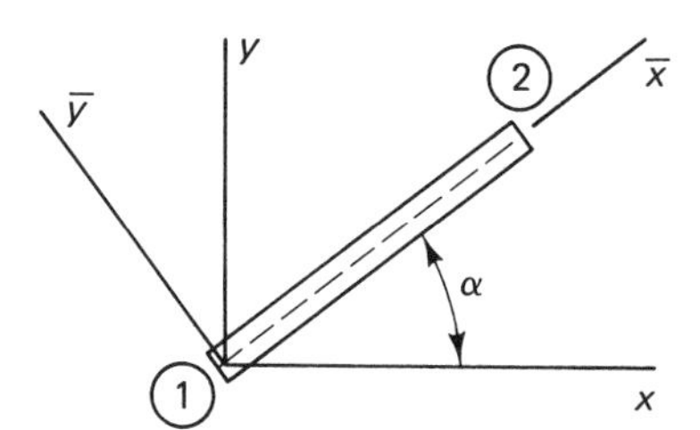

Fig. 2.16. Inclined plane-frame structural member.

formation between member and system components for this problem can be established from our previous considerations for axially loaded members. Remembering that the moments m_1 and m_2 can be regarded as vectors normal to the x-y plane, we see that these, as well as the corresponding angles θ_1 and θ_2, will be unaffected by the transformation. Hence, the transformation equations relating member-force components with system-force components may be written as

$$\begin{aligned} \{\bar{f}\} &= [R]\{f\} \\ \{f\} &= [R]^T\{\bar{f}\} \end{aligned} \tag{66}$$

where $\{\bar{f}\}$ and $\{f\}$ are given by

$$\{\bar{f}\} = \begin{Bmatrix} \bar{f}_{1x} \\ \bar{f}_{1y} \\ \bar{m}_1 \\ \bar{f}_{2x} \\ \bar{f}_{2y} \\ \bar{m}_2 \end{Bmatrix}, \qquad \{f\} = \begin{Bmatrix} f_{1x} \\ f_{1y} \\ m_1 \\ f_{2x} \\ f_{2y} \\ m_2 \end{Bmatrix}$$

and the transformation matrix $[R]$ is given from our earlier work as

$$[R] = \begin{bmatrix} \lambda & \mu & 0 & 0 & 0 & 0 \\ -\mu & \lambda & 0 & 0 & 0 & 0 \\ 0 & 0 & 1 & 0 & 0 & 0 \\ 0 & 0 & 0 & \lambda & \mu & 0 \\ 0 & 0 & 0 & -\mu & \lambda & 0 \\ 0 & 0 & 0 & 0 & 0 & 1 \end{bmatrix} \tag{67}$$

with $\lambda = \cos\alpha$, $\mu = \sin\alpha$, as before.

Similarly, we have for the displacement components

$$\begin{aligned} \{\bar{U}\} &= [R]\{U\} \\ \{U\} &= [R]^T\{\bar{U}\} \end{aligned} \tag{68}$$

where $\{\bar{U}\}$ and $\{U\}$ are given by

$$\{\bar{U}\} = \begin{Bmatrix} \bar{U}_1 \\ \bar{V}_1 \\ \bar{\theta}_1 \\ \bar{U}_2 \\ \bar{V}_2 \\ \bar{\theta}_2 \end{Bmatrix}, \qquad \{U\} = \begin{Bmatrix} U_1 \\ V_1 \\ \theta_1 \\ U_2 \\ V_2 \\ \theta_2 \end{Bmatrix}$$

Using the relations above with the member stiffness equation

$$\{\bar{f}\} = [\bar{K}]\{\bar{U}\} \tag{69}$$

we may easily obtain, as before, the relation

$$[K] = [R]^T[\bar{K}][R] \tag{70}$$

relating the member stiffness matrix to system axes. The member stiffness matrix $[\bar{K}]$ is given by equation (65), so that we find, after matrix multiplication, the following result given in equation (71).

$$[K] = \frac{E}{l}\begin{bmatrix} A\lambda^2 + \frac{12I}{l^2}\mu^2 & & & & & \text{Symmetric} \\ \left(A - \frac{12I}{l^2}\right)\lambda\mu & A\mu^2 + \frac{12I}{l^2}\lambda^2 & & & & \\ \frac{6I}{l}\mu & -\frac{6I}{l}\lambda & 4I & & & \\ -\left(A\lambda^2 + \frac{12I}{l^2}\mu^2\right) & -\left(A - \frac{12I}{l^2}\right)\lambda\mu & -\frac{6I}{l}\mu & A\lambda^2 + \frac{12I}{l^2}\mu^2 & & \\ -\left(A - \frac{12I}{l^2}\right)\lambda\mu & -\left(A\mu^2 + \frac{12I}{l^2}\lambda^2\right) & \frac{6I}{l}\lambda & \left(A - \frac{12I}{l^2}\right)\lambda\mu & A\mu^2 + \frac{12I}{l^2}\lambda^2 & \\ \frac{6I}{l}\mu & -\frac{6I}{l}\lambda & 2I & -\frac{6I}{l}\mu & \frac{6I}{l}\lambda & 4I \end{bmatrix} \tag{71}$$

The expression above gives the stiffness matrix of an inclined member in terms of system axes. We use this result in the same way that we used the analogous one for axial-loaded inclined members. For example, if we are considering a frame made up of several inclined members, we first transform the stiffness matrix of each of these to common system axes using the result above and then add these using the direct stiffness method (which ensures balance of internal forces and moments with external forces and moments) to obtain the structural stiffness matrix. Next, we assign boundary conditions and solve for unknown displacements in terms of known forces using the resulting reduced stiffness equations. We may then use these displacements with the remaining part of the structural stiffness equations to find the unknown reaction forces.

It should be emphasized that the sign convention for the applied forces F_x, F_y and moment M at a joint is the same as that assumed for the internal end forces and moment shown in Fig. 2.14; namely, applied forces are positive when acting in the positive coordinate directions and applied moments are positive when acting clockwise. Similarly, the joint displacements U, V are positive in the positive coordinate directions and the joint rotation θ is positive when clockwise.

The boundary conditions to be employed with frame problems are similar to those described earlier for truss problems. At each coordinate location we must specify three boundary conditions, F_x or U, F_y or V, and M or θ. In order to solve for any displacements of the structure, we must also specify a sufficient number of displacement components (U, V, θ) so as to eliminate the possibility of rigid-body motion of the structure. Otherwise, no unique solution for unknown displacements can be obtained.

Once the joint displacements are determined, these may be used with the individual member stiffness matrices (referred to system axes) to find the internal forces and bending moments acting at the ends of the members. Having these, the stresses in the individual members may then be calculated using formulas from solid mechanics.

In general, these stresses will consist, at any point P in a member, of a normal stress σ and a shear stress τ, as indicated in Fig. 2.17. The normal stress consists, in turn, of an axial stress σ_a and a bending stress σ_b such that

$$\sigma = \sigma_a + \sigma_b \tag{72}$$

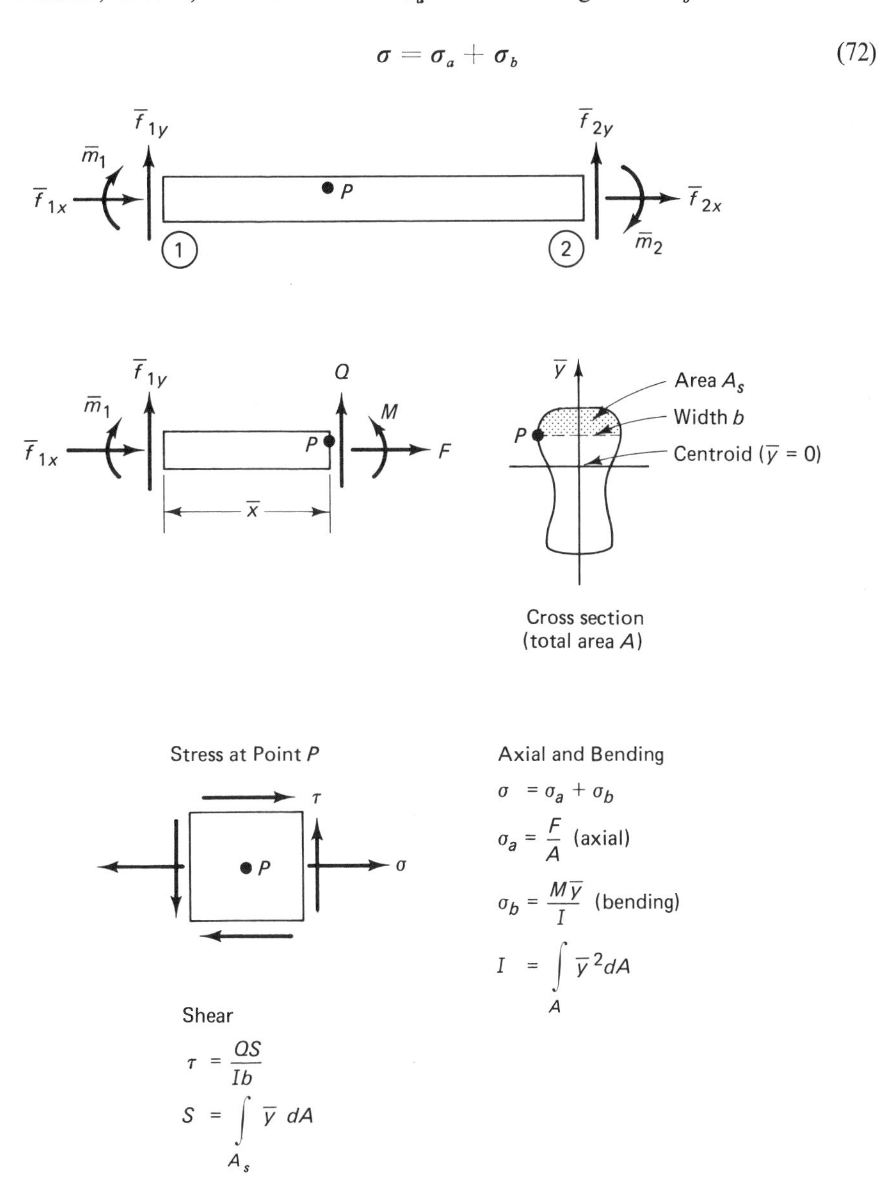

Fig. 2.17. Stresses in member subjected to axial, bending, and shear loading.

with σ_a and σ_b expressible in terms of the axial force F and moment M as

$$\sigma_a = \frac{F}{A}, \qquad \sigma_b = \frac{-M\bar{y}}{I} \tag{73}$$

where, as in the stiffness equation for the member, A and I denote, respectively, the cross-sectional area and moment of inertia of this area about its centroid, and $\bar{y}$ denotes vertical distance from the centroid of the section. The shear stress is similarly expressible in terms of the shear force Q by the equation

$$\tau = \frac{QS}{Ib} \tag{74}$$

where, as indicated in Fig. 2.17, S is the first moment of the area A_s above the plane on which the shear stress is calculated, I is as defined above, and b denotes the width of material on the shear plane.

EXAMPLE 2.4-1. The simple frame shown in Fig. 2.18 consists of members built in at locations 1 and 3 and connected at 2 with a rigid joint. The members are aluminum ($E = 10 \times 10^6$ lb/in^2) hollow tubes having 3 in outside diameter and $\frac{1}{8}$ in wall thickness. We wish to determine the deflections at 2 and the maximum stress existing within each member.

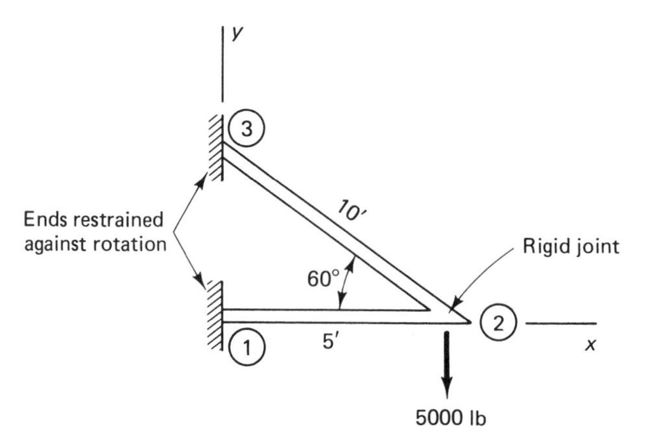

Fig. 2.18

Note that this problem is identical to the truss problem considered in Example 2.2-1 except that the ends 1 and 3 are now restrained against rotation and the joint 2 is rigid, rather than pinned.

To solve this problem, we first construct the stiffness matrix of the entire frame by addition of the individual member stiffness matrices. For both members, the sectional load-carry area A and the moment of inertia I are given by

$$A = \frac{\pi}{4}(D_o^2 - D_i^2) = 1.129 \text{ in}^2$$

$$I = \frac{\pi}{64}(D_o^4 - D_i^4) = 1.169 \text{ in}^4$$

where $D_o = 3$ in and $D_i = 2.75$ in denote outside and inside diameters of the members. Noting the length $l = l_1 = 60$ in of member 1–2, we may thus use equation (71) to obtain the stiffness matrix of this member ($\lambda = 1$, $\mu = 0$) as

$$[K]_{1-2} = k\begin{bmatrix} 1 & 0 & 0 & -1 & 0 & 0 \\ 0 & 0.003 & -0.104 & 0 & -0.003 & -0.104 \\ 0 & -0.104 & 4.141 & 0 & 0.104 & 2.070 \\ -1 & 0 & 0 & 1 & 0 & 0 \\ 0 & -0.003 & 0.104 & 0 & 0.003 & 0.104 \\ 0 & -0.104 & 2.070 & 0 & 0.104 & 4.141 \end{bmatrix}$$

where $k = EA/l_1 = 1.882 \times 10^5$ lb/in.

In a similar way, we may also write for member 2–3 ($\lambda = -0.500$, $\mu = 0.8660$) with $l = l_2 = 120$ in,

$$[K]_{2-3} = k\begin{bmatrix} 0.125 & -0.216 & 0.022 & -0.125 & 0.216 & 0.022 \\ -0.216 & 0.375 & 0.013 & 0.216 & -0.375 & 0.013 \\ 0.022 & 0.013 & 2.070 & -0.022 & -0.013 & 1.035 \\ -0.125 & 0.216 & -0.022 & 0.125 & -0.216 & -0.022 \\ 0.216 & -0.375 & -0.013 & -0.216 & 0.375 & -0.013 \\ 0.022 & 0.013 & 1.035 & -0.022 & -0.013 & 2.070 \end{bmatrix}$$

where the scalar k, defined above, has been factored out for convenience in combining this matrix with $[K]_{1-2}$.

Using the direct stiffness method, we now expand matrix $[K]_{1-2}$ by including seventh, eighth, and ninth rows and columns of zeros to show nondependence on the forces and displacements at coordinate 3. Similarly, we expand matrix $[K]_{2-3}$ by including an additional first three rows and columns of zeros to indicate nondependence on the forces and displacements at coordinate 1. Adding these expanded matrices together, we then obtain the stiffness matrix for the structure in the form

$$[K] = k\begin{bmatrix} 1 & 0 & 0 & -1 & 0 & 0 & 0 & 0 & 0 \\ 0 & 0.003 & -0.104 & 0 & -0.003 & -0.104 & 0 & 0 & 0 \\ 0 & -0.104 & 4.141 & 0 & 0.104 & 2.070 & 0 & 0 & 0 \\ -1 & 0 & 0 & 1.125 & -0.216 & 0.022 & -0.125 & 0.216 & 0.022 \\ 0 & -0.003 & 0.104 & -0.216 & 0.378 & 0.117 & 0.216 & -0.375 & 0.013 \\ 0 & -0.104 & 2.070 & 0.022 & 0.117 & 6.211 & -0.022 & -0.013 & 1.035 \\ 0 & 0 & 0 & -0.125 & 0.216 & -0.022 & 0.125 & -0.216 & -0.022 \\ 0 & 0 & 0 & 0.216 & -0.375 & -0.013 & -0.216 & 0.375 & -0.013 \\ 0 & 0 & 0 & 0.022 & 0.013 & 1.035 & -0.022 & -0.013 & 2.070 \end{bmatrix}$$

The boundary conditions are expressible (Fig. 2.18) as

$$U_1 = V_1 = \theta_1 = 0$$
$$U_3 = V_3 = \theta_3 = 0$$
$$F_{2x} = M_2 = 0, \qquad F_{2y} = -5000 \text{ lb}$$

Using these with the structural stiffness equation

$$\{F\} = [K]\{U\},$$

where $[K]$ is as described above, we may extract the following reduced matrix equations:

$$\begin{Bmatrix} F_{1x} \\ F_{1y} \\ M_1 \\ F_{3x} \\ F_{3y} \\ M_3 \end{Bmatrix} = k \begin{bmatrix} -1 & 0 & 0 \\ 0 & -0.003 & -0.104 \\ 0 & 0.104 & 2.070 \\ -0.125 & 0.216 & -0.022 \\ 0.216 & -0.375 & -0.013 \\ 0.022 & 0.013 & 1.035 \end{bmatrix} \begin{Bmatrix} U_2 \\ V_2 \\ \theta_2 \end{Bmatrix}$$

and

$$\begin{Bmatrix} F_{2x} \\ F_{2y} \\ M_2 \end{Bmatrix} = k \begin{bmatrix} 1.125 & -0.216 & 0.022 \\ -0.216 & 0.378 & 0.117 \\ 0.022 & 0.117 & 6.211 \end{bmatrix} \begin{Bmatrix} U_2 \\ V_2 \\ \theta_2 \end{Bmatrix}$$

Solving this last equation for U_2, V_2, θ_2 with $F_{2x} = M_2 = 0$, $F_{2y} = -5000$, we find

$$U_2 = -0.0153 \text{ in}, \qquad V_2 = -0.0795 \text{ in}, \qquad \theta_2 = 0.00154 \text{ rad}$$

The first equation then gives

$$F_{1x} = 2880 \text{ lb}, \qquad F_{1y} = 20 \text{ lb}, \qquad M_1 = -956 \text{ in-lb}$$
$$F_{3x} = -2880 \text{ lb}, \qquad F_{3y} = 4980 \text{ lb}, \qquad M_3 = 42 \text{ in-lb}$$

To find the internal forces acting at the ends of member 1–2, we multiply the displacements above by the member stiffness matrix $[K]_{1-2}$. Since $U_1 = V_1 = \theta_1 = 0$, we have the equation

$$\begin{Bmatrix} f_{1x} \\ f_{1y} \\ m_1 \\ f_{2x} \\ f_{2y} \\ m_2 \end{Bmatrix} = \begin{bmatrix} -1 & 0 & 0 \\ 0 & -0.003 & -0.104 \\ 0 & 0.104 & 2.070 \\ 1 & 0 & 0 \\ 0 & 0.003 & 0.104 \\ 0 & 0.104 & 4.141 \end{bmatrix} \begin{Bmatrix} U_2 \\ V_2 \\ \theta_2 \end{Bmatrix}$$

where f_{1x}, f_{1y}, m_1, etc., denote the internal forces and moments. Carrying out the multiplication, we thus find

$$f_{1x} = 2880 \text{ lb}, \qquad f_{1y} = 20 \text{ lb}, \qquad m_1 = -956 \text{ in-lb}$$
$$f_{2x} = -2880 \text{ lb}, \qquad f_{2y} = -20 \text{ lb}, \qquad m_2 = -356 \text{ in-lb}$$

These internal end forces and moments are shown in Fig. 2.19(a) together

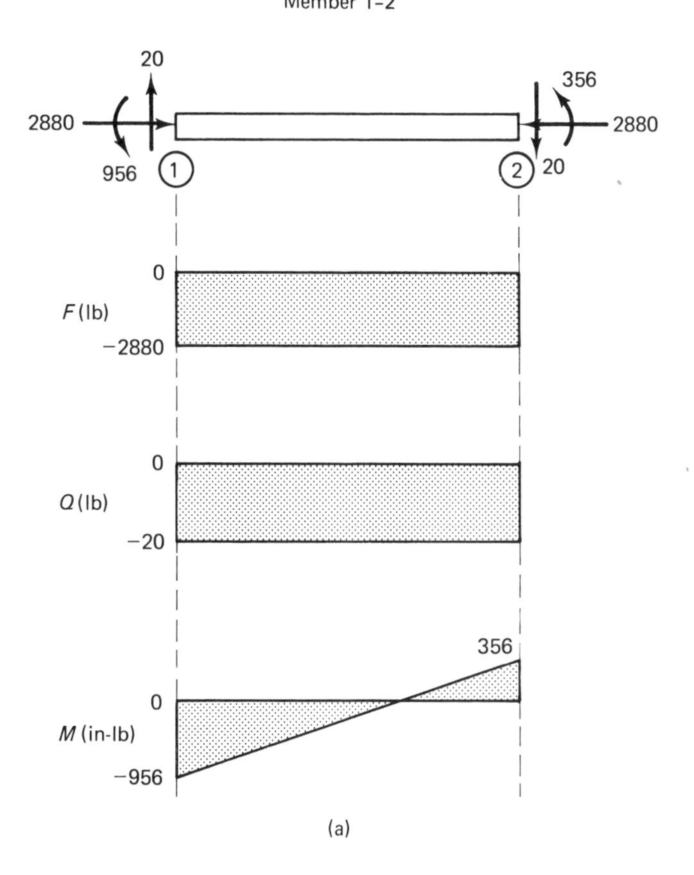

(a)

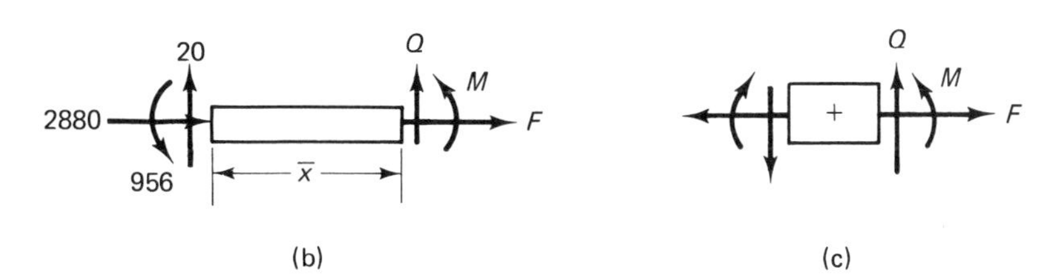

(b) (c)

Fig. 2.19

with the distribution of the axial force F, shear force Q, and moment M along the member, as determined from application of the equilibrium equations to the free-body diagram of Fig. 2.19(b). The sign convention for F, Q, and M are as indicated in Fig. 2.19(b) and (c).

In a similar way, we may multiply the displacements at joint 2 by the stiffness matrix $[K]_{2-3}$ to obtain the internal forces and moments acting at the ends of member 2–3. This operation yields

$$f_{2x} = 2880 \text{ lb}, \qquad f_{2y} = -4980 \text{ lb}, \qquad m_2 = 342 \text{ in-lb}$$
$$f_{3x} = -2880 \text{ lb}, \qquad f_{3y} = 4980 \text{ lb}, \qquad m_3 = 42 \text{ in-lb}$$

These components refer, of course, to the system axes and, in this case, are not the same as the axial and transverse force components acting on the member, since it is inclined [Fig. 2.20(a)]. To obtain these components we may resolve the system axes components using the first of equation (66). We find

$$\bar{f}_{2x} = -5750 \text{ lb}, \qquad \bar{f}_{2y} = -4 \text{ lb}, \qquad \bar{m}_2 = 342 \text{ in-lb}$$
$$\bar{f}_{3x} = 5750 \text{ lb}, \qquad \bar{f}_{3y} = 4 \text{ lb}, \qquad \bar{m}_3 = 42 \text{ in-lb}$$

These end force components are illustrated in Fig. 2.20(b) and the resulting distributions of the axial force, shear force, and moment along the member are shown in Fig. 2.20(c).

It remains to calculate the stresses in the individual members from equations (72) and (74). For the case of a circular tube having inside radius R_i and outside radius R_o, as considered here, the extreme bending stresses will occur at the outside radius where $\bar{y} = \pm R_o$ and at that point along the tube where the bending moment $M = M_m$ is numerically greatest. The axial stress is uniform over the cross section and, in the present problem, is constant along the length. Hence, the extreme normal stresses σ are given from equations (72) and (73) by

$$\sigma = \sigma_a + \sigma_b = \frac{F}{A} \pm \frac{M_m R_o}{I}$$

In contrast, the extreme shear stress will occur at the center plane of the tube where $\bar{y} = 0$ and at that point along the tube where the shear force $Q = Q_m$ is numerically greatest. The value of S in equation (74) is then given by

$$S = \tfrac{2}{3}(R_o^3 - R_i^3)$$

and the extreme shear stress is accordingly determined, with $b = 2(R_o - R_i)$, from equation (74) as

$$\tau = \frac{Q_m(R_o^3 - R_i^3)}{3I(R_o - R_i)}$$

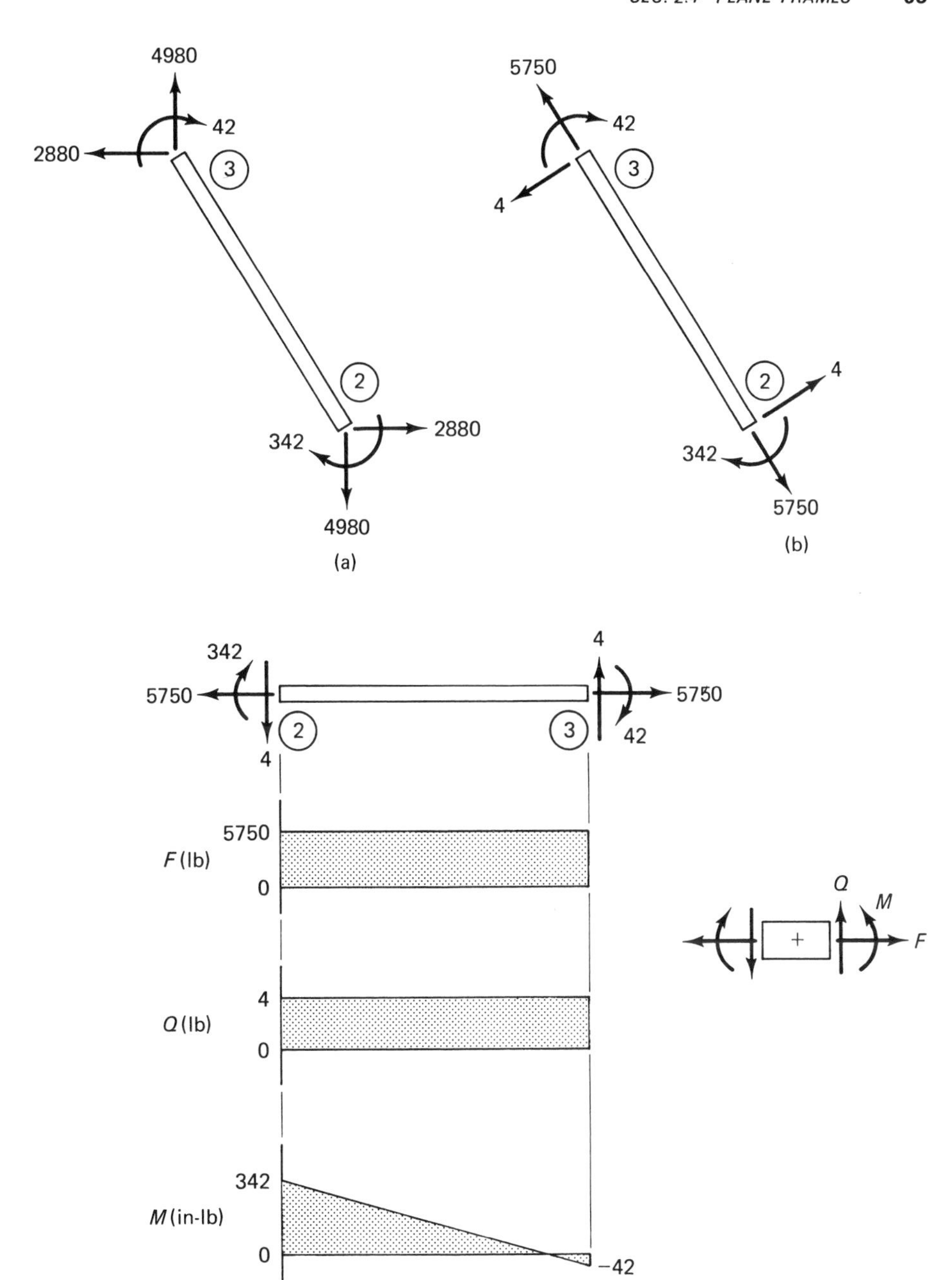
4980
42
2880
3
2
2880
342
4980
(a)
5750
42
3
4
2
4
342
5750
(b)
342
5750
2
4
4
5750
3
42
5750
F (lb)
0
4
Q (lb)
0
342
M (in-lb)
0
−42
(c)
Q
M
+
F

Fig. 2.20

Applying the formulas above to members 1–2 and 2–3, we find, using results from Figs. 2.19 and 2.20:

Member 1–2:

$$\sigma_a = \frac{-2880}{1.13} = -2550 \text{ lb/in}^2$$

$$\sigma_b = \pm\frac{(-956)(1.5)}{1.17} = \pm 1230 \text{ lb/in}^2$$

$$\sigma = \sigma_a + \sigma_b = -3780 \text{ lb/in}^2,\ -1320 \text{ lb/in}^2$$

$$\tau = \frac{(-20)(3.37 - 2.60)}{(3)(1.17)(0.125)} = -35 \text{ lb/in}^2$$

Member 2–3:

$$\sigma_a = \frac{5750}{1.13} = 5090 \text{ lb/in}^2$$

$$\sigma_b = \pm\frac{(342)(1.5)}{1.17} = \pm 438 \text{ lb/in}^2$$

$$\sigma = \sigma_a + \sigma_b = 5530 \text{ lb/in}^2,\ 4650 \text{ lb/in}^2$$

$$\tau = \frac{(4)(3.37 - 2.60)}{(3)(1.17)(0.125)} = 7 \text{ lb/in}^2$$

The extreme shear stresses in the members are seen to be very small when compared with the bending stresses. This is generally true for members that are long in comparison with their cross-sectional dimensions. For such members, which are typical of those used for structural purposes, the shear stress may therefore usually be neglected altogether and stresses in the member calculated from considerations of axial and bending stresses only.

It is interesting to compare the deflections and stresses calculated here with those found earlier in Example 2.2-1, where the member ends were assumed free to rotate. Examination shows, in fact, that the displacements U_2 and V_2 and the axial stresses in the members are essentially identical. However, the presence of the restraining moments in the present problem causes the total maximum longitudinal stress in member 1–2 to be increased by about 48% and that in member 2–3 by about 8%.

These stress increases would be much reduced if the severe foundational restraints at joints 1 and 3 of the structure were replaced by pinned connections, allowing free rotation. In this case, calculation shows the rigid joint at location 2 would then cause less than a 10% increase in the maximum stress in each member over that given by truss analysis. This result accordingly illustrates a general rule in structural mechanics, namely that for a structure that would act as a truss (members carry only tension or compression loadings) except for the presence of rigid joints connecting the members, truss analysis will provide a good approximate treatment provided that the members are relatively flexible, that is, long in comparison with their cross-sectional dimensions.

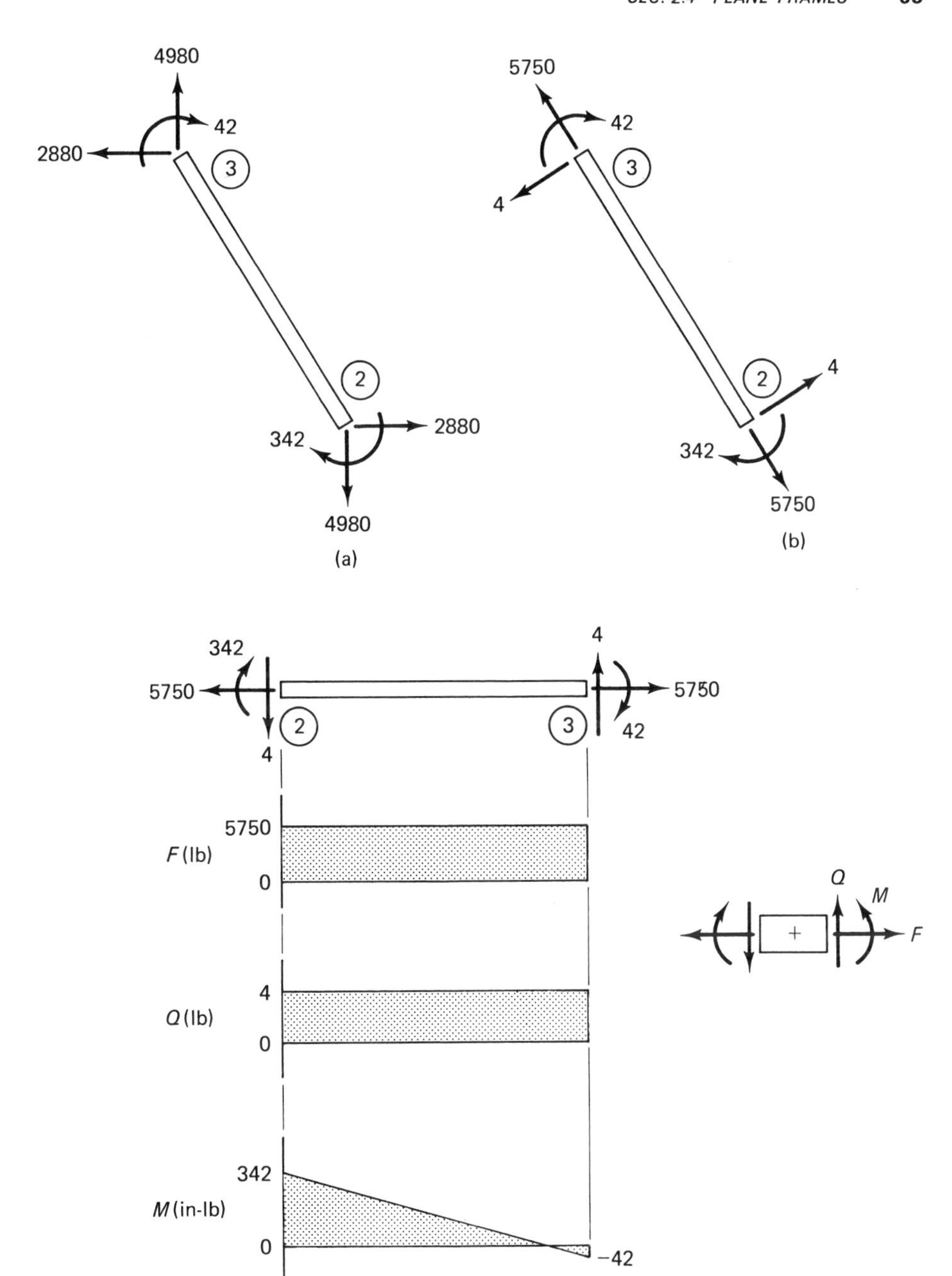
4980
42
2880
3
2
342
2880
4980
(a)
5750
42
4
3
2
4
342
5750
(b)
342
4
5750
5750
2
3
42
4
F (lb)
5750
0
Q
M
F
+
Q (lb)
4
0
342
M (in-lb)
0
−42
(c)

Fig. 2.20

Applying the formulas above to members 1–2 and 2–3, we find, using results from Figs. 2.19 and 2.20:

Member 1–2:

$$\sigma_a = \frac{-2880}{1.13} = -2550 \text{ lb/in}^2$$

$$\sigma_b = \pm\frac{(-956)(1.5)}{1.17} = \pm 1230 \text{ lb/in}^2$$

$$\sigma = \sigma_a + \sigma_b = -3780 \text{ lb/in}^2, \; -1320 \text{ lb/in}^2$$

$$\tau = \frac{(-20)(3.37 - 2.60)}{(3)(1.17)(0.125)} = -35 \text{ lb/in}^2$$

Member 2–3:

$$\sigma_a = \frac{5750}{1.13} = 5090 \text{ lb/in}^2$$

$$\sigma_b = \pm\frac{(342)(1.5)}{1.17} = \pm 438 \text{ lb/in}^2$$

$$\sigma = \sigma_a + \sigma_b = 5530 \text{ lb/in}^2, \; 4650 \text{ lb/in}^2$$

$$\tau = \frac{(4)(3.37 - 2.60)}{(3)(1.17)(0.125)} = 7 \text{ lb/in}^2$$

The extreme shear stresses in the members are seen to be very small when compared with the bending stresses. This is generally true for members that are long in comparison with their cross-sectional dimensions. For such members, which are typical of those used for structural purposes, the shear stress may therefore usually be neglected altogether and stresses in the member calculated from considerations of axial and bending stresses only.

It is interesting to compare the deflections and stresses calculated here with those found earlier in Example 2.2-1, where the member ends were assumed free to rotate. Examination shows, in fact, that the displacements U_2 and V_2 and the axial stresses in the members are essentially identical. However, the presence of the restraining moments in the present problem causes the total maximum longitudinal stress in member 1–2 to be increased by about 48% and that in member 2–3 by about 8%.

These stress increases would be much reduced if the severe foundational restraints at joints 1 and 3 of the structure were replaced by pinned connections, allowing free rotation. In this case, calculation shows the rigid joint at location 2 would then cause less than a 10% increase in the maximum stress in each member over that given by truss analysis. This result accordingly illustrates a general rule in structural mechanics, namely that for a structure that would act as a truss (members carry only tension or compression loadings) except for the presence of rigid joints connecting the members, truss analysis will provide a good approximate treatment provided that the members are relatively flexible, that is, long in comparison with their cross-sectional dimensions.

It may finally be noticed that the internal x- and y-components of force and the internal moment acting on member 1–2 at joint 1 and member 2–3 at joint 3 are identical to the reaction loads F_{1x}, F_{1y}, etc., calculated using the first of the foregoing reduced matrix equations for the structure. The reason, of course, is that the support reactions act on only the end of a single member. If two or more members came together at the foundation support, this would no longer be so, and the support reactions would then equal the sum of the corresponding internal forces of the members at the foundation support. #

EXAMPLE 2.4-2. Show that the equilibrium equations applied to joint 2 of the structure of Example 2.4-1 will yield, with the member stiffness equations, the same reduced matrix equation for determining unknown displacements as determined in that example using the direct stiffness method.

Free-body diagrams of the members and joint 2 are shown in Fig. 2.21. Applying the equilibrium equations

$$\sum F_x = 0, \qquad \sum F_y = 0, \qquad \sum M = 0$$

to the free-body diagram of joint 2, we find

$$\begin{aligned} F_{2x} &= f_{2x}^1 + f_{2x}^2 \\ F_{2y} &= f_{2y}^1 + f_{2y}^2 \\ M_2 &= m_2^1 + m_2^2 \end{aligned}$$

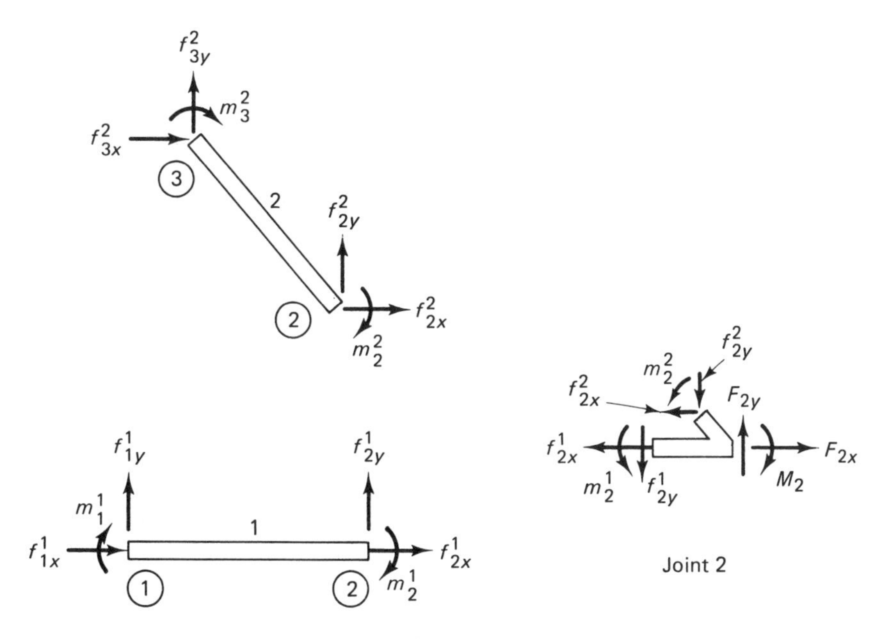

Fig. 2.21

Using the stiffness equations for the individual members given in Example 2.4-1, we have (with $U_1 = V_1 = \theta_1 = U_3 = V_3 = \theta_3 = 0$)

$$\begin{aligned}
f_{2x}^1 &= kU_2 \\
f_{2y}^1 &= k(0.003V_2 + 0.104\theta_2) \\
m_2^1 &= k(0.104V_2 + 4.141\theta_2) \\
f_{2x}^2 &= k(0.125U_2 - 0.216V_2 + 0.022\theta_2) \\
f_{2y}^2 &= k(-0.216U_2 + 0.375V_2 + 0.013\theta_2) \\
m_2^2 &= k(0.022U_2 + 0.013V_2 + 2.070\theta_2)
\end{aligned}$$

Substituting these into the equilibrium equations above and arranging the results in matrix form, we have

$$\begin{Bmatrix} F_{2x} \\ F_{2y} \\ M_2 \end{Bmatrix} = k\begin{bmatrix} 1.125 & -0.216 & 0.022 \\ -0.216 & 0.378 & 0.117 \\ 0.022 & 0.117 & 6.211 \end{bmatrix}\begin{Bmatrix} U_2 \\ V_2 \\ \theta_2 \end{Bmatrix}$$

which is identical to that found in Example 2.4-1 using the direct stiffness method. The remaining reduced matrix equation for determining reaction forces and moments at joints 1 and 3 can be determined by applying a similar procedure to free-body diagrams of these joints. #

EXAMPLE 2.4-3. The simple offshore structure shown in Fig. 2.22(a) consists of a single vertical pile driven into the seafloor and extending upward above the water level to support a deck and equipment. For purposes of analysis, that part of the pile beneath the groundline is replaced by an equivalent free-standing pile, fixed at its base and having stiffness properties representative of the actual pile at the groundline [Fig. 2.22(b)]. Determine the reduced matrix equation for calculating displacements at joints 2 and 3 in terms of known forces acting there.

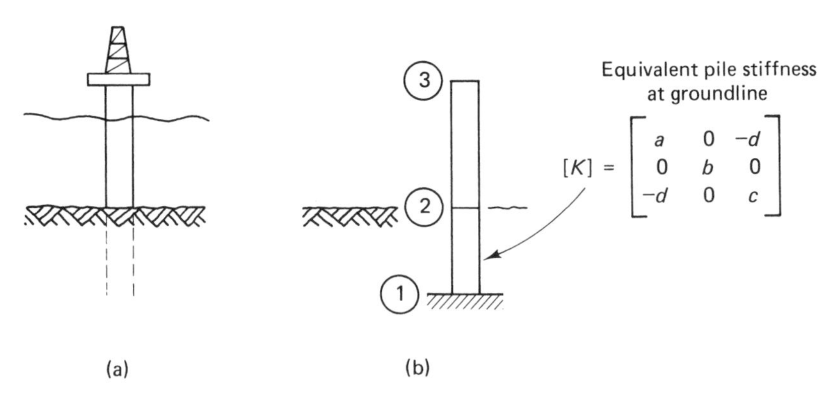

Fig. 2.22

The force–deflection relations at the groundline for the equivalent pile are expressible, from the stiffness matrix given in Fig. 2.22(b), as

$$\begin{Bmatrix} f_{2x} \\ f_{2y} \\ m_2 \end{Bmatrix} = \begin{bmatrix} a & 0 & -d \\ 0 & b & 0 \\ -d & 0 & c \end{bmatrix} \begin{Bmatrix} U_2 \\ V_2 \\ \theta_2 \end{Bmatrix}$$

The stiffness equations for the pile 2–3 are similarly expressible as

$$\begin{Bmatrix} f_{2x} \\ f_{2y} \\ m_2 \\ f_{3x} \\ f_{3y} \\ m_3 \end{Bmatrix} = \begin{bmatrix} k_{22} & k_{23} & k_{24} & \cdots & k_{27} \\ k_{32} & k_{33} & k_{34} & \cdots & \cdot \\ k_{42} & k_{43} & k_{44} & \cdots & \cdot \\ \cdot & \cdot & \cdot & \cdots & \cdot \\ \cdot & \cdot & \cdot & \cdots & \cdot \\ k_{72} & k_{73} & k_{74} & \cdots & k_{77} \end{bmatrix} \begin{Bmatrix} U_2 \\ V_2 \\ \theta_2 \\ U_3 \\ V_3 \\ \theta_3 \end{Bmatrix}$$

where the elements k_{22}, k_{23}, etc., are given by equation (71).

Using the direct stiffness method, these may be combined to obtain the matrix equation

$$\begin{Bmatrix} F_{2x} \\ F_{2y} \\ M_2 \\ F_{3x} \\ F_{3y} \\ M_3 \end{Bmatrix} = \begin{bmatrix} k_{22}+a & k_{23} & k_{24}-d & \cdots & k_{27} \\ k_{32} & k_{33}+b & k_{34} & \cdots & \cdot \\ k_{42}-d & k_{43} & k_{44}+c & \cdots & \cdot \\ \cdot & \cdot & \cdot & \cdots & \cdot \\ \cdot & \cdot & \cdot & \cdots & \cdot \\ k_{72} & \cdot & \cdot & \cdots & k_{77} \end{bmatrix} \begin{Bmatrix} U_2 \\ V_2 \\ \theta_2 \\ U_3 \\ V_3 \\ \theta_3 \end{Bmatrix}$$

which may be inverted and solved for the displacements and rotations when the forces and moments are specified. #

Inclusion of Uniformly Distributed Loads

The foregoing treatment did not allow for loadings along the length of a structural member. To examine this situation, we may return to the beam member of Fig. 2.14 and generalize the subsequent treatment to include distributed loads. To fix ideas, we consider the member of Fig. 2.23 having a uniformly distributed loading w_0 applied along its length.

Carrying out an analysis similar to that done earlier, we find for the present case that the member stiffness equation, analogous to that of equation (65)

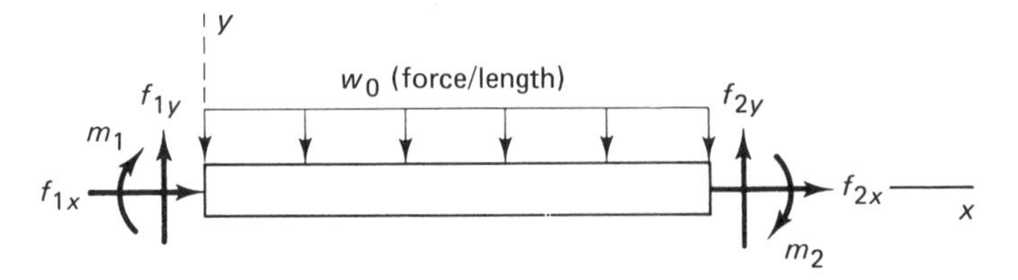

Fig. 2.23. Member subjected to uniformly distributed loading.

can be expressed as

$$\begin{Bmatrix} f_{1x} \\ f_{1y} - \frac{w_0 l}{2} \\ m_1 + \frac{w_0 l^2}{12} \\ f_{2x} \\ f_{2y} - \frac{w_0 l}{2} \\ m_2 - \frac{w_0 l^2}{12} \end{Bmatrix} = \begin{bmatrix} \frac{EA}{l} & & & & & \\ 0 & \frac{12EI}{l^3} & & \text{Symmetric} & & \\ 0 & -\frac{6EI}{l^2} & \frac{4EI}{l} & & & \\ \frac{EA}{l} & 0 & 0 & \frac{EA}{l} & & \\ 0 & -\frac{12EI}{l^3} & \frac{6EI}{l^2} & 0 & \frac{12EI}{l^3} & \\ 0 & -\frac{6EI}{l^2} & \frac{2EI}{l} & 0 & \frac{6EI}{l^2} & \frac{4EI}{l} \end{bmatrix} \begin{Bmatrix} U_1 \\ V_1 \\ \theta_1 \\ U_2 \\ V_2 \\ \theta_2 \end{Bmatrix} \tag{75}$$

Inspection shows that the only difference between this stiffness equation and that of equation (65) rests in the additional terms in the force matrix. Hence, in solving problems involving distributed loadings, we may proceed as before, provided that we treat these additional terms as equivalent applied loads at the ends of the member. Using the superscript e to denote these equivalent forces and moments, we have

$$f_{1y}^e = -\frac{w_0 l}{2}, \qquad m_1^e = \frac{w_0 l^2}{12}$$

$$f_{2y}^e = -\frac{w_0 l}{2}, \qquad m_2^e = -\frac{w_0 l^2}{12}$$

For example, for the members shown in Fig. 2.24, the distributed loadings are replaced by equivalent joint loadings at the ends of the members. The solution for the joint displacements then follows the same procedure as discussed earlier for structures having only joint loadings.

In using the equivalent loads it is convenient to denote the sum of the actual and equivalent loadings at a joint as F_x^T, F_y^T, M^T. If the actual applied joint loads F_x, F_y, M are known, the total forces and moments can be used in the appropriate reduced matrix equation for the structure to determine the

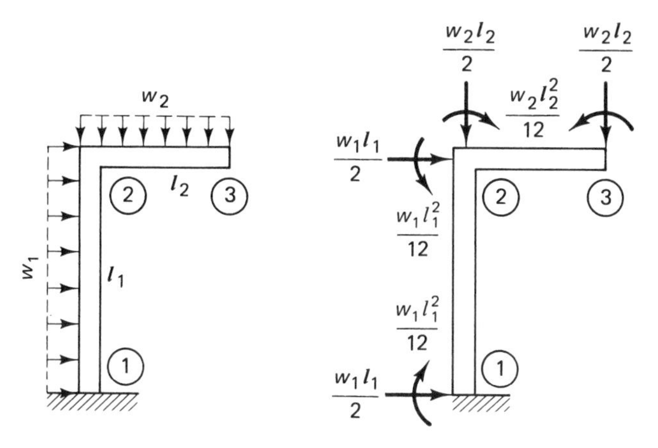

Fig. 2.24. Illustration of equivalent loads.

displacements. If the actual loading is a reaction loading, as with joint 1 in Fig. 2.24, the unknown total force and moment can be determined from the appropriate reduced matrix equation, once the displacements are known, and the actual reaction loadings then determined by solving equations of the form

$$
\begin{aligned}
F_x^T &= F_x + F_x^e \\
F_y^T &= F_y + F_y^e \\
M^T &= M + M^e
\end{aligned}
\tag{76}
$$

where F_x^e, F_y^e, and M^e denote the equivalent joint loads.

Similarly, for a member having a distributed loading, the sum of the actual and equivalent internal forces and moments acting at the ends of the member are conveniently denoted by f_x^T, f_y^T, m^T. Once the joint displacements are known, these total forces and moments may be determined, as before, by simply multiplying the end displacements by the appropriate member stiffness matrix. The actual internal end forces and moments f_x, f_y, m are then determined by solving equations of the form

$$
\begin{aligned}
f_x^T &= f_x + f_x^e \\
f_y^T &= f_y + f_y^e \\
m^T &= m + m^e
\end{aligned}
\tag{77}
$$

where f_x^e, f_y^e, m^e denote equivalent loadings of the distributed loading on the member.

It is worth noticing that, when all displacements and rotations are set equal to zero in equation (75), the actual internal forces and moments are determined as

$$
\begin{aligned}
f_{1x} &= 0, \qquad f_{1y} = -f_{1y}^e, \qquad m_1 = -m_1^e \\
f_{2x} &= 0, \qquad f_{2y}^e = -f_{2y}^e, \qquad m_2 = -m_2^e
\end{aligned}
$$

Since the actual internal forces and moments are equal to the reaction forces and moments associated with the fixed ends, these equations thus suggest a simple way to determine the equivalent loads for loadings along the member other than uniformly distributed loads. Specifically, determine the reaction forces and moments when the ends of the beam are fixed against displacement and rotation. These will be equal in magnitude and opposite in sign to the equivalent loads.

EXAMPLE 2.4-4. A proposed navigational aid is to be supported by a single steel column 22.5 ft high, having an outside diameter of 5 ft and a wall thickness of 1 in. The column is embedded in a concrete footing as shown in Fig. 2.25(a).

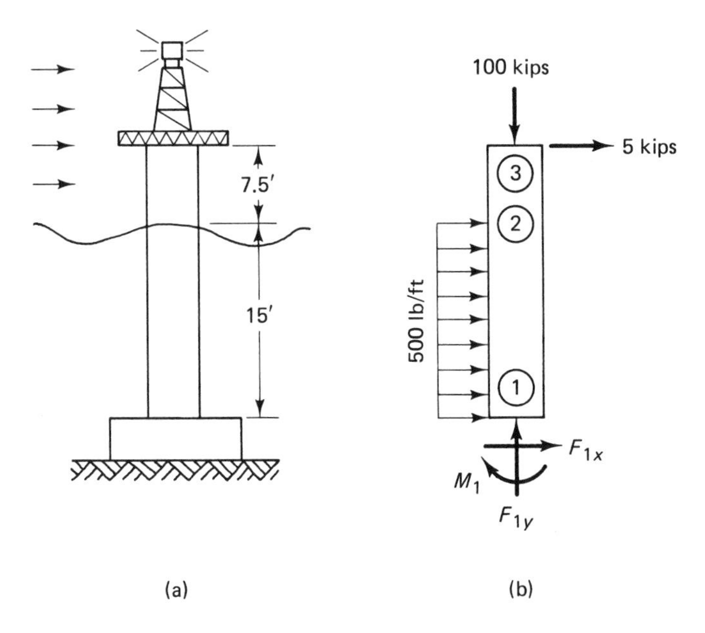

Fig. 2.25

Assumed horizontal wind and wave forces are as shown in Fig. 2.25(b). In addition, the weight of the column is assumed to be concentrated at its upper end and this weight and that of the deck and its parts provide a downward force of 100 kips, as indicated. For these loads we wish to determine the deck deflection and the maximum stress within the columns.

To solve this problem, we first imagine the support column divided into two members 1–2 and 2–3, as shown in Fig. 2.25(b). The stiffness matrices of these two members are obtainable from equation (71). Using units of pounds and feet, we have $E = 4.32 \times 10^9$ lb/ft², $I = 3.891$ ft⁴, $A = 1.287$ ft². For member 1–2 ($\lambda = 0$, $\mu = 1$, $l = l_1 = 15$ ft) we thus find

$$[K]_{1-2} = E\begin{bmatrix} 0.014 & 0 & 0.104 & -0.014 & 0 & 0.104 \\ 0 & 0.086 & 0 & 0 & -0.086 & 0 \\ 0.104 & 0 & 1.038 & -0.104 & 0 & 0.519 \\ -0.014 & 0 & -0.104 & 0.014 & 0 & -0.104 \\ 0 & -0.086 & 0 & 0 & 0.086 & 0 \\ 0.104 & 0 & 0.519 & -0.104 & 0 & 1.038 \end{bmatrix}$$

and for member 2–3 ($\lambda = 0$, $\mu = 1$, $l = l_2 = 7.5$ ft) we find

$$[K]_{2-3} = E\begin{bmatrix} 0.111 & 0 & 0.415 & -0.111 & 0 & 0.415 \\ 0 & 0.172 & 0 & 0 & -0.172 & 0 \\ 0.415 & 0 & 2.075 & -0.415 & 0 & 1.038 \\ -0.111 & 0 & -0.415 & 0.111 & 0 & -0.415 \\ 0 & -0.172 & 0 & 0 & 0.172 & 0 \\ 0.415 & 0 & 1.038 & -0.415 & 0 & 2.075 \end{bmatrix}$$

The joint loadings equivalent to the 0.5 kip/ft = 500 lb/ft distributed wave loading are, from the discussion above,

$$F_{1x} = F_{2x} = \frac{w_0 l_1}{2} = 3750 \text{ lb}$$

$$M_1 = -M_2 = \frac{w_0 l_1^2}{12} = 9375 \text{ ft-lb}$$

Combining these with the actual loadings

$$F_{3x} = 5000 \text{ lb}, \qquad F_{3y} = -100{,}000 \text{ lb}$$

we have the force boundary conditions at joints 2 and 3 given as

$$F_{2x}^T = 3750 \text{ lb}, \qquad F_{2y}^T = 0, \qquad M_2^T = -9375 \text{ ft-lb}$$
$$F_{3x}^T = 5000 \text{ lb}, \qquad F_{3y}^T = -100{,}000 \text{ lb}, \qquad M_3^T = 0$$

The displacement boundary conditions at joint 1 are

$$U_1 = V_1 = \theta_1 = 0$$

The reduced matrix equations are determined from the direct stiffness method as

$$\begin{Bmatrix} F_{1x}^T \\ F_{1y}^T \\ M_1^T \end{Bmatrix} = E\begin{bmatrix} -0.014 & 0 & 0.104 \\ 0 & -0.086 & 0 \\ -0.104 & 0 & 0.519 \end{bmatrix}\begin{Bmatrix} U_2 \\ V_2 \\ \theta_2 \end{Bmatrix}$$

and

$$\begin{Bmatrix} F^T_{2x} \\ F^T_{2y} \\ M^T_2 \\ F^T_{3x} \\ F^T_{3y} \\ M^T_3 \end{Bmatrix} = E \begin{bmatrix} 0.125 & 0 & 0.311 & -0.111 & 0 & 0.415 \\ 0 & 0.258 & 0 & 0 & -0.172 & 0 \\ 0.311 & 0 & 3.113 & -0.415 & 0 & 1.038 \\ -0.111 & 0 & -0.415 & 0.111 & 0 & -0.415 \\ 0 & -0.172 & 0 & 0 & 0.172 & 0 \\ 0.415 & 0 & 1.038 & -0.415 & 0 & 2.075 \end{bmatrix} \begin{Bmatrix} U_2 \\ V_2 \\ \theta_2 \\ U_3 \\ V_3 \\ \theta_3 \end{Bmatrix}$$

Using the known values of the total loads in this second equation and solving for the displacements, we find

$$U_2 = 7.74 \times 10^{-4}\text{ ft}, \qquad V_2 = -2.70 \times 10^{-4}\text{ ft}, \qquad \theta_2 = 8.37 \times 10^{-5}\text{ rad}$$
$$U_3 = 1.44 \times 10^{-3}\text{ ft}, \qquad V_3 = -4.05 \times 10^{-4}\text{ ft}, \qquad \theta_3 = 9.20 \times 10^{-5}\text{ rad}$$

The total forces and total moment at the foundation are then determined from the first reduced equation as

$$F^T_{1x} = -8750\text{ lb}, \qquad F^T_{1y} = 100{,}000\text{ lb}, \qquad M^T_1 = -159{,}000\text{ ft-lb}$$

The actual reaction forces and moment at joint 1 are determined by solving equation (76) for F_x, F_y, and M. We have

$$\begin{aligned} F_{1x} &= F^T_{1x} - 3750 = -12{,}500\text{ lb} \\ F_{1y} &= F^T_{1y} = 100{,}000\text{ lb} \\ M_1 &= M^T_1 - 9375 = -168{,}000\text{ ft-lb} \end{aligned}$$

Consider now the internal forces and moments acting on the ends of member 1–2. The total forces and moments are determined by multiplying the stiffness matrix for this member by the displacements and rotations at joints 1 and 2. We find

$$f^T_{1x} = -8750\text{ lb}, \qquad f^T_{1y} = 100{,}000\text{ lb}, \qquad m^T_1 = -159{,}000\text{ ft-lb}$$
$$f^T_{2x} = 8750\text{ lb}, \qquad f^T_{2y} = -100{,}000\text{ lb}, \qquad m^T_2 = 28{,}100\text{ ft-lb}$$

The actual internal forces and moments are determined by solving equation (77) for f_x, f_y, and m. We have

$$\begin{aligned} f_{1x} &= f^T_{1x} - 3750 = -12{,}500\text{ lb} \\ f_{1y} &= f^T_{2y} = 100{,}000\text{ lb} \\ m_1 &= m^T_1 - 9375 = -168{,}000\text{ ft-lb} \end{aligned}$$

$$f_{2x} = f_{2x}^T - 3750 = 5000 \text{ lb}$$
$$f_{2y} = f_{2y}^T = -100{,}000 \text{ lb}$$
$$m_2 = m_2^T + 9375 = 37{,}500 \text{ ft-lb}$$

These internal end forces and moments are shown on the free-body diagram of member 1–2 given in Fig. 2.26(a).

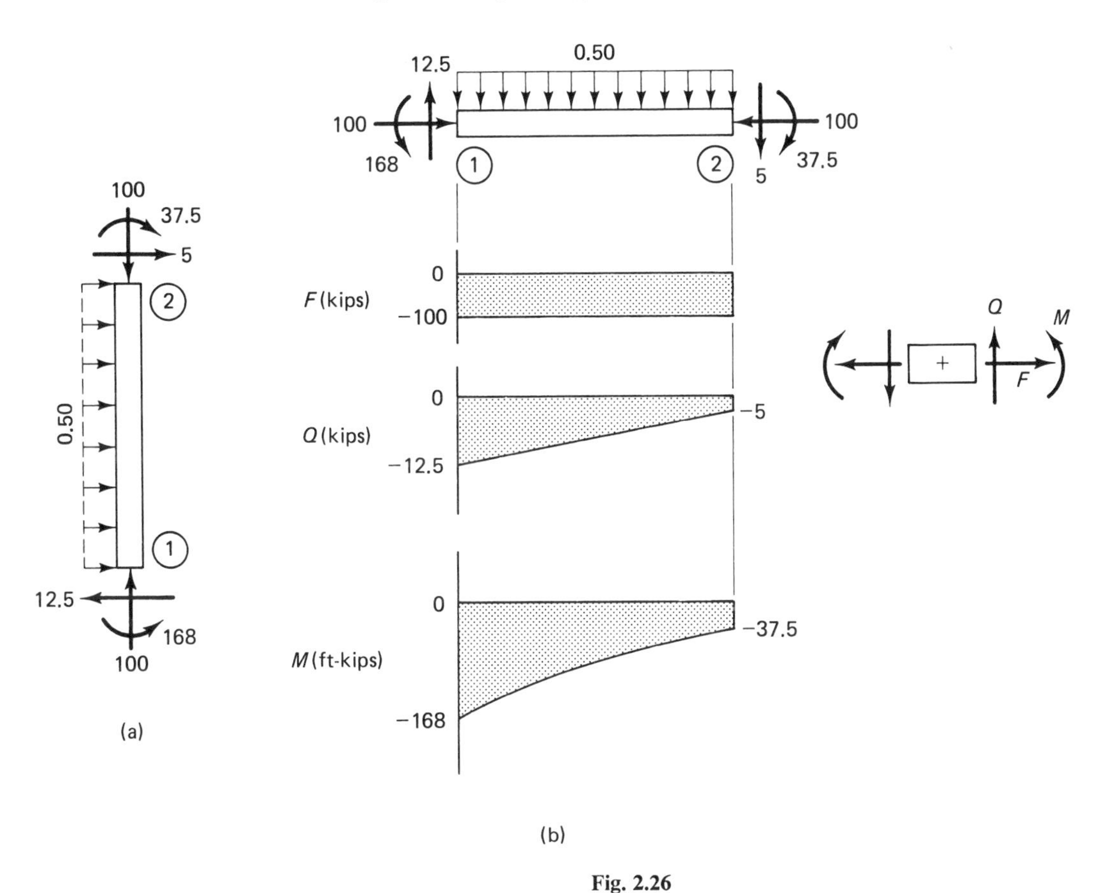

Fig. 2.26

Also shown in Fig. 2.26(b) are the variations of axial force, shear force, and moment along the beam, as determined from statics (see Example 2.4-1). The maximum moment is seen to occur at the joint 1 and to equal −168 ft-kips. The extreme bending stresses existing there are thus

$$\sigma_b = \pm\frac{MR_0}{I} = \pm\frac{(168)(2.5)}{3.89} = \pm 108 \text{ kips/ft}^2$$

The axial force along the member is seen to be constant and equal to −100 kips.

The axial stress is thus

$$\sigma_a = \frac{F}{A} = \frac{-100}{1.29} = -77.5 \text{ kips/ft}^2$$

The maximum normal stress $\sigma = \sigma_b + \sigma_a$ is thus seen to occur on the compressive side of the bending and to be determined as

$$\sigma = -186 \text{ kips/ft}^2(-1290 \text{ lb/in}^2)$$

Finally, consider the internal forces and moments acting on member 2–3. Multiplying the column matrix made up of the end displacements of this member by its stiffness matrix, we find

$$f^T_{2x} = -5000 \text{ lb}, \qquad f^T_{2y} = 100{,}000 \text{ lb}, \qquad m_2 = -37{,}500 \text{ ft-lb}$$

$$f^T_{3x} = 5000 \text{ lb}, \qquad f_{3y} = -100{,}000 \text{ lb}, \qquad m_3 = 0$$

Since no equivalent loads act on this member, these total internal end forces and moments are the actual ones, and we may immediately construct the free-body diagram of the member and the axial force, shear force, and moment diagrams, as shown in Fig. 2.27.

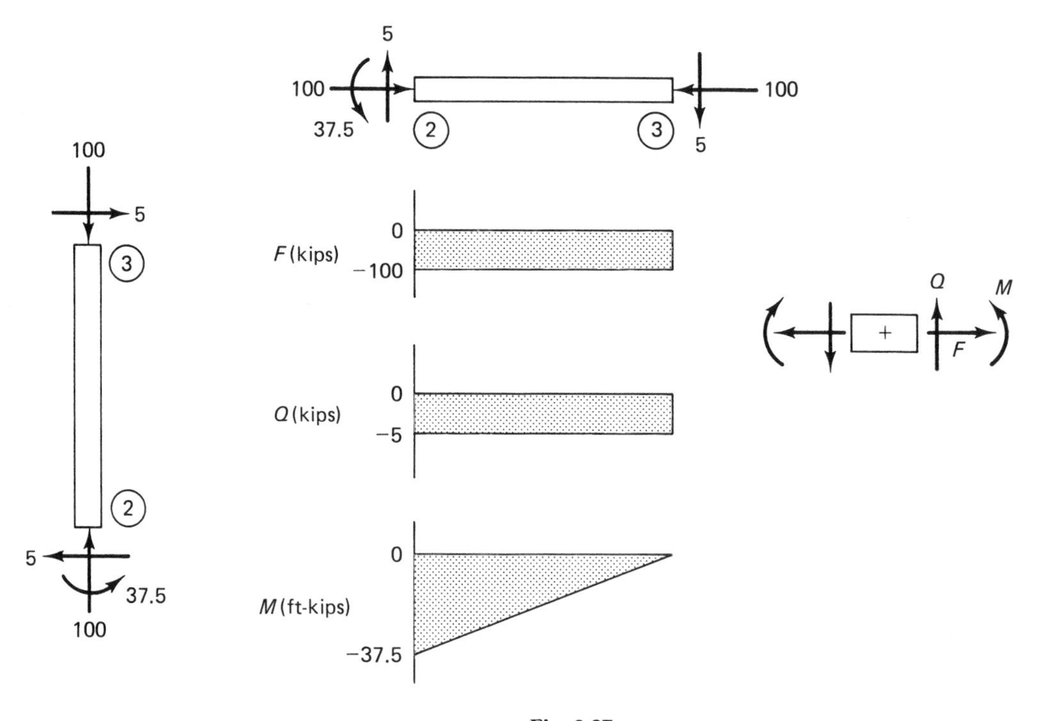

Fig. 2.27

The extreme bending stresses are seen to occur at joint 2 and to be given by

$$\sigma_b = \pm\frac{MR_0}{I} = \pm\frac{(37.5)(2.5)}{3.89} = \pm 24.1 \text{ kips/ft}^2$$

and the axial stress is

$$\sigma_a = \frac{F}{A} = \frac{-100}{1.29} = -77.5 \text{ kips/ft}^2$$

The maximum normal stress $\sigma = \sigma_b + \sigma_a$ in this member is thus seen to occur on the compressive side of the bending at joint 2 and to be determined as

$$\sigma = -102 \text{ kips/ft}^2(-705 \text{ lb/in})$$

Examining the solution above, we thus find the horizontal deck deflection and the maximum longitudinal stress in the column to be

$$U_3 = 0.017 \text{ in}, \qquad \sigma = -1290 \text{ lb/in}^2$$

the maximum stress occurring at the base of the column on the compressive side of the bending. #

Additional Lateral Loadings

A treatment similar to that given above for uniformly distributed loads is also possible for other loadings along the length of the beam. The actual loadings and the equivalent end loadings for three cases of interest are shown in Fig. 2.28 on page 76 and are available for use in the same manner as that discussed above for a uniformly distributed loading.

2.5 SPACE FRAMES

The most general frame structure consists of a three-dimensional assemblage of members, each of which is subjected to axial and transverse loadings and twisting and bending moments as shown in Fig. 2.29. In this figure, the forces and moments are referred to member axes $\bar{x}, \bar{y}, \bar{z}$, with $\bar{y}$ and $\bar{z}$ in the plane of the cross section, and $\bar{x}$ along the member axis. The forces are assumed positive when they act in the direction of the member axes and the moments are assumed positive according to the right-hand rule. Associated with these forces and moments are corresponding displacements and angular rotations, assumed positive in the same sense as the forces and moments.

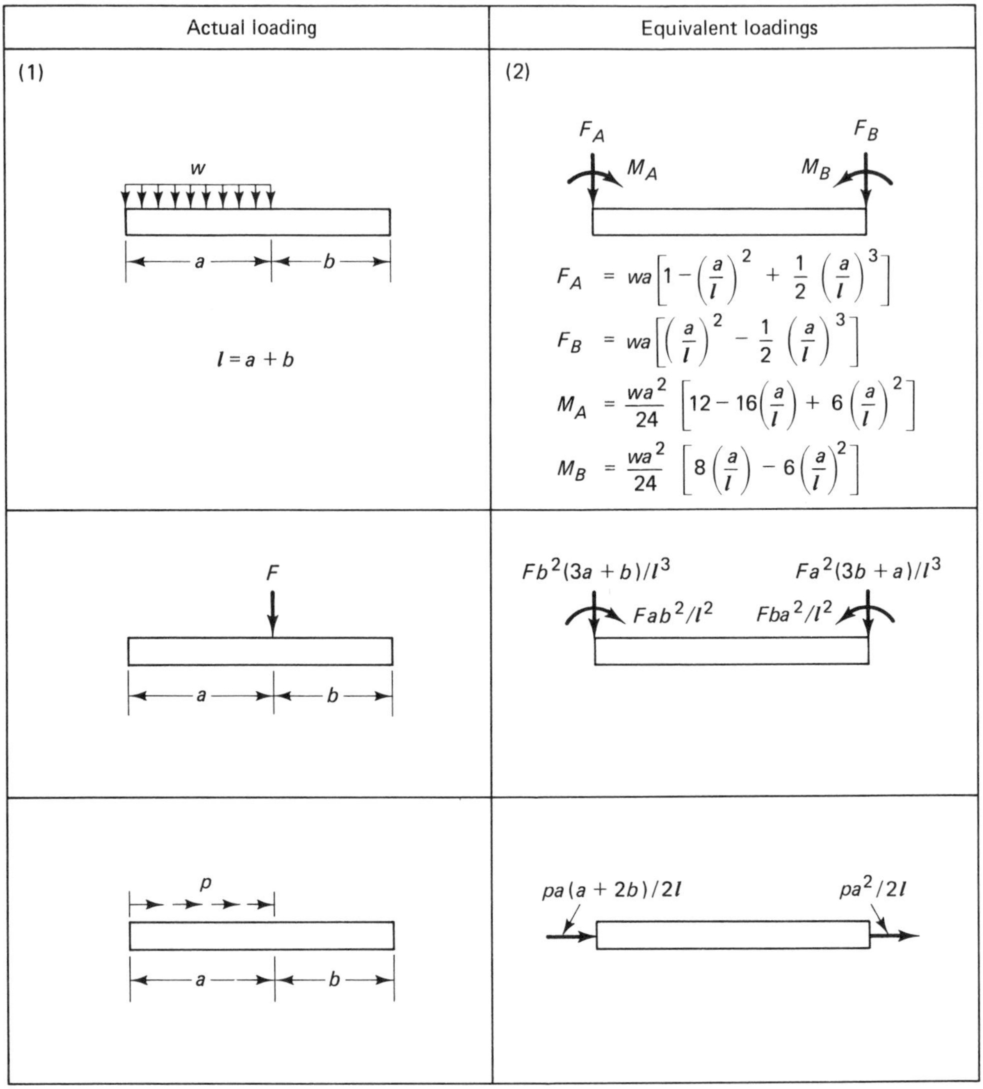

Actual loading	Equivalent loadings
(1) w a b $l = a + b$	(2) F_A M_A M_B F_B $F_A = wa\left[1 - \left(\frac{a}{l}\right)^2 + \frac{1}{2}\left(\frac{a}{l}\right)^3\right]$ $F_B = wa\left[\left(\frac{a}{l}\right)^2 - \frac{1}{2}\left(\frac{a}{l}\right)^3\right]$ $M_A = \frac{wa^2}{24}\left[12 - 16\left(\frac{a}{l}\right) + 6\left(\frac{a}{l}\right)^2\right]$ $M_B = \frac{wa^2}{24}\left[8\left(\frac{a}{l}\right) - 6\left(\frac{a}{l}\right)^2\right]$
F a b	$Fb^2(3a + b)/l^3$ $Fa^2(3b + a)/l^3$ Fab^2/l^2 Fba^2/l^2
p a b	$pa(a + 2b)/2l$ $pa^2/2l$

Fig. 2.28. Actual and equivalent loadings for (a) partial uniform transverse loading, (b) concentrated transverse force, and (c) partial uniform longitudinal loading.

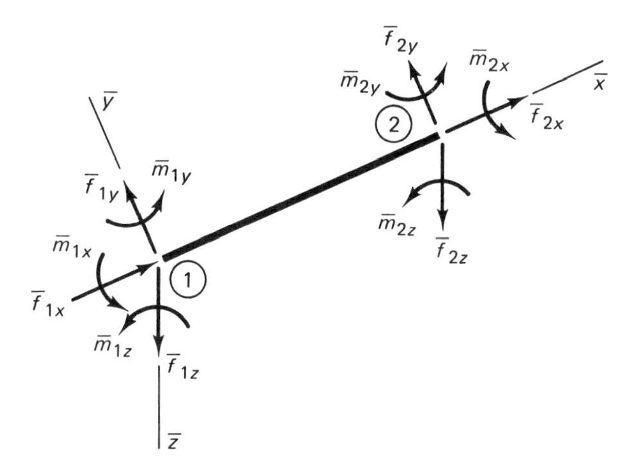

Fig. 2.29. General loading on structural member.

If we write force and corresponding displacement matrices as

$$\{\bar{f}\} = \begin{Bmatrix} \bar{f}_{1x} \\ \bar{f}_{1y} \\ \bar{f}_{1z} \\ \bar{m}_{1x} \\ \bar{m}_{1y} \\ \bar{m}_{1z} \\ \bar{f}_{2x} \\ \cdot \\ \cdot \\ \cdot \end{Bmatrix}, \qquad \{\bar{U}\} = \begin{Bmatrix} \bar{U}_1 \\ \bar{V}_1 \\ \bar{W}_1 \\ \bar{\theta}_{1x} \\ \bar{\theta}_{1y} \\ \bar{\theta}_{1z} \\ \bar{U}_{2x} \\ \cdot \\ \cdot \\ \cdot \end{Bmatrix} \tag{78}$$

the force–deflection equation for the member may be written as

$$\{\bar{f}\} = [\bar{K}]\{\bar{U}\}$$

where $[\bar{K}]$ is the member stiffness matrix whose components may be determined in a manner similar to that used for the two-dimensional matrix of equation (65). Figure 2.30 on page 78 illustrates this matrix for the special case of a member having circular cross section. The more general case of a member having arbitrary cross section has been discussed by Willems and Lucas (1968).

Having the member stiffness $[\bar{K}]$ for each member, the stiffness $[K]$ of the structure can, of course, be determined by transforming each to system axes and adding using the direct stiffness method. The transformation from member to system axes is given, as earlier, by

$$[K] = [R]^T[\bar{K}][R] \tag{79}$$

$$
[K] = \begin{bmatrix}
\frac{EA}{l} & & & & & & & & & & & \\
0 & \frac{12EI}{l^3} & & & & & & & & & & \\
0 & 0 & \frac{12EI}{l^3} & & & & & & \text{Symmetric} & & & \\
0 & 0 & 0 & \frac{GJ}{l} & & & & & & & & \\
0 & 0 & -\frac{6EI}{l^2} & 0 & \frac{4EI}{l} & & & & & & & \\
0 & \frac{8EI}{l^2} & 0 & 0 & 0 & \frac{4EI}{l} & & & & & & \\
-\frac{EA}{l} & 0 & 0 & 0 & 0 & 0 & \frac{EA}{l} & & & & & \\
0 & -\frac{12EI}{l^3} & 0 & 0 & 0 & -\frac{6EI}{l^2} & 0 & \frac{12EI}{l^3} & & & & \\
0 & 0 & -\frac{12EI}{l^3} & 0 & \frac{6EI}{l^2} & 0 & 0 & 0 & \frac{12EI}{l^3} & & & \\
0 & 0 & 0 & -\frac{GJ}{l} & 0 & 0 & 0 & 0 & 0 & \frac{GJ}{l} & & \\
0 & 0 & -\frac{6EI}{l^2} & 0 & \frac{2EI}{l} & 0 & 0 & 0 & \frac{6EI}{l^2} & 0 & \frac{4EI}{l} & \\
0 & \frac{6EI}{l^2} & 0 & 0 & 0 & \frac{2EI}{l} & 0 & -\frac{6EI}{l^2} & 0 & 0 & 0 & \frac{4EI}{l}
\end{bmatrix}
$$

Fig. 2.30. General member stiffness matrix for a member having circular cross section. Here l denotes the length of the member, E denotes Young's modulus, G denotes the shear modulus, and I, J, and A are defined in terms of the outside diameter D_o and the inside diameter D_i as

$$I = \frac{\pi}{64}(D_o^4 - D_i^4), \quad J = \frac{\pi}{32}(D_o^4 - D_i^4), \quad A = \frac{\pi}{4}(D_o^2 - D_i^2)$$

where $[R]$ denotes the transformation matrix. This is expressible in terms of submatrices $[L]$ and $[0]$ as

$$[R] = \begin{bmatrix} [L] & [0] & [0] & [0] \\ [0] & [L] & [0] & [0] \\ [0] & [0] & [L] & [0] \\ [0] & [0] & [0] & [L] \end{bmatrix}$$

with $[L]$ and $[0]$ defined by

$$[L] = \begin{bmatrix} \lambda_{\bar{x}} & \mu_{\bar{x}} & \nu_{\bar{x}} \\ \lambda_{\bar{y}} & \mu_{\bar{y}} & \nu_{\bar{y}} \\ \lambda_{\bar{z}} & \mu_{\bar{z}} & \nu_{\bar{z}} \end{bmatrix}, \qquad [0] = \begin{bmatrix} 0 & 0 & 0 \\ 0 & 0 & 0 \\ 0 & 0 & 0 \end{bmatrix}$$

where $\lambda_{\bar{x}}$ denotes the cosine of the angle between the x and $\bar{x}$ axes, $\mu_{\bar{x}}$ denotes the cosine of the angle between the y and $\bar{x}$ axes, $\nu_{\bar{x}}$ denotes the cosine of the angle between the z and $\bar{x}$ axes, etc.

Once the stiffness matrix of the structure is established, boundary conditions can be applied and displacements and stresses determined as in the earlier work. The level of computational difficulty is, however, increased considerably for space structures and the only reasonable attack on such problems is through the use of a fully automated computer analysis. A number of these programs are now available commercially and their use involves mainly the specification of the sectional characteristics of the individual members; the x, y, z coordinates of their ends relative to system axes; and the appropriate boundary conditions on the structure. From these conditions, the stiffness matrix of the structure can be constructed and joint displacements and internal end forces and moments acting on the individual members calculated automatically.

Although most realistic structures fall within the category of space structures, we may often avoid the full three-dimensional analysis described above by treating parts of the structure as plane frames. This is illustrated in the following example.

EXAMPLE 2.5-1. Consider the steel structure shown in two-dimensions in Fig. 2.31 and determine the displacements of joint 4 using both three- and two-dimensional analyses. All four faces of the structure are identical to the side face shown. The loadings on both side faces are also identical. The lengths L, outside diameters D, and wall thicknesses t of the members are as indicated. The bottom joints are assumed fixed against displacement and rotation.

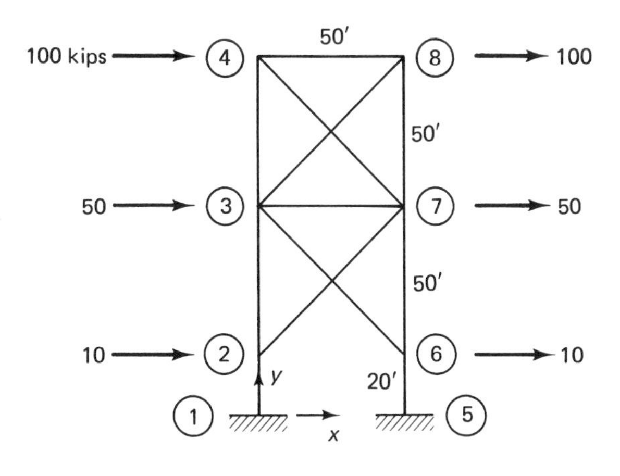

Member	L (ft)	D (ft)	t (in)
Vertical	20	3.9	1.0
Vertical	50	4.0	1.5
Horizontal	50	2.0	0.5
Diagonal	70.7	2.0	0.5

Fig. 2.31

The displacements of joint 4 are listed in Table 2.2 for the full three-dimensional analysis of the structure and for a two-dimensional analysis of the side frame shown in Fig. 2.31. It can be seen that the results from the two analyses

Table 2.2

Analysis	U (ft)	V (ft)	W (ft)
3-D	8.68×10^{-2}	5.91×10^{-3}	5.60×10^{-5}
2-D	8.71×10^{-2}	6.00×10^{-3}	0

agree very well within about 0.4% of one another. The slightly lower horizontal and vertical displacements U, V given by the three-dimensional analysis are the result of the additional lateral stiffness provided by the diagonal members in the front and back faces of the structures. The stiffness of these members is, of course, ignored in the two-dimensional analysis. The error committed is, however, negligible and the simpler two-dimensional analysis is fully adequate for this structure and loadings. #

REFERENCES

Martin, H. C. (1966). *Introduction to Matrix Methods of Structural Analysis*, McGraw-Hill Book Company, New York.

Willems, N., and W. M. Lucas, Jr. (1968). *Matrix Analysis for Structural Engineers*, Prentice-Hall, Inc., Englewood Cliffs, N.J.

PROBLEMS

1. A 20-ft cylindrical member of 5-in diameter has a force $P = 50$ kips applied as shown in Fig. P.1. Assuming that $U_1 = U_3 = 0$, determine the displacement U_2 and the reaction forces F_1 and F_3.
Ans. $U_2 = 3.82 \times 10^{-3}$ in, $F_1 = -37.5$ kips, $F_3 = -12.5$ kips.

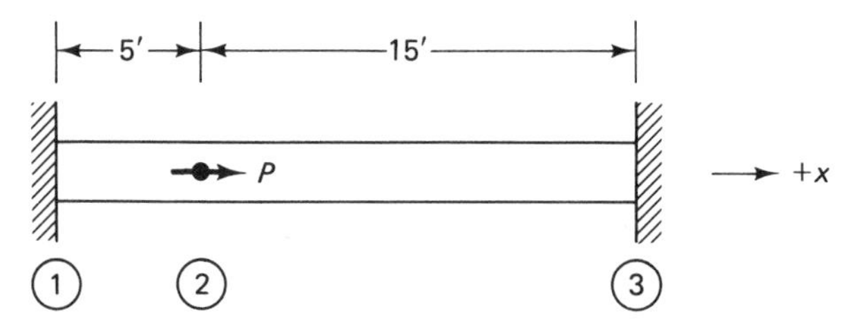

Fig. P.1

2. A 15-ft steel column of 5-in diameter has downward forces $P_1 = 30$ kip, $P_2 = 20$ kip applied as shown in Fig. P.2.

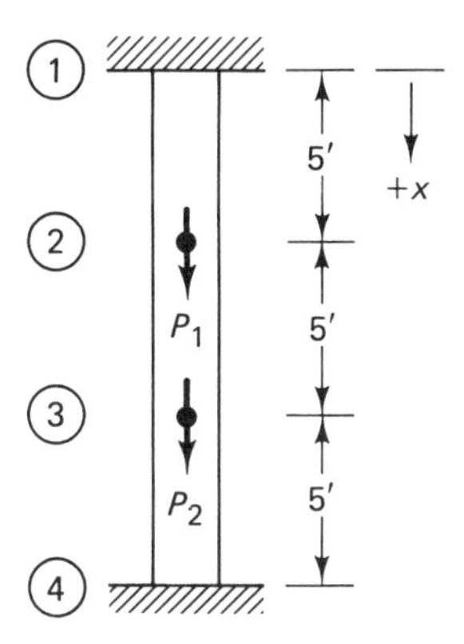

Fig. P.2

(a) Determine the stiffness matrix for the structure 1–2–3–4.
(b) Determine the unknown deflections and reactions.
(c) Determine the stress in members 1–2, 2–3, and 3–4.
Assume that $E = 3 \times 10^7$ lb/in² and $U_1 = U_4 = 0$.

Ans. (a) $$[K] = \begin{bmatrix} k & -k & 0 & 0 \\ -k & 2k & -k & 0 \\ 0 & -k & 2k & -k \\ 0 & 0 & -k & k \end{bmatrix}, \qquad k = \frac{EA}{l} = 9817 \text{ kips/in.}$$

(b) $U_2 = 2.72 \times 10^{-3}$ in, $U_3 = 2.38 \times 10^{-3}$ in, $F_1 = -26.7$ kips, $F_4 = -23.3$ kips.

(c) $\sigma_{1-2} = 1358$ lb/in², $\sigma_{2-3} = 169.8$ lb/in², $\sigma_{3-4} = -1189$ lb/in².

3. The simple truss shown in Fig. P.3 is subjected to a force $P = 1000$ lb at joint 2.

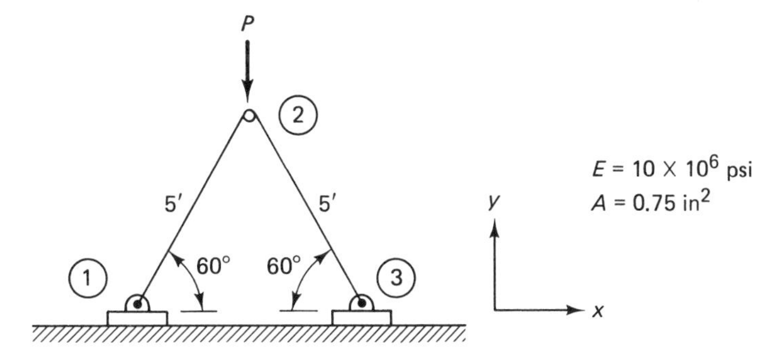

Fig. P.3

(a) Determine the general stiffness matrix for the truss.

(b) Assuming joints 1 and 3 fixed against displacement, determine the unknown displacements and reactions.

(c) Determine the internal forces in the two bars.

Ans. (b) $U_2 = 0$, $V_2 = -5.33 \times 10^{-3}$ in, $F_{1x} = -F_{3x} = 288$ lb, $F_{1y} = F_{3y} = 500$ lb.

(c) $S_{1-2} = S_{2-3} = -578$ lb.

4. Consider the support truss 1–2–3–4–5 shown in Fig. P.4 and determine reduced equations for calculating unknown displacements and reaction forces. Assume joints 1 and 3 fixed against displacement but joints 2, 4, and 5 free to displace.

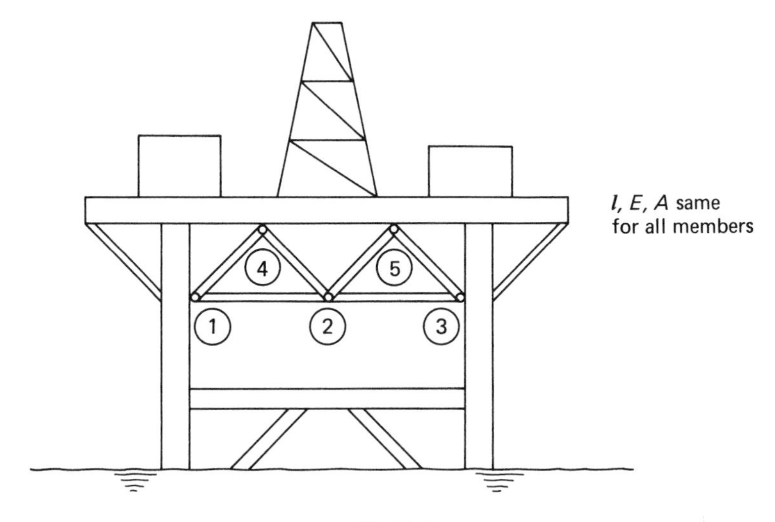

Fig. P.4

5. For the frame shown in Fig. P.5, determine the reduced matrix equation

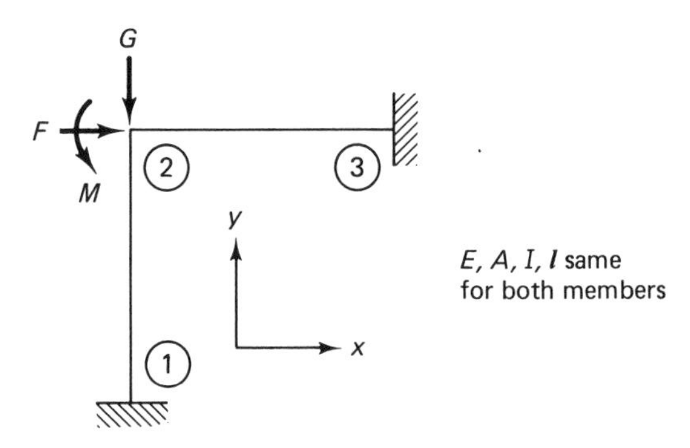

Fig. P.5

connecting known loads F, G, M with unknown displacements at joint 2. Assume joints 1 and 3 fixed against displacement and rotation.

$$\textit{Ans.}\quad \begin{Bmatrix} F \\ -G \\ -M \end{Bmatrix} = \frac{E}{l}\begin{bmatrix} \frac{12I}{l^2}+A & 0 & -\frac{6I}{l} \\ 0 & \frac{12I}{l^2}+A & -\frac{6I}{l} \\ -\frac{6I}{l} & -\frac{6I}{l} & 8I \end{bmatrix}\begin{Bmatrix} U_2 \\ V_2 \\ \theta_2 \end{Bmatrix}$$

6. For the structure of Problem 5, assume that

$$E = 3 \times 10^7 \text{ lb/in}^2, \qquad I = 1.3 \text{ in}^4$$
$$A = 4 \text{ in}^2, \qquad l = 12 \text{ ft}$$
$$G = 50 \text{ kips}, \qquad F = M = 0$$

and determine U_2, V_2, θ_2.

7. For the structure of Problem 5 and the data of Problem 6, determine the internal end forces acting on members 1–2 and 2–3. Sketch shear and moment diagrams for each member and determine the maximum longitudinal stress (axial plus bending) in each, assuming both members have solid square cross sections with 2-in sides.

8. For the frame and loading shown in Fig. P.8:
   (a) Write the reduced matrix equations for determining the unknown deflections and reaction forces.
   (b) Write force boundary conditions for the structure, replacing distributed loads with equivalent loads.

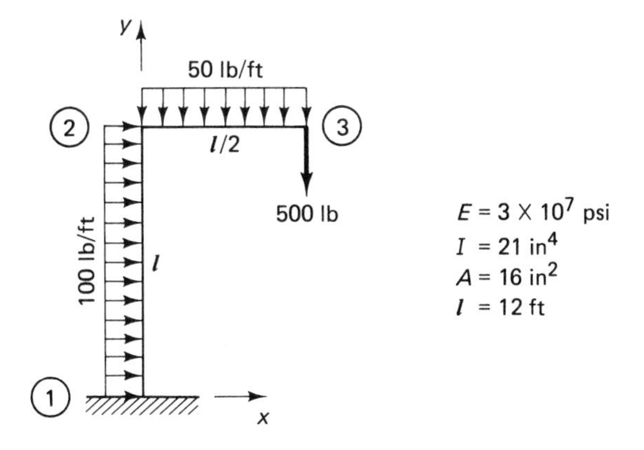

Fig. P.8

9. In Problem 8, if

$$u_2 = 1.481 \text{ in}, \qquad v_2 = -2.400 \times 10^4 \text{ in}, \qquad \theta_2 = 0.01728 \text{ rad}$$

determine the internal end forces acting on member 1–2.
Ans. $f_{1x} = -1200$ lb, $f_{1y} = 800$ lb, $m_1 = -11{,}100$ ft-lb, $f_{2x} = 0$, $f_{2y} = -800$ lb, $m_2 = 3{,}900$ ft-lb.

10. Use the results of Problem 9 to construct the shear and moment diagram for member 1–2 of Fig. P.8. Determine the extreme stresses (axial plus bending) in the member assuming its cross section a solid square of sides 4 in.
Ans. $\sigma = -12{,}740$ lb/in², $+12{,}640$ lb/in².

11. Assemble the 12 × 12 reduced stiffness matrix for determining unknown joint displacements of the structure shown in Fig. P.11, assuming known forces at joints 2, 3, 5, 6 and assuming joints 1 and 4 fixed against displacement and rotation.

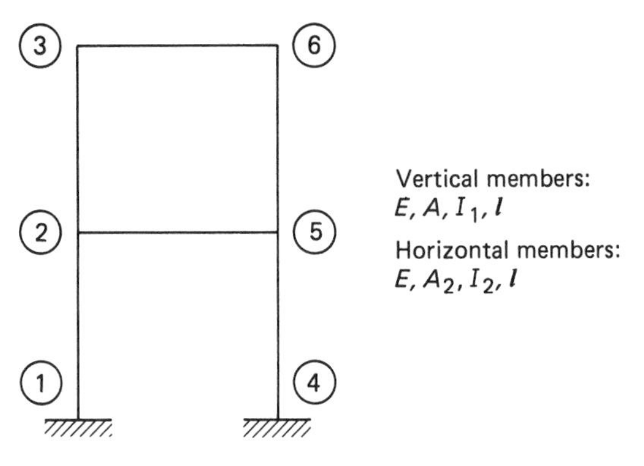

Fig. P.11

12. The simple offshore structure shown in two dimensions in Fig. P.12(a) consists of four cylindrical support legs of 4-ft outside diameter and 1-in wall thickness, connected at their upper ends by four horizontal cylindrical braces of 2-ft outside diameter and 0.50-in wall thickness, and supporting a deck and equipment weighing 150 kips. The members are made of steel ($E = 3 \times 10^7$ lb/in²) and all four sides of the structure are identical.

During storm conditions, the deck experiences a total horizontal wind force of 10 kips and a total horizontal wave force of 40 kips. As an approximation, assume this total force to act at the top of the structure and to be divided equally between the two side faces, as shown in Fig. P.12(b). The weight of the structure is also estimated to be 50 kips and this is accounted for by simply adding it to the deck weight, giving an effective total deck weight of 200 kips and a weight per side face of 100 kips, as also indicated

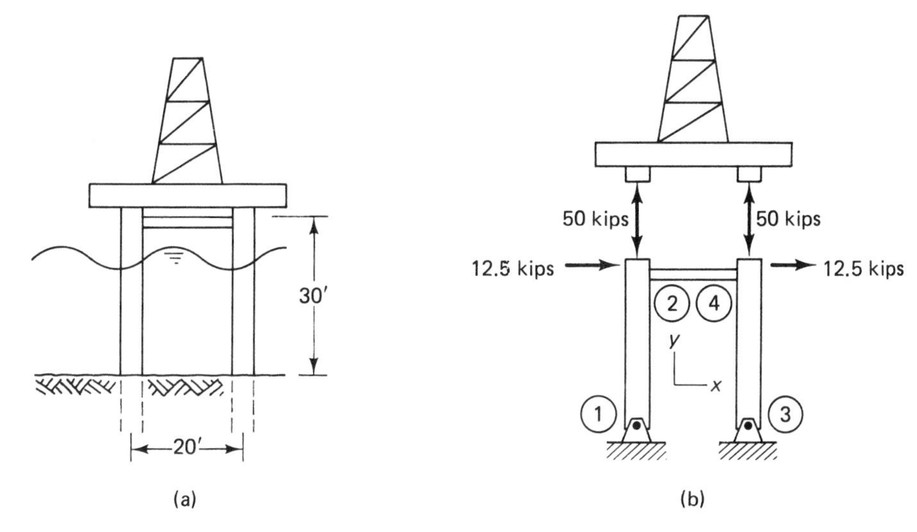

Fig. P.12

in Fig. P.12(b).

Assuming the foundation support to be approximated by pinned connections at joints 1 and 3 (fixed against displacement but free to rotate), determine:

(a) The unknown displacements of the structure

(b) The extreme longitudinal stresses in each member of the side face shown in Fig. P.12(b)

Ans. (a) $\theta_1 = \theta_3 = 3.041 \times 10^{-3}$ rad
$U_2 = U_4 = 1.015$ in
$V_2 = -1.016 \times 10^{-3}$ in, $V_4 = -7.111 \times 10^{-3}$ in
$\theta_2 = \theta_4 = 2.379 \times 10^{-3}$ rad

(b) $\sigma_{1-2} = -2730$ lb/in², $+2560$ lb/in² (at 2)
$\sigma_{3-4} = -3240$ lb/in², $+2060$ lb/in² (at 4)
$\sigma_{2-4} = \pm 21{,}200$ lb/in² (at 2, 4)

13. Repeat Problem 12 for the case where the pinned joint 3 is replaced by a roller joint fixed against vertical displacement but free to rotate and experience horizontal displacement.

Ans. (a) $\theta_1 = 1.076 \times 10^{-2}$ rad, $U_2 = 3.716$ in.
$V_2 = 1.016 \times 10^{-3}$ in, $\theta_2 = 9.440 \times 10^{-3}$ rad
$U_3 = 5.404$ in, $\theta_3 = -4.682 \times 10^{-3}$ rad
$U_4 = 3.719$ in, $V_4 = -7.111 \times 10^{-3}$ in
$\theta_4 = -4.682 \times 10^{-3}$ rad

(b) $\sigma_{1-2} = -5380$ lb/in², $+5210$ lb/in² (at 2)
$\sigma_{3-4} = -593$ lb/in² (at 3, 4)
$\sigma_{2-4} = +42{,}700$ lb/in², $-42{,}030$ lb/in² (at 2)

14. Assemble the 18×18 reduced stiffness matrix for determining unknown displacements of the steel structure shown in Fig. P.14, assuming known forces at joints 2, 3, 4 and 6, 7, 8 and assuming joints 1 and 5 fixed against displacement and rotation. Vertical members above the seafloor all have $l = 50$ ft, $A = 1.52$ ft^2, $I = 2.86$ ft^4. Horizontal and diagonal members have $l = 50$ ft, $A = 0.256$ ft^2, $I = 0.1229$ ft^4. Properties of equivalent piles beneath the seafloor are $l = 15$ ft, $A = 0.160$ ft^2, $I = 2.30$ ft^4. All members have $E = 3 \times 10^7$ lb/in^2.

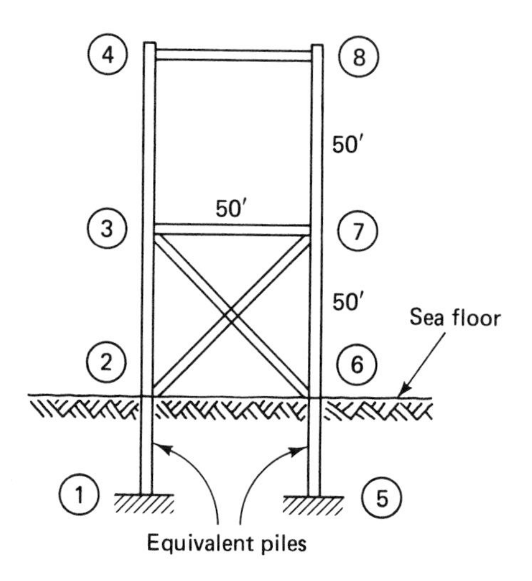

Fig. P.14

Flare bridge of an offshore platform in the North Sea being battered by a severe storm. (Courtesy of the Phillips Petroleum Company and the American Petroleum Institute.)

3

ENVIRONMENTAL LOADINGS

BEFORE THE RESPONSE of a proposed offshore structure can be analyzed, it is necessary to have quantitative estimates of all the significant loadings that the structure is likely to experience in the ocean environment. For structural engineering purposes, this environment may be characterized mainly by overwater wind, by surface waves, and by currents that exist during severe storm conditions (Fig. 3.1).

Overwater wind during storm conditions is significant in the design of offshore structures because of the large forces it can induce on the upper exposed parts of the structures. Wind speed during hurricane conditions in the Gulf of Mexico can, for example, exceed 100 mph, causing, in turn, horizontal forces on a typical offshore structure of 100 kips or more.

Surface waves during storm conditions are also of major importance in the design of offshore structures because of the large forces produced on submerged parts of the structure by the accompanying water motion. Wave heights (difference between maximum and minimum elevations of the water surface at any

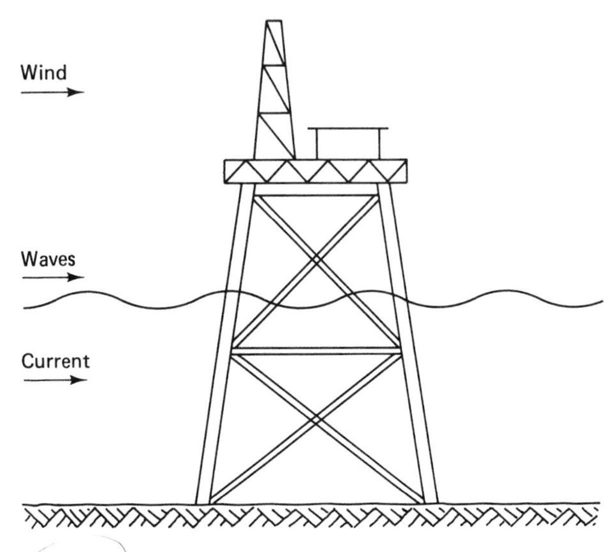

Fig. 3.1. Environmental loadings on an offshore structure.

instant) in the Gulf of Mexico during storm conditions can exceed 50 ft and these can have associated witn them water motions that induced horizontal forces of several hundred kips or more on a typical structure.

Finally, currents at a particular site can contribute significantly to the total forces exerted on the submerged parts of the structure. Currents refer, of course, generally to the motion of water that arises from sources other than surface waves. Tidal currents, for example, arise from the astronomical forces exerted on the water by the moon and sun, wind-drift currents from the drag of local wind on the water surface, river currents from the discharge of rivers, and ocean currents from the drag of large-scale wind systems on the ocean. During storm conditions, currents at the surface of 2 ft/sec or more are not uncommon, giving rise to horizontal structural forces that equal 10%, or more, of the wave-induced forces.

The present chapter is concerned with details of these environmental elements and with suitable engineering estimates of the forces they induce on offshore structures.

3.1 WIND SPEEDS

The forces exerted on a structure by wind depend on the size and shape of the structural members in the path of the wind and on the speed at which the wind is blowing. The greatest wind speed to be expected at a particular site can be estimated from analysis of local daily weather records. Because of wind fluctuations over any measuring time, such records necessarily contain averaged wind-speed measurements over a finite interval of time. In the United States, this averaging time is customarily chosen as the time required for a horizontal column of air 1 mile long to pass the measuring station. The *fastest mile of wind* is then the highest wind speed so measured in a single day, and the *annual extreme fastest mile of wind* is the largest of the daily maximums recorded during a single year.

Figure 3.2 shows, for selected coastal regions of the United States, statistical projections of the annual extreme fastest-mile wind speeds 30 ft above the earth that can be expected, on the average, every 100 years. Either these or those based on a 50-year recurrence interval are generally used for structural design purposes, the 50-year speeds being about 10% less than the 100-year ones. For most permanent land structures, the 50-year recurrence interval is generally considered adequate, but for offshore structures, where there exists the possibility of unusually high loss of life and property in case of failure, the 100-year interval is generally recommended.

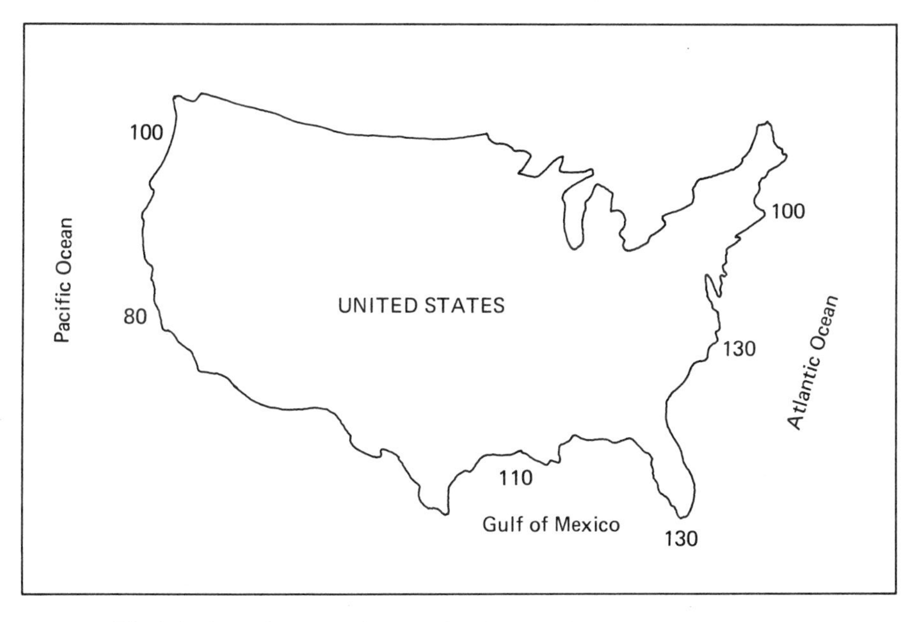

Fig. 3.2. Annual extreme fastest-mile wind speeds (mph) 30 ft above the earth, having a 100-year mean recurrence interval. (After Thom, 1968.)

It should be emphasized that the 100-year wind speeds shown in Fig. 3.2 are statistical projections and that, accordingly, there exists a finite chance that they will be exceeded during any given period of time. It is only *on the average* that they will have a return period of 100 years. The chance that these wind speeds will be exceeded at least once during the first m years is tabulated in Table 3.1 for selected values of m and compared with corresponding results for the 50-year recurrence interval.

Table 3.1 CHANCE THAT 100-YEAR AND 50-YEAR WINDS WILL BE EXCEEDED IN THE FIRST m YEARS

m (years)	100-Year Wind Chance (%)	50-Year Wind Chance (%)
1	1	2
5	5	10
10	10	18
20	18	33
50	39	64
100	63	87

It can be seen, for example, that for a 20-year design life of a structure, there is an 18% chance that the 100-year wind speed will be exceeded at least once. This, of course, does not necessarily mean that there exists an 18% chance that a structure designed on the basis of the 100-year wind speed will fail, since the design itself will, if proper, allow for some overload through a factor of safety.

The wind speeds shown in Fig. 3.2 refer to values 30 ft above the earth. To determine the wind speeds at other elevations, a one-seventh power law has generally been found to be adequate for elevations to about 600 ft. Thus, if V denotes the wind speed at an elevation y, and V_0 denotes the wind speed at the 30-ft elevation, then

$$V = V_0\left(\frac{y}{30}\right)^{1/7} \tag{1}$$

where y is measured in feet.

In the absence of more specific information, it is generally assumed for offshore structures that the overwater wind speeds are about 10% greater than those for nearby coastal stations. For example, in the Gulf of Mexico in the vicinity of the Gulf coast states, the 100-year overwater wind speed can be expected to be about 1.1 times the 110-mph coastal wind speed, i.e., about 120 mph at 30 ft above the earth.

In addition to the sustained wind-speed values discussed above, it is also necessary in structural design to consider the effects of wind fluctuations. These fluctuations are referred to as *gusts* and are usually accounted for by multiplying the basic wind speeds given above by a *gust factor*. Measurements show that the gust factor depends on the averaging time used for the basic wind and on the smaller durations considered in connection with the gusts. Since the averaging time for the basic wind depends on the fastest mile of wind, the gust factor can also be regarded as depending on the fastest mile of wind as well as on the durations considered for the gusts.

Table 3.2 gives typical gust factors for basic wind speeds of 60 and 120 mph, corresponding to averaging times of 60 and 30 sec, respectively, and for various shorter durations over which average gust speeds are measured.

Table 3.2 GUST FACTORS FOR VARIOUS DURATIONS OF MEASUREMENT

Fastest Mile (mph)	Duration (sec)					
	60	30	20	10	5	0.5
60	1.0	1.08	1.12	1.18	1.24	1.37
120	—	1.0	1.04	1.10	1.12	1.29

Source: Data from Bretschneider (1969).

It can be seen that the gust factors range from 1.0 to about 1.4, depending on the fastest mile of basic wind and the duration considered. Not all durations are, however, effective in loading the structure. In fact, gusts with durations less than the time needed for the structure to respond to the loading will have very little effect on the structure and can be neglected. For example, a small sign board is likely to respond to gusts of 1 sec or greater duration; hence, for a basic wind speed of 120 mph, we find by interpolation from Table 3.2 that a gust factor of about 1.3 would be appropriate for this case. Similarly, a large building is likely to respond only to gusts of 10 sec or greater duration so that, in a 120-mph basic wind, a gust factor of about 1.1 would be appropriate. Offshore structures can generally be regarded as lying somewhere between these two extreme structures, and therefore for a basic wind speed of 120 mph, a gust factor of about 1.2 is appropriate, giving a design wind speed of (1.2)(120) = 144 mph.

3.2 WIND FORCES

The wind force acting on an ocean structure is the sum of the wind forces acting on its individual parts. For any part such as a structural member, storage tank, deck house, etc., the wind force arises from the viscous drag of the air on the body and from the difference in pressure on the windward and leeward sides. The net force on the object is known experimentally to be described by an equation of the form

$$F = \tfrac{1}{2}\rho C A V^2 \tag{2}$$

where ρ denotes the density of the air, A denotes a characteristic area of the body, V denotes the wind speed, and C denotes a dimensionless force coefficient dependent on the shape of the body and on the viscosity μ of the air through the value of the Reynolds number $N_R = \rho V D/\mu$, D being a characteristic dimension of the body.

Because the density and viscosity of the air do not vary greatly with ordinary changes in atmospheric temperature and pressure at sea level, it is customary to take $\rho = 2.38 \times 10^{-3}$ slug/ft³ and $\mu = 3.75 \times 10^{-7}$ lb-sec/ft², corresponding to standard temperature and pressure conditions of 60°F and 14.7 lb/in². Thus, with the wind force F expressed in units of pounds, the characteristic area A in units of ft² and the wind speed V in units of MPH, equation (2) may be written as

$$F = 0.00256CAV^2 \tag{3}$$

and the Reynolds number N_R upon which C depends may be expressed as $N_R = 9310VD$, with the characteristic length D being expressed in units of feet.

Figure 3.3 shows the variation of the force coefficient with the Reynolds number for the case of wind normal to the surface of a long, thin rectangular beam of length L and width D and for the case of wind normal to the axis of a long circular cylinder of length L and diameter D. For these cases, the magnitude of the wind force is given by equation (3), with $A = LD$, and its direction is in the direction of the wind.

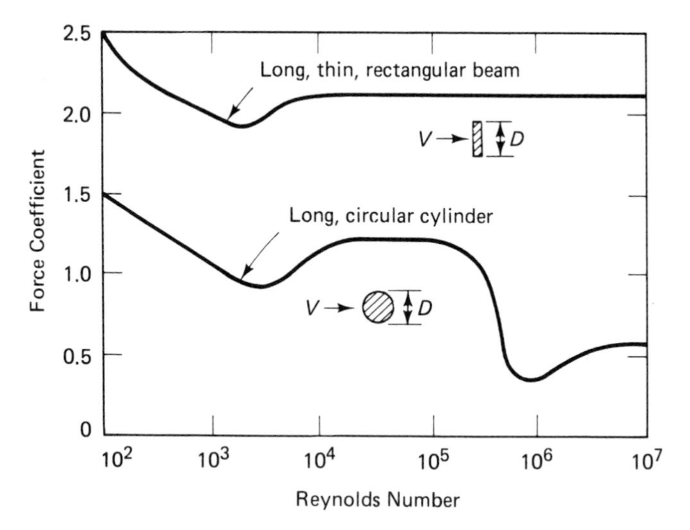

Fig. 3.3. Variation of force coefficient with Reynolds number for a long, thin rectangular beam and a long circular cylinder. (After ASCE, 1961.)

Typical design winds and structural dimensions generally yield Reynolds numbers of the order of 10^6 or greater. Hence, for engineering purposes, the force coefficient may usually be assumed constant and equal to about 2.1 for a long thin rectangular beam and about 0.6 for a long circular cylinder.

Force coefficients for thin rectangular beams and circular cylinders of finite length are generally smaller than those given above for long members because of wind flow around the ends. Typical values used in engineering calculations are listed in Table 3.3 for both beams and cylinders as well as for flat sides of buildings and the overall projected area of an offshore platform.

Table 3.3 WIND-FORCE COEFFICIENTS

Object	Force Coefficient
Beams	1.5
Cylinders	0.5
Sides of buildings	1.5
Projected area of platform	1.0

Source: Data from API (1980).

The discussion above considered only cases where the wind direction was normal to the surface of a member. In cases where the member is inclined to the direction of the wind, the resulting force will act essentially in the direction normal to the member and its magnitude will be given approximately by equation (2) with the normal velocity component of the wind replacing the total wind velocity.

Consider, for example, the member shown in Fig. 3.4. If α denotes the angle between the wind direction and the normal to the member surface, the

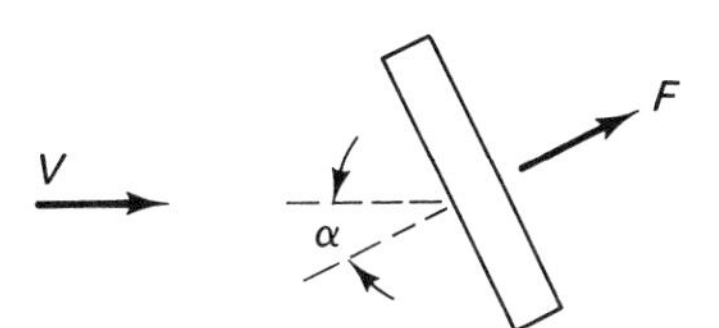

Fig. 3.4. Wind force on an inclined member.

velocity component normal to the surface will be $V \cos \alpha$ and the equation for the force is expressible as

$$F = \tfrac{1}{2}\rho C A V^2 \cos^2 \alpha \tag{4a}$$

where A denotes the area presented to the normal velocity component. For a circular cylinder of length L and diameter D, or a flat plate of length L and width D, $A = LD$.

An alternative method for calculating the force on an inclined member that is sometimes employed is to project the area A of the member onto the wind direction and use this projected area $A \cos \alpha$ in equation (2) together with the full wind velocity. In this case, we have

$$F = \tfrac{1}{2}\rho C A V^2 \cos \alpha \tag{4b}$$

which is more conservative than equation (4a) since it gives a greater estimate of the force.

By applying equation (4a) or (4b) to each exposed element of a structure and resolving each force into horizontal (x) and vertical (y) components, the total horizontal and vertical forces F_x^T, F_y^T exerted on the structure by the wind can be determined by summation. The moments of the individual force components about a fixed point can also be calculated assuming the force components uniformly distributed over each element and the total moments M_1^T, M_2^T resulting, respectively, from the horizontal and vertical forces can similarly be determined by summation. The horizontal and vertical distances d_x, d_y from the fixed point used for the moment calculations to the location of the resultant total forces F_x^T, F_y^T are then finally determined by the equations

$$d_x = \frac{M_2^T}{F_y^T}, \qquad d_y = \frac{M_1^T}{F_x^T} \tag{5}$$

EXAMPLE 3.2-1. Consider the offshore platform shown in Fig. 3.5 and estimate the total horizontal wind force exerted on the structure for a 150-mph design wind using the total projected area of the exposed structure normal to the wind. The total area of all members of the platform and deck tower on the windward face, when projected normal to the direction of the wind, is 700 ft². On a side face, it is 400 ft². The area of the windward face of the deck house is 200 ft².

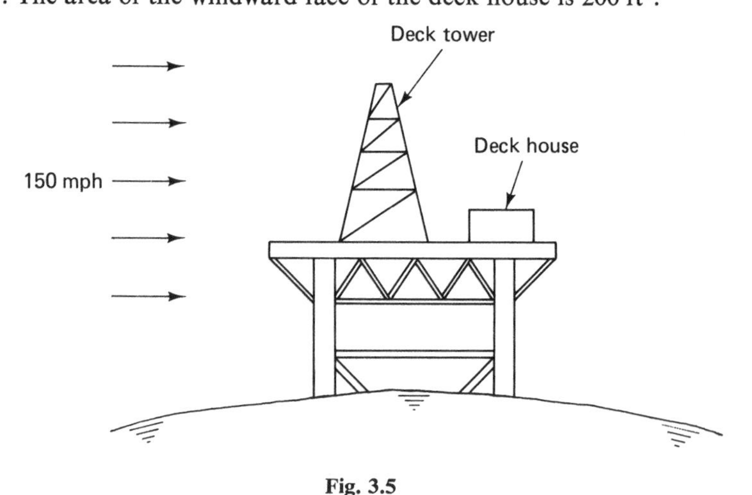

Fig. 3.5

The total projected area of the platform and deck tower is, accounting for all four faces of the structure,

$$A = 2 \times 700 + 2 \times 400 = 2200 \text{ ft}^2$$

From equation (3), with $C = 1$, the horizontal wind force is estimated as

$$F = 127 \text{ kips}$$

The area of the deck house is 200 ft², so that, with $C = 1.5$, we also have

$$F = 17.3 \text{ kips}$$

The total horizontal force is thus estimated as

$$F = 127 + 17 = 144 \text{ kips}$$

This calculation is conservative insofar as horizontal forces are concerned since it assumes the force on members inclined to the direction of the wind to be horizontal rather than normal to the members, as is more nearly the case. #

EXAMPLE 3.2-2. Consider again the offshore structure of Fig. 3.5 and calculate the horizontal and vertical forces exerted on the inclined truss member shown below in Fig. 3.6. The member has a length of 35 ft and a diameter of 2 ft and is inclined 60° to the direction of the wind.

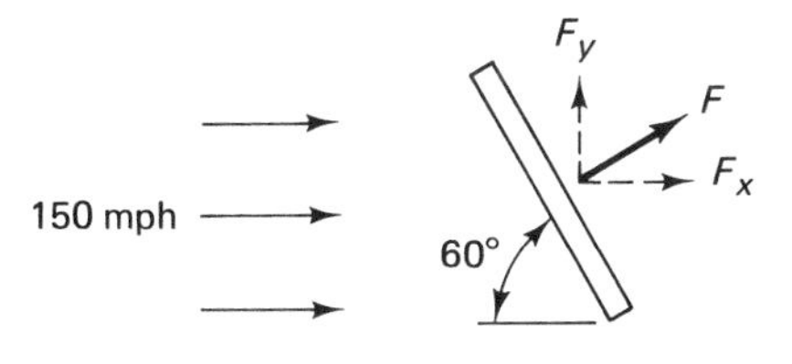

Fig. 3.6

Using the first of the methods described above, equation (4a), we have (with $C = 0.5$, $\alpha = 30°$) $F = 1.51$ kips with horizontal and vertical components $F_x = F \cos \alpha$, $F_y = F \sin \alpha$ given by

$$F_x = 1.31 \text{ kips}, \qquad F_y = 0.76 \text{ kip}$$

If the second of the methods discussed above is used, equation (4b), we obtain $F = 1.75$ kips and

$$F_x = 1.52 \text{ kips}, \qquad F_y = 0.88 \text{ kip}$$

In addition to the horizontal and vertical forces acting on this member, we may also calculate the moments of these forces about a fixed point, say, the base of the member. The horizontal and vertical forces act at the midlength of the member so that the moment M_1 resulting from the horizontal force is, using the first of the methods above for calculating forces,

$$M_1 = (17.5 \sin 60)(1.31) = 19.8 \text{ ft-kips}$$

and the moment M_2 resulting from the vertical forces is

$$M_2 = (17.5 \cos 60)(0.76) = 6.7 \text{ ft-kips}$$

both moments being clockwise. #

3.3 OCEAN SURFACE WAVES

Ocean surface waves refer generally to the moving succession of irregular humps and hollows on the ocean surface. They are generated primarily by the drag of the wind on the water surface and hence are greatest at any offshore site when storm conditions exist there.

For engineering purposes, it is customary to analyze the effects of surface waves on structures either by use of a single design wave chosen to represent extreme storm conditions in the area of interest, or by use of a statistical representation of the waves during extreme storm conditions. In either case, it is necessary to relate the surface-wave data to the water velocity, acceleration, and pressure beneath the waves. This is achieved by the use of an appropriate wave theory.

Airy Wave Theory

A relatively simple theory of wave motion, known as Airy wave theory, was given by G. B. Airy in 1842. This description assumes a sinusoidal wave form whose height H is small in comparison with the wave length λ and the water depth h (Fig. 3.7). Although not strictly applicable to typical design waves used in offshore structural engineering, the Airy theory is valuable for preliminary calculations and for revealing the basic characteristics of wave-induced water motion. It also serves as a basis for the statistical representation of waves and induced water motion during storm conditions (see Chapter 6).

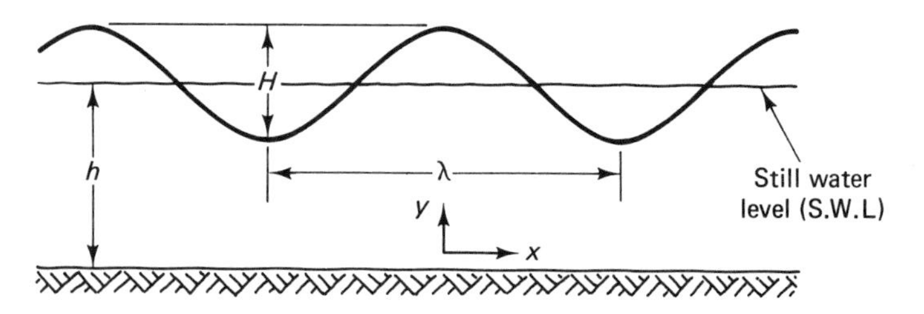

Fig. 3.7. Definition of wave parameters.

With axes x and y as indicated in Fig. 3.7 and a surface-wave form described by

$$\eta = \frac{H}{2} \cos (kx - \omega t) \tag{6}$$

the horizontal velocity u and vertical velocity v of the water particles at place (x, y) and time t are expressible according to the appropriate hydrodynamics equations and Airy theory (see, for example, Kinsman, 1965, or McCormick, 1973) as

$$u = \frac{\omega H}{2} \frac{\cosh ky}{\sinh kh} \cos (kx - \omega t) \tag{7}$$

$$v = \frac{\omega H}{2} \frac{\sinh ky}{\sinh kh} \sin (kx - \omega t) \tag{8}$$

where k and ω denote, respectively, the wavenumber and frequency, defined in terms of the wave length λ and period T (time required for one complete oscillation at any fixed place) by

$$k = \frac{2\pi}{\lambda}, \qquad \omega = \frac{2\pi}{T} \tag{9}$$

These are related through the Airy theory by the equation

$$\omega^2 = gk \tanh kh \tag{10}$$

where g denotes the acceleration of gravity.

It will be noticed that the value of the term $(kx - \omega t)$ in the expressions above remains unchanged at time $t + \Delta t$ when we advance forward a distance $\Delta x = (\omega/k)\Delta t$, i.e.,

$$kx - \omega t = k(x + \Delta x) - \omega(t + \Delta t)$$

The surface wave described by equation (6) can therefore easily be seen to represent a fixed wave form propagating to the right with speed c given by (Fig. 3.8)

$$c = \frac{\omega}{k} = \frac{\lambda}{T} \tag{11}$$

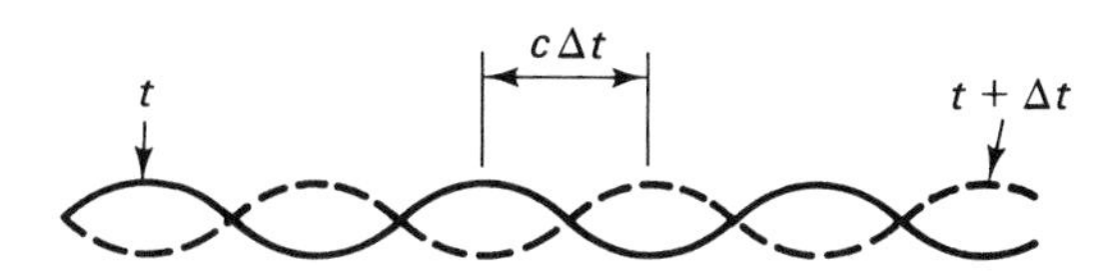

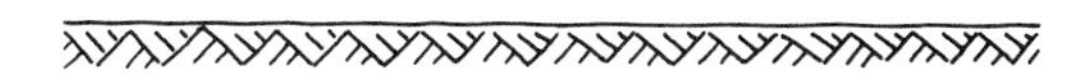

Fig. 3.8. Propagating wave.

Using equation (10), we thus have the Airy wave–speed relation

$$c = \left(\frac{g}{k} \tanh kh\right)^{1/2} \tag{12}$$

For the small wave heights considered by the Airy theory, the horizontal acceleration a_x and vertical acceleration a_y of the water particles at place (x, y) and time t are expressible, approximately, as $a_x = \partial u/\partial t$ and $a_y = \partial v/\partial t$, so that, on using equations (7) and (8), we have

$$a_x = \frac{\omega^2 H}{2} \frac{\cosh ky}{\sinh kh} \sin(kx - \omega t) \tag{13}$$

$$a_y = -\frac{\omega^2 H}{2} \frac{\sinh ky}{\sinh kh} \cos(kx - \omega t) \tag{14}$$

The gage pressure p (difference between actual pressure and atmospheric pressure) at any place (x, y) and time t resulting from the overhead wave and from the hydrostatic contribution is similarly expressible from the Airy theory by the equation

$$p = \rho g \frac{H}{2} \frac{\cosh ky}{\cosh kh} \cos(kx - \omega t) + \rho g(h - y) \tag{15}$$

where ρ denotes the mass density of water.

In using the foregoing wave relations, certain simplifications are possible for relatively deep water, where kh is large, and for relatively shallow water, where kh is small. These simplifications involve approximations of the hyperbolic functions as summarized in Table 3.4.

Table 3.4 APPROXIMATIONS FOR HYPERBOLIC FUNCTIONS

Function	Definition	Approximation (Large α)	Approximation (Small α)
$\sinh \alpha$	$\dfrac{e^{\alpha} - e^{-\alpha}}{2}$	$\frac{1}{2}e^{\alpha}$	α
$\cosh \alpha$	$\dfrac{e^{\alpha} + e^{-\alpha}}{2}$	$\frac{1}{2}e^{\alpha}$	1
$\tanh \alpha$	$\dfrac{e^{\alpha} - e^{-\alpha}}{e^{\alpha} + e^{-\alpha}}$	1	α

For deep water, say $kh > \pi$ or $h/\lambda > \frac{1}{2}$, we have, for example, the simplified frequency equation $\omega^2 = gk$ and the simplified velocity components

$$u = \frac{\omega H}{2} e^{ky'} \cos (kx - \omega t)$$
$$v = \frac{\omega H}{2} e^{ky'} \sin (kx - \omega t) \tag{16}$$

where $y' = y - h$.

For shallow water, say $kh < \pi/10$ or $h/\lambda < \frac{1}{20}$, we have $\omega^2 = ghk^2$ and

$$u = \frac{\omega H}{2kh} \cos (kx - \omega t)$$
$$v = \frac{\omega H}{2h} y \sin (kx - \omega t) \tag{17}$$

Similar relations may also be developed for the accelerations and pressure.

EXAMPLE 3.3-1. Consider water of depth 50 ft and a wave of height 4 ft and period 6 sec. Determine the wavenumber, wave speed, and variation of horizontal water velocity with distance below a wave crest.

We determine the wavenumber by solution of equation (10). The frequency is $\omega = 2\pi/T = 1.047\ \text{sec}^{-1}$ and for determination of the wavenumber, we write equation (10) as

$$k = \frac{\omega^2}{g \tanh kh}$$

We may obtain the solution of this equation by iteration. Choosing tank $kh = 1$ initially, we may solve the equation to get $k = 0.0341\ \text{ft}^{-1}$. For $h = 50$ ft, this then gives $\tanh kh = 0.936$. Using this value in the right-hand side of the equation, we then find a second value of $k = 0.0364\ \text{ft}^{-1}$. Repeating the process for third, fourth, etc., values of k, we obtain the solution when the value of k used in the right-hand side is identical to that found on the left-hand side. The calculations are summarized in Table 3.5, where the solution for k is seen to be $k = 0.0360\ \text{ft}^{-1}$. This corresponds to a wave length $\lambda = 2\pi/k = 175$ ft.

Table 3.5

k (Assumed) (ft^{-1})	kh	$\tanh kh$	k (Calculated) (ft^{-1})
0.0341	1.705	0.936	0.0364
0.0364	1.822	0.949	0.0359
0.0359	1.797	0.946	0.0360
0.0360	1.801	0.947	0.0360

From equation (12), we next determine the wave speed as $c = 29.1$ ft/sec. Finally, the horizontal water velocity below the wave crest is given by equation (7), with $kx - \omega t = 0$, as

$$u = 0.711 \cosh ky$$

Taking $ky = (0.0360)(50 + 2) = 1.87$, we find the maximum velocity at the wave crest to be $u_{\max} = 2.37$ ft/sec. At the seafloor, we have $ky = 0$ and a horizontal velocity of 0.711 ft/sec. Intermediate values may similarly be calculated from the relation above, and variation with depth determined as illustrated in Fig. 3.9 #

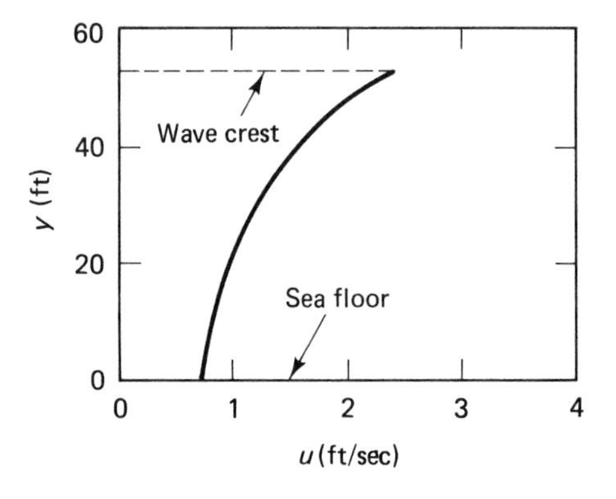

Fig. 3.9

Stokes Wave Theory

An extension of the Airy theory to waves of finite height was made by G. G. Stokes in 1847. His method was to expand the wave solution in series form and determine the coefficients of the individual terms so as to satisfy the appropriate hydrodynamic equations for finite-amplitude waves. Stokes carried the analysis forward to third order of accuracy in the wave steepness H/λ. This solution has been discussed in detail by Skjelbreia (1959) and Wiegel (1964), among others. An extension of the method to fifth order has been made by Skjelbreia and Hendrickson (1961). This work, commonly referred to as the Stokes fifth-order wave theory, is now widely employed in ocean engineering calculations for finite amplitude waves. Because of the slowness of convergence of the series in shallow water, the theory is considered most nearly valid in water where the relative depth h/λ is greater than about $\frac{1}{10}$. This condition is usually met for storm waves considered in the design of fixed offshore platforms.

The equations for the Stokes fifth-order wave theory are given here using the same notation as used above for the Airy theory. The presentation generally follows that of Skjelbreia and Hendrickson.

For a wave of height H, wavenumber k, and frequency ω propagating in the positive x-direction, the free-surface water deflection η from the still-water level is, according to the Stokes fifth-order theory, described by the equation

$$\eta = \frac{1}{k} \sum_{n=1}^{5} F_n \cos n(kx - \omega t) \tag{18}$$

where F_1, F_2, etc., are given by

$$\begin{aligned} F_1 &= a \\ F_2 &= a^2 F_{22} + a^4 F_{24} \\ F_3 &= a^3 F_{33} + a^5 F_{35} \\ F_4 &= a^4 F_{44} \\ F_5 &= a^5 F_{55} \end{aligned} \tag{19}$$

with F_{22}, F_{24}, etc., denoting wave-profile parameters dependent on kh and a denoting a wave-height parameter related to the wave height through the equation

$$kH = 2[a + a^3 F_{33} + a^5(F_{35} + F_{55})] \tag{20}$$

The horizontal water velocity u and the vertical water velocity v (at place x, time t, and distance y above the seafloor) caused by the free-surface wave propagating over water of depth h are expressible as

$$u = \frac{\omega}{k} \sum_{n=1}^{5} G_n \frac{\cosh nky}{\sinh nkh} \cos n(kx - \omega t) \tag{21}$$

$$v = \frac{\omega}{k} \sum_{n=1}^{5} G_n \frac{\sinh nky}{\sinh nkh} \sin n(kx - \omega t) \tag{22}$$

where G_1, G_2, etc., are given by

$$\begin{aligned} G_1 &= aG_{11} + a^3 G_{13} + a^5 G_{15} \\ G_2 &= 2(a^2 G_{22} + a^4 G_{24}) \\ G_3 &= 3(a^3 G_{33} + a^5 G_{35}) \\ G_4 &= 4a^4 G_{44} \\ G_5 &= 5a^5 G_{55} \end{aligned} \tag{23}$$

with G_{11}, G_{13}, etc., denoting wave-velocity parameters dependent on kh.

Explicit expressions for the parameters F_{22}, F_{24}, G_{11}, etc., have been given by Skjelbreia and Hendrickson ($F_{22} = B_{22}$, $F_{24} = B_{24}$, etc., and $G_{11} = A_{11} \sinh kh$, $G_{24} = A_{24} \sinh 2kh$, etc., in their notation) and explicit values worked out for various values of $h/\lambda = kh/2\pi$. Tables 3.6 and 3.7 give illustrative approximate values.

Table 3.6 SELECTED VALUES OF WAVE PROFILE PARAMETERS

h/λ	F_{22}	F_{24}	F_{33}	F_{35}	F_{44}	F_{55}
0.10	3.892	−28.61	13.09	−138.6	44.99	163.8
0.15	1.539	1.344	2.381	6.935	4.147	7.935
0.20	0.927	1.398	0.996	3.679	1.259	1.734
0.25	0.699	1.064	0.630	2.244	0.676	0.797
0.30	0.599	0.893	0.495	1.685	0.484	0.525
0.35	0.551	0.804	0.435	1.438	0.407	0.420
0.40	0.527	0.759	0.410	1.330	0.371	0.373
0.50	0.507	0.722	0.384	1.230	0.344	0.339
0.60	0.502	0.712	0.377	1.205	0.337	0.329

Source: Based on Skjelbreia and Hendrickson (1961).

Table 3.7 SELECTED VALUES OF VELOCITY PARAMETERS

h/λ	G_{11}	G_{13}	G_{15}	G_{22}	G_{24}	G_{33}	G_{35}	G_{44}	G_{55}
0.10	1.000	−7.394	−12.73	2.996	−48.14	5.942	−121.7	7.671	0.892
0.15	1.000	−2.320	−4.864	0.860	−0.907	0.310	2.843	−0.167	−0.257
0.20	1.000	−1.263	−2.266	0.326	0.680	−0.017	1.093	−0.044	0.006
0.25	1.000	−0.911	−1.415	0.154	0.673	−0.030	0.440	−0.005	0.005
0.30	1.000	−0.765	−1.077	0.076	0.601	−0.020	0.231	0.002	0.001
0.35	1.000	−0.696	−0.925	0.038	0.556	−0.012	0.152	0.002	0.000
0.40	1.000	−0.662	−0.850	0.020	0.528	−0.006	0.117	0.001	0.000
0.50	1.000	−0.635	−0.790	0.006	0.503	−0.002	0.092	0.000	0.000
0.60	1.000	−0.628	−0.777	0.002	0.502	−0.001	0.086	0.000	0.000

Source: Based on Skjelbreia and Hendrickson (1961).

In addition to the foregoing relations, it is also necessary to have the frequency relation connecting wave frequency with wavenumber. This relation is given by the equation

$$\omega^2 = gk(1 + a^2C_1 + a^4C_2) \tanh kh \tag{24}$$

where C_1 and C_2 are frequency parameters. Illustrative values are listed in Table 3.8 for various ratios of h/λ.

The wave speed c is determined as in the Airy theory from the relation $c = \omega/k$, which for the Stokes fifth-order solution is expressible as

$$c = \left[\frac{g}{k}(1 + a^2C_1 + a^4C_2) \tanh kh\right]^{1/2} \tag{25}$$

Table 3.8 FREQUENCY AND PRESSURE PARAMETERS

h/λ	C_1	C_2	C_3	C_4
0.10	8.791	383.7	−0.310	−0.060
0.15	2.646	19.82	−0.155	0.257
0.20	1.549	5.044	−0.082	0.077
0.25	1.229	2.568	−0.043	0.028
0.30	1.107	1.833	−0.023	0.010
0.35	1.055	1.532	−0.012	0.004
0.40	1.027	1.393	−0.007	0.002
0.50	1.008	1.283	−0.001	~0
0.60	1.002	1.240	−0.001	~0

Source: Based on Skjelbreia and Hendrickson (1961).

Once the coefficients of the expressions for the velocity components u and v are determined, the horizontal acceleration a_x and vertical acceleration a_y of the water particles can be determined from the equations

$$a_x = \frac{\partial u}{\partial t} + u\frac{\partial u}{\partial x} + v\frac{\partial u}{\partial y}$$

$$a_y = \frac{\partial v}{\partial v} + u\frac{\partial v}{\partial x} + v\frac{\partial v}{\partial y}$$

Writing the velocity coefficients as

$$U_n = G_n\frac{\cosh nky}{\sinh nkh}$$
$$V_n = G_n\frac{\sinh nky}{\sinh nkh} \tag{26}$$

and making use of appropriate trigonometric identities, these expressions can be put in the following explicit forms:

$$a_x = \frac{kc^2}{2}\sum_{n=1}^{5} R_n \sin n(kx - \omega t) \tag{27}$$

$$a_y = \frac{-kc^2}{2}\sum_{n=1}^{5} S_n \cos n(kx - \omega t) \tag{28}$$

where the coefficients R_n and S_n are given in terms of U_n and V_n by

$$\begin{aligned}
R_1 &= 2U_1 - U_1U_2 - V_1V_2 - U_2U_3 - V_2V_3 \\
R_2 &= 4U_2 - U_1^2 + V_1^2 - 2U_1U_3 - 2V_1V_3 \\
R_3 &= 6U_3 - 3U_1U_2 + 3V_1V_2 - 3U_1U_4 - 3V_1V_4 \\
R_4 &= 8U_4 - 2U_2^2 + 2V_2^2 - 4U_1U_3 + 4V_1V_3 \\
R_5 &= 10U_5 - 5U_1U_4 - 5U_2U_3 + 5V_1V_4 + 5V_2V_3
\end{aligned} \tag{29}$$

and

$$
\begin{aligned}
S_0 &= -2U_1V_1 \\
S_1 &= 2V_1 - 3U_1V_2 - 3U_2V_1 - 5U_2V_3 - 5U_3V_2 \\
S_2 &= 4V_2 - 4U_1V_3 - 4U_3V_1 \\
S_3 &= 6V_3 - U_1V_2 + U_2V_1 - 5U_1V_4 - 5U_4V_1 \\
S_4 &= 8V_4 - 2U_1V_3 + 2U_3V_1 + 4U_2V_2 \\
S_5 &= 10V_5 - 3U_1V_4 + 3U_4V_1 - U_2V_3 + U_3V_2
\end{aligned}
\tag{30}
$$

The gage pressure in the water resulting from the overhead wave and the hydrostatic contribution can also be determined from the velocity components by substitution into the relation

$$p = \rho\frac{\omega}{k}u - \tfrac{1}{2}\rho(u^2 + v^2) - \frac{\rho g}{k}(a^2C_3 + a^4C_4 + ky') \tag{31}$$

where $y' = y - h$ and where C_3 and C_4 denote pressure parameters dependent on kh or h/λ. Illustrative values are given in Table 3.8.

EXAMPLE 3.3-2. Consider a wave of height $H = 35$ ft and length $\lambda = 375$ ft in 75 ft of water and determine expressions for the surface-wave profile and the horizontal velocity distribution under the wave crest using the Stokes fifth-order wave theory.

We first determine the wave-height parameter a from equation (20). This equation may be written in the form

$$a = \frac{kH}{2} - a^3F_{33} - a^5(F_{35} + F_{55})$$

and solved by iteration. Noting that $h/\lambda = 0.20$, $k = 0.0168\ \text{ft}^{-1}$, $kH/2 = 0.293$, we have from Table 3.6, $F_{33} = 0.996$, $F_{35} = 3.679$, $F_{55} = 1.734$, and we find, choosing $a = 0.293$ initially, after several iterations the solution $a = 0.267$.

Next we determine the wave-profile relation from equation (18). Substituting the appropriate parameters, we find the surface-wave deflection η (in feet) expressible as

$$\eta = 15.89 \cos\theta + 4.36 \cos 2\theta + 1.43 \cos 3\theta + 0.38 \cos 4\theta + 0.14 \cos 5\theta$$

where $\theta = kx - \omega t$. At the crest ($\theta = 0$) and trough ($\theta = \pi$), the maximum and minimum deflections η_c, η_T are found to be

$$\eta_c = 22.2\ \text{ft}, \qquad \eta_T = -12.8\ \text{ft}$$

The entire surface-wave profile over a half-cycle is shown in Fig. 3.10, for $t = 0$, and contrasted with that of the Airy theory.

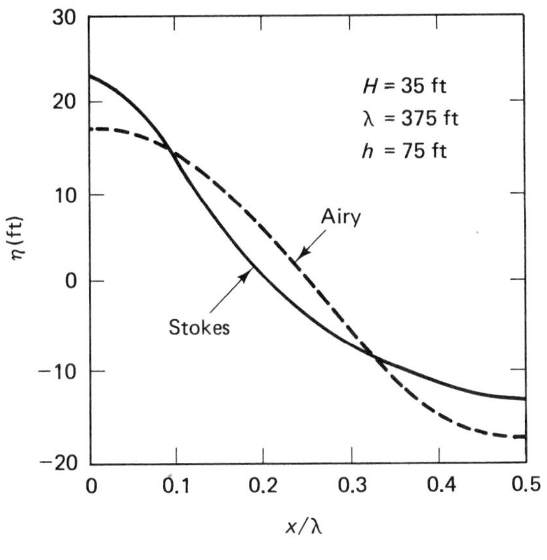

Fig. 3.10

Next consider the equation for the horizontal water velocity. From equation (24), we find the wave frequency as $\omega = 0.723\ \text{sec}^{-1}$. Substituting this into equation (21) and evaluating the coefficients, we find, for locations under the wave crest the expression

$$u = 10.32 \frac{\cosh ky}{\sinh kh} + 2.33 \frac{\cosh 2ky}{\sinh 2kh} + 0.150 \frac{\cosh 3ky}{\sinh 2kh} - 0.039 \frac{\cosh 4ky}{\sinh 4kh} + 0.002 \frac{\cosh 5ky}{\sinh 5kh}$$

At the crest itself, we have $y = 97.2$ ft, and we thus find the water particle velocity there to be $u_c = 22.1$ ft/sec. The actual distribution under the wave crest is shown in Fig. 3.11 on page 108 together with results given by the Airy theory. #

Cnoidal Wave Theory

It has already been pointed out that the Stokes wave theory is most nearly valid for water whose relative depth h/λ is greater than about $\frac{1}{10}$. For shallower water, cnoidal wave theory is generally regarded as being more satisfactory. This theory was first presented by Korteweg and deVries in 1895 and has since been developed further by several writers. Wiegel (1960) has summarized these developments and has given a presentation of the theory convenient for practical application.

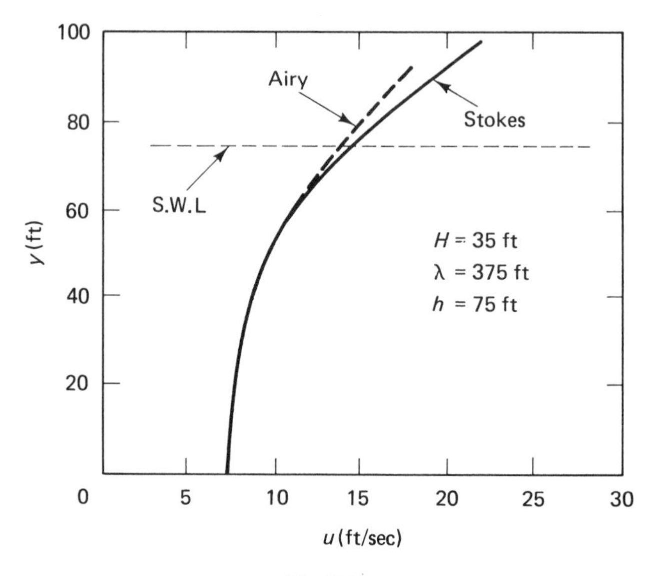

Fig. 3.11

Cnoidal theory is expressed in terms of tabulated elliptic functions and integrals (see, for example, the tables by Milne-Thomson, 1950). The theory, in its common first-order form, assumes chiefly that the ratio of wave height to water depth is sufficiently small that its square can be neglected. This may be contrasted with Airy theory, where the quantity itself, rather than its square, is assumed very small.

The theory presented here follows essentially the presentation of Wiegel (1960). The notation used is the same as that employed above for the Airy and Stokes theories.

Cnoidal waves are periodic waves with surface profiles described in terms of the wavenumber k and frequency ω by the equation

$$\eta = \eta_T + H\mathrm{cn}^2(kx - \omega t, m) \tag{32}$$

where η denotes the deflection of the water surface from its still-water level, at place x and time t, η_T the value of η at the wave trough, H the wave height and cn the Jacobian elliptic function of modulus m, with $0 \leq m \leq 1$.

The modulus m depends on the wave height H, the wave length λ, and the water depth h through the relation

$$mK^2 = \frac{3}{16}\frac{H\lambda^2}{h^3} \tag{33}$$

where K denotes a parameter (known as the complete elliptical integral) which is itself dependent on m. Values of m, K, and $H\lambda^2/h^3$ are tabulated in Table 3.9.

Table 3.9 PARAMETERS FOR CNOIDAL THEORY

m	$H\lambda^2/h^3$	K	E
0	0	1.571	1.571
0.100	1.38	1.612	1.531
0.200	7.94	1.660	1.489
0.300	4.71	1.714	1.445
0.400	6.74	1.778	1.399
0.500	9.16	1.854	1.351
0.600	12.17	1.950	1.298
0.700	16.09	2.075	1.242
0.800	21.74	2.257	1.178
0.900	31.90	2.578	1.105
0.950	42.85	2.908	1.060
0.990	72.13	3.696	1.016
1.000	∞	∞	1.000

The wavenumber k and frequency ω depend on the wave length λ and wave period T through the relations

$$k = \frac{2K}{\lambda}, \qquad \omega = \frac{2K}{T} \tag{34}$$

The frequency is also related to the wavenumber by the equation

$$\omega^2 = ghk^2\left[1 + \frac{H}{mh}\left(\frac{1}{2} - \frac{E}{K}\right)\right]^2 \tag{35}$$

where g denotes the acceleration of gravity and E denotes a parameter (known as the complete elliptic integral of the second kind) dependent on the modulus m. Values of E are also tabulated in Table 3.9. When the wave length is specified, the wavenumber can be determined from the first of equations (34) and the frequency and period then determined from equations (34) and (35).

The value of η_T appearing in equation (32) is expressible in terms of the wave height as

$$\frac{\eta_T}{H} = \frac{K(1 - m) - E}{mK} \tag{36}$$

Since K and E depend only on the modulus m, this ratio is also seen to be dependent only on m.

It remains to specify numerically the surface deflection η associated with the quantities given above. From equation (32), we have

$$\frac{\eta - \eta_T}{H} = \text{cn}^2(\theta, m) \tag{37}$$

where $\theta = kx - \omega t$. Values of this ratio for various values of θ and m are tabulated in Table 3.10. These values, together with those listed in Table 3.9,

Table 3.10 APPROXIMATE VALUES OF $(\eta - \eta_T)/H$

θ	$m = 0$	$m = 0.2$	$m = 0.4$	$m = 0.6$	$m = 0.8$	$m = 1.0$
0	1.000	1.000	1.000	1.000	1.000	1.000
0.2	0.960	0.960	0.960	0.960	0.960	0.960
0.4	0.848	0.850	0.852	0.852	0.854	0.856
0.6	0.681	0.687	0.694	0.699	0.706	0.712
0.8	0.487	0.500	0.516	0.530	0.545	0.560
1.0	0.292	0.317	0.342	0.368	0.394	0.420
1.2	0.131	0.162	0.194	0.229	0.266	0.305
1.4	0.029	0.053	0.085	0.123	0.166	0.216
1.6	0.001	0.003	0.019	0.049	0.094	0.151
1.8	0.052	0.016	0.000	0.009	0.044	0.104
2.0	0.175	0.062	0.028	0.001	0.013	0.071

may be used with linear interpolation when more complete tables of elliptic functions and integrals are unavailable.

With the exception of the case $m = 1$, the tabulated values in Table 3.10 repeat periodically with period $2K$. It can be seen that the tabulated values range only over a half-period, or so. For values of θ outside this range, we may replace θ by $2K - \theta$ and use this new value of θ in Table 3.10 to get the appropriate value for $(\eta - \eta_T)/H$.

For relatively shallow waters where cnoidal wave theory is considered most appropriate, the velocity of the water will be essentially horizontal and described by the equation

$$u = \left(\frac{g}{h}\right)^{1/2} \eta \tag{38}$$

Under this same shallow-water condition, the horizontal acceleration of the water can be determined from the equation

$$a_x = \frac{\partial u}{\partial t} + u\frac{\partial u}{\partial x}$$

Using appropriate identities for derivatives of the elliptic functions, this may be expressed as

$$a_x = \pm 2kH(c - u)\left(\frac{g}{h}\right)^{1/2} A \tag{39}$$

where $c = \omega/k$ denotes the wave speed, and A is defined by

$$A = \left[\frac{\eta - \eta_T}{H}\left(1 - \frac{\eta - \eta_T}{H}\right)\left(1 - m + m\frac{\eta - \eta_T}{H}\right)\right]^{1/2} \tag{40}$$

The positive sign for a_x is chosen for $0 \leq \theta \leq K$ and the negative sign for $K < \theta \leq 2K$.

The gage pressure in the water resulting from the wave and the hydrostatic contribution is finally expressible in terms of distance y from the seafloor as

$$p = \rho g(h + \eta - y) \tag{41}$$

where ρ denotes the mass density of the water.

It is worth noting that when $m = 1$ in the theory above, $\text{cn}\,(\theta) = \text{sech}\,(\theta)$ and the wave profile will be nonperiodic and lie entirely above the still-water level. In this limiting case, the wave is then known as a *solitary wave*.

EXAMPLE 3.3-3. Consider a wave of height 10 ft and length 425 ft propagating in water of depth 40 ft. Determine, using cnoidal wave theory, the period, the wave profile, and the horizontal velocity at the wave crest and trough.

We first determine the modulus m from Table 3.9. We have

$$\frac{H\lambda^2}{h^3} = 28.2$$

and, by interpolation, we find m, and also K and E, from Table 3.9 as

$$m = 0.86, \qquad K = 2.46, \qquad E = 1.13$$

The wavenumber k is next determined from equation (34) as $k = 0.0116\ \text{ft}^{-1}$ and the frequency is found from equation (35) as $\omega = 0.419\ \text{sec}^{-1}$. The period of the wave is thus

$$T = \frac{2K}{\omega} = 11.7\ \text{sec}$$

The wave profile is described by equation (37) with $H = 10$ ft and η_T determined from equation (36) as $\eta_T = -3.71$ ft. This profile, as determined by interpolation from Table 3.10, is shown in Fig. 3.12 and compared with that given by the Airy theory. It can be seen that the crest of the wave is higher than that given by the Airy theory, in agreement with the general finite-wave features exhibited by the Stokes theory.

Finally, the velocity of the water particles under the crest and trough are determined from equation (37) as

$$u_c = 5.42\ \text{ft/sec}, \qquad u_T = -3.40\ \text{ft/sec}$$

These may be contrasted with velocities under the crest and trough of 4.93 ft/sec and -4.93 ft/sec, as determined by the Airy theory. #

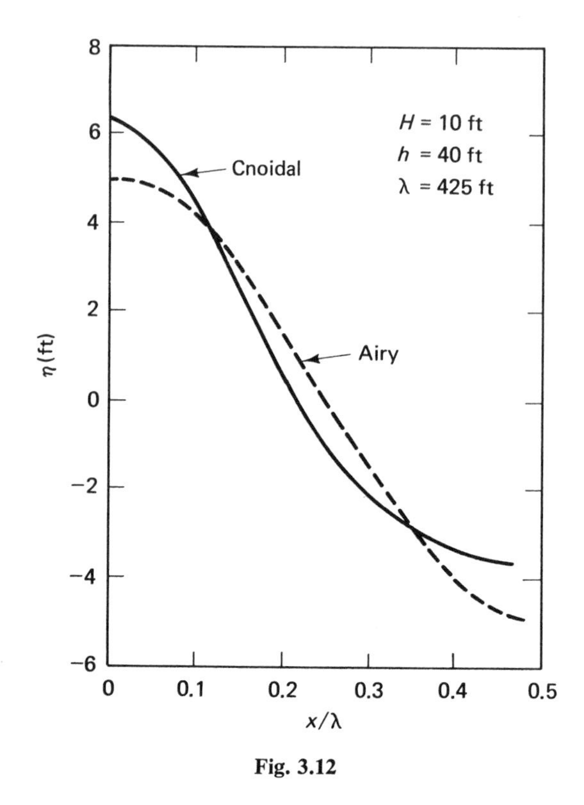

Fig. 3.12

Applicability of Wave Theories

We have previously noted that the Airy wave theory is useful for preliminary calculations involving assumed design waves even though the heights of such waves may exceed the restriction that they be small in comparison with the wave length and water depth. When more accurate calculations are needed, the Stokes theory should be used for design waves having lengths less than about 10 times the water depth. For longer waves, the cnoidal theory is generally recommended.

The question naturally arises as to range of values of the ratios of wave height to wave length and water depth to wave length for which the simpler Airy theory can be used to obtain a reasonably accurate estimate of the wave characteristics. The previous examples have indicated that a main feature of the more accurate Stokes and cnoidal theories is the increase in the amplitude of the crest of a wave over that given by Airy theory. This accordingly suggests a simple means for assessing the applicability of the Airy theory, namely that the theory can be considered applicable for waves whose crest amplitude differs from that determined from the more accurate theories by no more than some assigned percentage. In this way, a range of values of wave height to wave length can be established for which the Airy theory can be expected to give results of acceptable accuracy.

Figure 3.13 shows such ranges for an arbitrarily chosen allowable error in the crest amplitude of 10%. The division between the Airy and Stokes theories

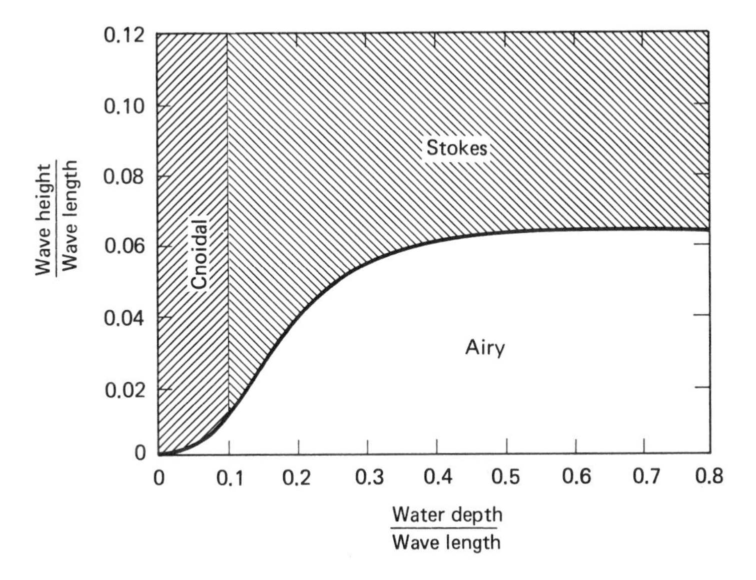

Fig. 3.13. Diagram showing the range of validity of the Airy wave theory, assuming tolerable errors of no more than 10%.

is calculated using the Stokes theory described earlier, with only the first two terms in the expression for the surface deflection retained, consistent with a first correction to the Airy theory. The division between the Airy and cnoidal theory is based on the results from the cnoidal theory given above. The division between the cnoidal and Stokes theories is based on the generally accepted view that cnoidal theory should be used for ratios of water depth to wave length less than one-tenth, except when Airy theory is acceptable.

In addition to providing a definite division between the theories, the diagram of Fig. 3.13 is also useful in assessing the error committed by using the Airy theory over the more accurate ones. For example, the division between the Airy and Stokes theories for a ratio of water depth to wave length of 0.5 occurs at a value of the ratio of wave height to wave length of about 0.63. This corresponds to a 10% error. If the ratio of wave height to wave length for a particular design wave is, say, only 0.56, the error committed in using the Airy theory is then approximately $10 \times 0.56/0.63 = 8.9\%$.

3.4 WAVE FORCES ON VERTICAL PILES

The force exerted on a fixed vertical cylindrical pile by surface waves was first considered by Morison et al. (1950) under the restriction that the diameter of the pile is small in comparison with the length of the waves encountered, say $\frac{1}{10}$ or less, so that the distortion of the waves by the pile is negligible. If f denotes the wave force per unit length acting on a fixed vertical pile of diameter D, the *Morison equation* developed in this work, and now widely employed in engineering calculations, may be expressed as

$$f = \frac{1}{2}\rho C_D D\,|u|\,u + \rho C_I \frac{\pi D^2}{4} a_x \tag{42}$$

where ρ denotes water density, C_D and C_I denote coefficients, and where u and a_x denote the horizontal velocity and acceleration of the water associated with the wave.

The first term on the right-hand side of this equation is referred to as the *drag term* and is seen to be proportional to the square of the water velocity, the absolute value sign being used to ensure that the sign of the drag component will coincide with that of the velocity. The second term is referred to as the *inertia term* and is seen to be proportional to the water acceleration. The coefficients C_D and C_I are accordingly known, respectively, as the *drag* and *inertia coefficients.*

The values of the drag and inertia coefficients can generally be expected from dimensional reasoning to vary with the maximum water velocity u_m of the wave motion and with the wave period T through the dimensionless numbers

$$N_R = \frac{\rho^{u_m D}}{\mu}, \qquad N_K = \frac{u_m T}{D} \tag{43}$$

where ρ and μ denotes the density and viscosity of the water. The first of these is the Reynolds number discussed earlier in connection with wind forces and representative of the effect of viscosity. The second is the *Keulegan–Carpenter number*, representative of the effect of the wave period. Unfortunately, however, only limited experimental data exist on the variation of drag and inertia coefficients with both these numbers, and usual engineering practice is simply to assume them constant, with values of the drag coefficient chosen within the range 0.6 to 1.0 and values of the inertia coefficient chosen within the range 1.5 to 2.0 (API, 1980).

The values of the velocity u and acceleration a_x in equation (42) are calculated from an appropriate wave theory and, together with chosen values of the drag and inertia coefficient, the equation yields, at any instant in the wave cycle, the force distribution along the pile. Since the wave-induced velocity and accleration of the water generally decay with depth, the wave force distribution

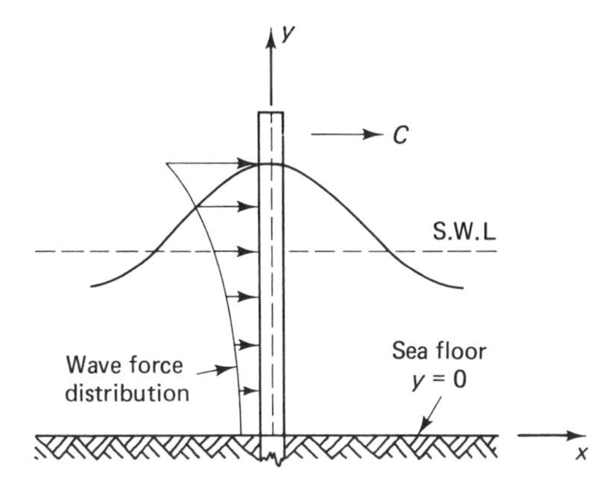

Fig. 3.14. Wave force on a vertical pile.

can also be expected to follow the same general behavior, as illustrated in Fig. 3.14.

The total horizontal force F exerted on that length of the pile ranging from $y = 0$ to $y = y$ is easily seen to be given by the relation

$$F = \int_0^y f(y)\, dy \tag{44}$$

Similarly, the total moment M about $y = 0$ of the force exerted on that part of the pile ranging from $y = 0$ to $y = y$ is given by

$$M = \int_0^y y f(y)\, dy \tag{45}$$

and the distance b above the base of the pile where the resultant force F acts is determined as $b = M/F$.

Forces Arising from Airy Waves

If we restrict attention to small-amplitude surface waves of height H, frequency ω and wavenumber k on water of depth h and use the Airy wave theory of Section 3.3, we may easily evaluate equation (44), with $x = 0$ taken as the pile location, to find

$$F = F_D + F_I \tag{46}$$

where F_D and F_I denote, respectively, the drag and inertia force contributions, as given for a uniform pile diameter by

$$F_D = \frac{\rho C_D D}{32k}(\omega H)^2\left(\frac{\sinh 2ky}{\sinh^2 kh} + \frac{2ky}{\sinh^2 kh}\right)|\cos \omega t| \cos \omega t \tag{47}$$

and

$$F_I = -\frac{\rho C_I}{2k}\frac{\pi D^2}{4}\omega^2 H \frac{\sinh ky}{\sinh kh}\sin \omega t \tag{48}$$

Similarly, for the moment given by equation (45), we find

$$M = M_D + M_I \tag{49}$$

where M_D and M_I denote, respectively, the moments arising from the drag and inertia forces as given by

$$M_D = \frac{\rho C_D D}{64k^2}(\omega H)^2 Q_1 \,|\cos \omega t|\cos \omega t \tag{50}$$

$$MI = -\frac{\rho C_I}{2k^2}\frac{\pi D^2}{4}\omega^2 H Q_2 \sin \omega t \tag{51}$$

with

$$Q_1 = \frac{2ky \sinh 2ky - \cosh 2ky + 2(ky)^2 + 1}{\sinh^2 kh} \tag{52}$$

$$Q_2 = \frac{ky \sinh ky - \cosh ky + 1}{\sinh kh} \tag{53}$$

It can be seen from the equations above that the drag and inertia force contributions to the total force and moment are out of phase by 90° with the maximum of either occuring when the other is zero. The total force and moment exerted on the entire pile is, of course, obtainable from these relations by choosing $y = h + \eta$, where η is the water deflection from the still-water level at the pile. The value of η at any instant ωt is determined from the general surface-wave relation $\eta = (H/2)\cos(kx - \omega t) = (H/2)\cos \omega t$.

When Airy theory is strictly applicable, the ratio η/h will be small and the total force and moment on the pile can then be determined by simply putting $y = h$ in the expressions above. The maximum values may then be determined by numerical investigation for various values of ωt. When, however, η/h is not negligibly small but Airy theory is employed as an approximation, the total force and moment are estimated better by choosing $y = h + (H/2)\cos \omega t$. Maximum values, may then be determined by numerical investigation taking into account this variation of the water elevation on the pile.

The relative significance of the drag and inertia forces in equation (46) may be estimated using equations (47) and (48). The ratio of the maximum drag force F_{DM} to the maximum inertia force F_{IM} is seen to be expressible as

$$\frac{F_{DM}}{F_{IM}} = \mu\frac{H}{D} \tag{54}$$

where μ is given, for $h + \eta \approx h$, by

$$\mu = \frac{1}{4\pi}\frac{C_D}{C_I}\left(\frac{\sinh 2kh}{\sinh^2 kh} + \frac{2kh}{\sinh^2 kh}\right)$$

Approximate values of μ are listed in Table 3.11 for various values of $h/\lambda = kh/2\pi$, assuming that $C_D/C_I = 0.5$. It can be seen that for values of h/λ greater than about 0.3, the ratio of the maximum drag and inertia forces will be less

Table 3.11 VALUES OF μ IN EQUATION (54)

h/λ	kh	μ
0.05	0.314	0.51
0.10	0.628	0.25
0.30	1.885	0.10
0.50	3.142	0.08
1.00	6.283	0.08
∞	∞	0.08

than 0.10 times the ratio of wave height to cylinder diameter. Thus, for this restriction on h/λ and for relatively large piles where H/D is equal to or less than unity, the drag force will be less than 10% of the inertia force and can be neglected in an approximate analysis.

If attention is restricted to the maximum total force, the error committed in neglecting the drag-force components will be even less than indicated above since the drag and inertia forces are out of phase by 90°. The maximum total force F_m is, accordingly, expressible approximately as

$$F_m = \sqrt{F_{Im}^2 + F_{Dm}^2} = F_{Im}\sqrt{1 + \left(\frac{\mu H}{D}\right)^2} \tag{55}$$

so that, for example, for a value $\mu H/D = 0.25$, the error involved in neglecting the drag force is only about 3%.

EXAMPLE 3.4-1. Consider water of depth 75 ft having waves of height 35 ft and length 375 ft and estimate the maximum force and associated base moment exerted on a 4-ft-diameter vertical pile that extends from the bottom through the water surface. Assume that $C_D = 1$, $C_I = 2$.

Although Airy theory is not strictly applicable for this problem because of the relatively large wave height, we use it and the associated force relations for an approximate estimate. The wavenumber k is determined as $k = 2\pi/\lambda = 2\pi/375 = 0.01676\ \text{ft}^{-1}$ and the frequency is determined from equation (10) as $\omega = 0.6773$ rad/sec. Since the ratio of wave height to pile diameter is considerably greater than unity, the drag force cannot be neglected according to the results of equations (54) and (55). The total drag and inertia forces (in pounds) are expressible from equations (47) and (48), with $\rho = 1.99$ slugs/ft³, $D =$ 4 ft, $H = 35$ ft, as

$$F_D = 3200\,[\sinh 2k(h + \eta) + 2k(h + \eta)]\,|\cos \omega t|\cos \omega t$$

$$F_I = -14{,}840 \sinh k(h + \eta) \sin \omega t$$

where η is given by

$$\eta = \frac{H}{2} \cos \omega t = 17.5 \cos \omega t$$

The total force is expressible as $F = F_D + F_I$ and numerical examination for various values of ωt reveals the maximum total force to occur at about $\omega t =$ 6.0 (345°) and to equal 50.2 kips. At this instant, the water surface η on the pile is found to be 16.9 ft above the still-water level, that is, at an elevation $y = h + \eta$ of 91.9 ft above the seafloor.

The associated maximum moment is determined from equation (49) with the values of ωt and y as 2943 ft/kips. The distance the resultant maximum force acts above the base of the pile is 2943/50.2 = 58.6 ft. #

Forces Arising from Stokes Waves

When finite-amplitude Stokes waves are considered, the calculation of wave forces from equation (44) is more difficult because of the series form of the wave solution. Taking the expressions for the horizontal velocity and acceleration from equations (21) and (27) and substituting into Morison's equation, we find, with $x = 0$ chosen as the pile location,

$$f = \frac{\rho C_D D^2}{2k^2} \sum_{m=1}^{4} \sum_{n=1}^{5-m} U_m U_n |\cos m\omega t| \cos n\omega t - \frac{\rho C_I \pi D \omega}{8K} \sum_{n=1}^{5} R_n \sin n\omega t \qquad (56)$$

where the coefficients U_n and R_n are defined by equations (26) and (29). The limits on the double summation are chosen such that products of $U_m U_n$ involving $m + n > 5$ are ignored, in accordance with the accuracy of the Stokes fifth-order theory. Substituting this result into equation (44), we then find the total force $F(y)$ acting on a segment of the pile of height y above the seafloor expressible in terms of drag and inertia forces, F_D and F_I, as

$$F(y) = F_D(y) + F_I(y) \qquad (57)$$

where

$$F_D = \frac{\rho C_D D \omega^2}{2k^3} \sum_{m=1}^{4} \sum_{n=1}^{5-m} A_{mn} |\cos m\omega t| \cos n\omega t \qquad (58)$$

$$F_I = -\frac{\rho C_I \pi D^2 \omega^2}{4k^2} \sum_{n=1}^{5} B_n \sin n\omega t \qquad (59)$$

Writing $S_n = \sinh nkh$ and introducing the velocity coefficient V_n defined by equation (26), the coefficient A_{mn} for $m \neq n$ and A_{nn} for $m = n$ are given by

$$A_{mn} = \frac{G_m G_n}{2S_m S_n}\left[\frac{S_{m+n}V_{m+n}}{(m+n)G_{m+n}} + \frac{S_{m-n}V_{m-n}}{(m-n)G_{m-n}}\right] \tag{60}$$

$$A_{nn} = \frac{G_n^2}{S_n^2}\frac{S_{2n}V_{2n}}{4nG_{2n}} + \frac{ky}{2} \tag{61}$$

and the coefficients B_n are given by

$$\begin{aligned}
B_1 &= V_1 - \frac{1}{6}\frac{G_1G_2}{G_3}\frac{S_3}{S_1S_2}V_3 - \frac{1}{10}\frac{G_3G_3}{G_5}\frac{S_5}{S_2S_3}V_5 \\
B_2 &= V_2 - \frac{1}{2}\frac{G_1^2}{S_1^2}ky - \frac{1}{4}\frac{G_1G_3}{G_4}\frac{S_4}{S_1S_3}V_4 \\
B_3 &= V_3 - \frac{3}{2}\frac{G_2}{S_2}V_1 - \frac{3}{10}\frac{G_1G_4}{G_5}\frac{S_5}{S_1S_4}V_5 \\
B_4 &= V_4 - \frac{1}{2}\frac{G_2^2}{S_2^2}ky - \frac{G_1G_3}{G_2}\frac{S_2}{S_1S_3}V_2 \\
B_5 &= V_5 - \frac{5}{2}\frac{G_2G_3}{G_1}\frac{S_1}{S_2S_3}V_1 - \frac{5}{6}\frac{G_1G_4}{G_3}\frac{S_3}{S_1S_4}V_3
\end{aligned} \tag{62}$$

To calculate the moment of the maximum wave force about the base of the pile, we may use equations (45) and (56) and obtain relations similar to those above for the force. The resulting equations are, however, rather cumbersome and it is generally better to use an approximate numerical method based on the force relations already obtained. In particular, if we imagine the pile divided into N segments, we may calculate the net force acting on each segment from equation (57) at the time associated with the maximum wave force, and then assume these forces uniformly distributed over the segment lengths. The moment about the base can then be determined simply as the sum of the moments of the individual distributed loads. Consider, for example, the pile shown in Fig. 3.15, and imagine it to be divided into two segments, with the lower segment of

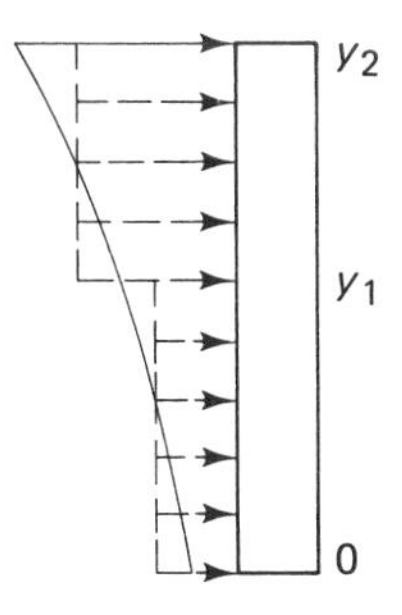

Fig. 3.15

height y_1 and the upper of height $y_2 - y_1$. The net force on the lower segment is given as $F_1 = F(y_1)$ and on the upper segment as $F_2 - F_1 = F(y_2) - F(y_1)$. Under the assumption that these forces are uniformly distributed, their resultants will act at the midlength of the segments and the total moment M about the base is thus expressible as

$$M = \tfrac{1}{2}F_1 y_1 + \tfrac{1}{2}(F_2 - F_1)(y_2 + y_1) \tag{63}$$

where the F's are calculated from equation (57) at the instant of interest. For the moment of the maximum wave force, we obviously must choose that instant at which the wave force is maximum.

The method described above provides a better and better estimate as more and more segments are considered. In the general case of N segments, we have

$$M = \frac{1}{2}\sum_{n=1}^{N}(F_n - F_{n-1})(y_n + y_{n-1}) \tag{64}$$

where F_0 and y_0 are zero.

EXAMPLE 3.4-2. Consider, as in the preceding example, water of depth 75 ft having waves of height 35 ft and length 375 ft, and determine the maximum total force and associated base moment exerted on a 4-ft-diameter vertical pile extending from the bottom through the water surface. Use Stokes fifth-order wave theory and assume that $C_D = 1$, $C_I = 2$.

The wave in this example is the same as that in the example of Stokes wave theory considered earlier in Example 3.3-2 so that those results apply here as well. Taking $\rho = 1.99$ slugs/ft^3, $D = 4$ ft, $\omega = 0.723$ sec^{-1}, and $k = 0.0167$ ft^{-1}, we thus find the total force on the pile by application of equation (57), with $y = h + \eta$, where η is described by equation (18) of the Stokes description. The calculations are conveniently carried out with the help of a digital computer. Figure 3.16 shows the variation of the total force with ωt, together with the variations of the individual drag and inertia parts. The maximum force on the pile is seen to occur at about $\omega t = 6.0$, and is given approximately as

$$F_{\max} = 59.4 \text{ kips}$$

At this instant, the drag force is 46.2 kips and the inertia force is 13.2 kips.

To calculate the corresponding moment of the wave force, we use equation (64) with the uppermost value of $y = h + \eta = 95$ ft, as determined from equation (18) for $\omega t = 6.0$. Table 3.12 shows the results obtained considering 2 segments ($N = 2$), 5 segments ($N = 5$), 10 segments ($N = 10$), and 20 segments ($N = 20$). For $N = 10$ and $N = 20$, the moment is essentially the same and the moment is accordingly determined as $M = 3830$ ft-kips. For $N = 5$, the difference is only about 1% and for $N = 2$, the difference is only about 8%. Thus, the moment relation of equation (61) is seen to provide a good approxi-

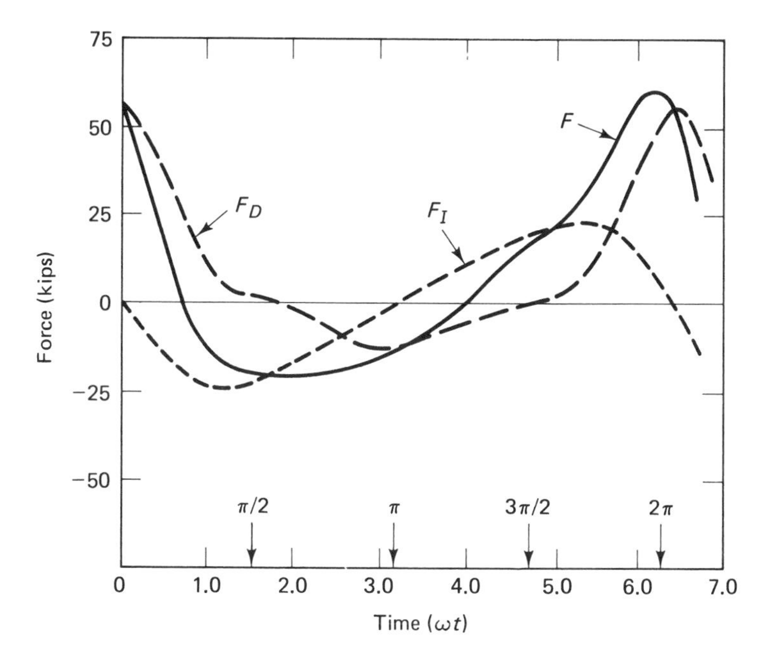

Fig. 3.16

Table 3.12

N	M (ft-kips)
2	3540
5	3790
10	3820
20	3830

mation even with a very small number of segments. The resultant maximum force is seen to act a distance $b = M/F = 64.5$ ft above the bottom. #

These results may be contrasted with those found in Example 3.4-1 using Airy theory and the associated force relations for an approximate estimate. There we found $F_{max} = 50.2$ kips, in contrast with 59.4 kips found here. The associated base moment of 2943 ft-kips estimated earlier may similarly be contrasted with the more precise value of 3830 ft-kips found here.

Forces Arising from Cnoidal Waves

In the case of shallow-water cnoidal waves, the resulting horizontal water velocity and acceleration do not vary with depth so that the wave force f per unit of pile length is therefore constant. The integration of equations (44) for the total force on a pile segment ranging from the seafloor to height y is accordingly equal simply to the product yf. The base moment is similarly determined from equation (45) as $y^2f/2$.

Effects of Relative Motion

The discussion above has been restricted to the case where the pile is fixed against all motion. Such a situation is commonly assumed in the engineering analysis of offshore structures even though the piles do, in fact, suffer some motion under the action of the waves.

When the motion of the pile is considered, the drag force is reduced by the relative motion and the inertia force is reduced by a factor proportional to the acceleration of the pile. The appropriate form of the Morison equation, giving the force per unit of length acting on an element of the pile, can be obtained from equation (42) by substituting $|u - \dot{x}|(u - \dot{x})$ for $|u|u$ and adding the term

$$-\rho(C_I - 1)\frac{\pi D^2}{4}\ddot{x}$$

to the right-hand side, where $\dot{x}$ and $\ddot{x}$ denote the horizontal velocity and acceleration of the pile element. The form of this last term is derived from fluid mechanics and represents the force associated with water acceleration from the pile motion. It is especially important in the dynamic analysis of offshore structures (considered in Chapter 6) where the inertia of the members must be considered.

As an illustration, we may consider a unit length of vertical pile. If F denotes the total horizontal force acting on the segment other than that associated with the water motion, and f denotes the water force represented by the modified Morison equation, the equation of motion of the segment in the horizontal direction is

$$m\ddot{x} = F + f \tag{65}$$

where m denotes the mass of the pile segment and f is expressible, taking into account the foregoing modifications of the Morison equation, from equation (42) as

$$f = \frac{1}{2}\rho C_D D\,|u - \dot{x}|\,(u - \dot{x}) + \rho C_I\frac{\pi D^2}{4}a_x - \rho(C_I - 1)\frac{\pi D^2}{4}\ddot{x}$$

Combining the two equations above we have

$$(m + m')\ddot{x} = F + \frac{1}{2}\rho C_D D\,|u - \dot{x}|\,(u - \dot{x}) + \rho C_I \frac{\pi D^2}{4} a_x \tag{66}$$

where $m' = \rho(C_I - 1)\pi D^2/4$.

From these considerations, we thus see that the effect of the pile segment accelerating in the water is equivalent to increasing the mass of the pile segment (of unit length) by the amount m'. This mass m' is known as the *added mass per unit length* of the pile and the total mass $m + m'$ is known as the *virtual mass per unit length* of the pile.

Effects of Pile Diameter

As noted earlier, the Morison equation is restricted to cases where the pile diameter is small in comparison with the length of the waves encountered, so that the effects of the pile on the waves can be neglected. When this is not the case, the distortion of the wave must be considered.

Fortunately, in cases where the diameter of the pile is comparable with the wave length, the ratio of wave height to pile diameter is usually small and drag forces are negligible in comparison with inertia forces in accordance with equation (54). Thus, in discussing cases where pile-diameter effects must be considered, it is usually sufficient to restrict attention to the inertia force only. This problem has been considered by MacCamy and Fuchs (1954) and, more recently, by Mogridge and Jamieson (1976), assuming small-amplitude wave theory and a rigid pile. The theory is generally known as the diffraction theory of wave forces and results from it show that the effect of the pile is to modify the value of the inertia coefficient in the Morison equation and introduce a phase lag. The equation for the force per unit of length of a vertical pile or cylinder arising from an *incident* surface wave described by

$$\eta = \frac{H}{2}\cos(kx - \omega t)$$

is thus expressible, with $x = 0$ taken at the center of the pile, as

$$f = -\rho C_I \frac{\pi D^2}{4}\omega^2 \frac{H}{2}\frac{\cosh ky}{\sinh kh}\sin(\omega t - \epsilon) \tag{67}$$

where ϵ denotes the phase angle and y is measured from the seafloor, as in our earlier discussions.

Because of the distortion of the wave by the pile, the inertia coefficient C_I and the phase angle ϵ will vary with the ratio of pile diameter to wave length. The inertia-coefficient and phase-angle variations are shown in Fig. 3.17.

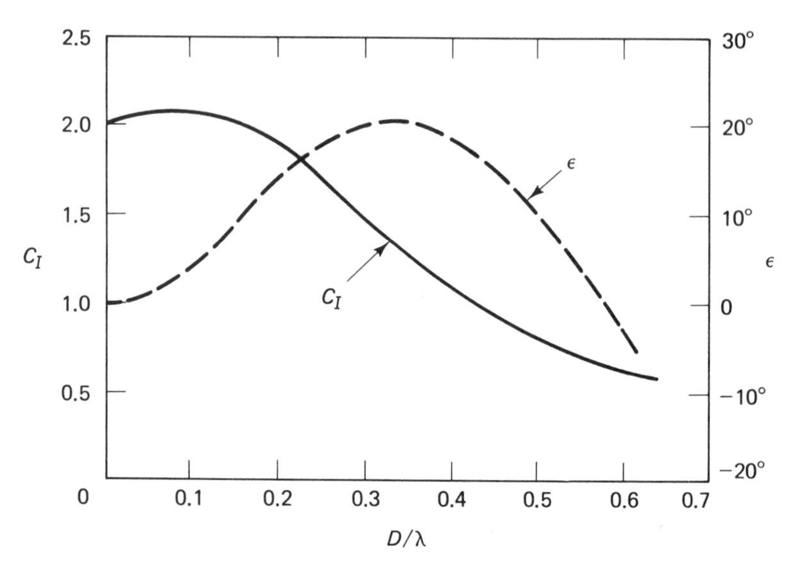

Fig. 3.17. Variation of inertia coefficient and phase angle with ratio of pile diameter to wave length. (After Mogridge and Jamieson, 1976.)

3.5 WAVE FORCES ON ARBITRARILY ORIENTED CYLINDERS

The application of the Morison equation to an arbitrarily oriented cylinder is of interest in connection with the determination of wave force on cross bracings of ocean structures and on inclined (battered) piles. Various approximate methods for applying the Morison equation to these cases have been discussed by Wade and Dwyer (1976). The most consistent generalization is that given by Chakrabarti et al. (1975). This involves resolving the water velocity and acceleration into components normal and tangential to the cylinder axis and using only the normal components in the Morison equation to calculate the wave force per unit length of the cylinder. Such a generalization is of course consistent with equation (42) for a vertical cylinder, since only normal components are involved there. The direction of the wave force on the inclined cylinder is normal to the cylinder axis, but may conveniently be resolved into horizontal and vertical components for engineering applications.

To illustrate, consider a fixed cylinder arbitrarily inclined to axes x, y, z with the y-axis vertical, as shown in Fig. 3.18. Assuming that the wave is propagating in the $+x$-direction, the resulting water motion will have horizontal and vertical velocities u and v and horizontal and vertical accelerations a_x and a_y, as in our previous discussions.

With polar coordinates θ, ϕ defining the orientation of the cylinder axis, as indicated in Fig. 3.18, the magnitude ν of the water velocity normal to the

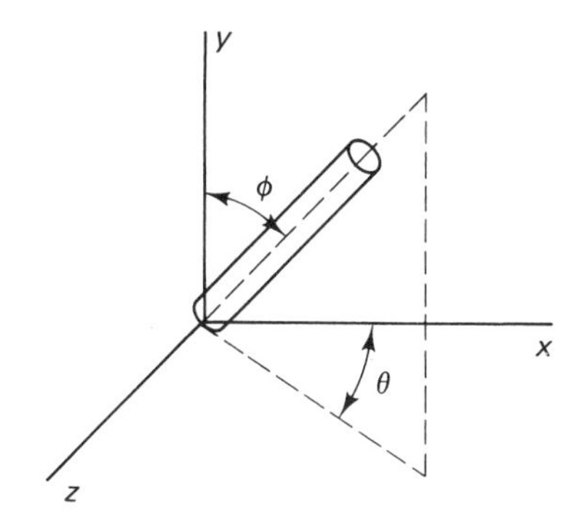

Fig. 3.18. Arbitrarily oriented cylindrical member.

cylinder axis is given by

$$\nu = [u^2 + v^2 - (c_x u + c_y v)^2]^{1/2} \tag{68}$$

and its components in the x, y, and z directions are given, respectively, by

$$\begin{aligned} u_n &= u - c_x(c_x u + c_y v) \\ v_n &= v - c_y(c_x u + c_y v) \\ w_n &= -c_z(c_x u + c_y v) \end{aligned} \tag{69}$$

where

$$\begin{aligned} c_x &= \sin\phi \cos\theta \\ c_y &= \cos\phi \\ c_z &= \sin\phi \sin\theta \end{aligned} \tag{70}$$

The components of the normal water acceleration in the x, y, and z directions are given, respectively, by

$$\begin{aligned} a_{nx} &= a_x - c_x(c_x a_x + c_y a_y) \\ a_{ny} &= a_y - c_y(c_x a_x + c_y a_y) \\ a_{nz} &= -c_z(c_x a_x + c_y a_y). \end{aligned} \tag{71}$$

With these relations, the components of the force per unit of cylinder length acting in the x, y, and z directions are given, respectively, by the generalized Morison equations

$$\begin{aligned} f_x &= \frac{1}{2}\rho C_D D \nu u_n + \rho C_I \frac{\pi D^2}{4} a_{nx} \\ f_y &= \frac{1}{2}\rho C_D D \nu v_n + \rho C_I \frac{\pi D^2}{4} a_{ny} \\ f_z &= \frac{1}{2}\rho C_D D \nu w_n + \rho C_I \frac{\pi D^2}{4} a_{nz} \end{aligned} \tag{72}$$

and the force per unit length f normal to the member is given by

$$f = \pm(f_x^2 + f_y^2 + f_z^2)^{1/2} \tag{73}$$

the sign of f being chosen consistent with the sign of the components f_x, f_y, and f_z.

For a small member such as a cross bracing on a structure, where the water motion does not vary appreciably from one end of the member to the other, average values of u, v, a_x, and a_y may be used in the equations above and the total force on the member described simply by

$$F_x = f_x L, \qquad F_y = f_y L, \qquad F_z = f_z L \tag{74}$$

where L denotes the length of the member. In the more general case where the water velocities and accelerations vary appreciably over the length of the member, the total forces must be calculated by numerical integration of the relations

$$F_x = \int_s f_x\, ds, \qquad F_y = \int_s f_y\, ds, \qquad F_z = \int_s f_z\, ds \tag{75}$$

where s denotes distance along the member axis, and the limits on the integrals are chosen so as to include all of the pile on which the wave force acts.

EXAMPLE 3.5-1. Consider the front view of an offshore structure as shown in Fig. 3.19 and determine the force exerted on member 1–2 for relatively uniform wave-induced water motion over the member described by

$$u = 14 \text{ ft/sec}, \qquad v = 4 \text{ ft/sec}$$

$$a_x = 4 \text{ ft/sec}^2, \qquad a_y = -6 \text{ ft/sec}^2$$

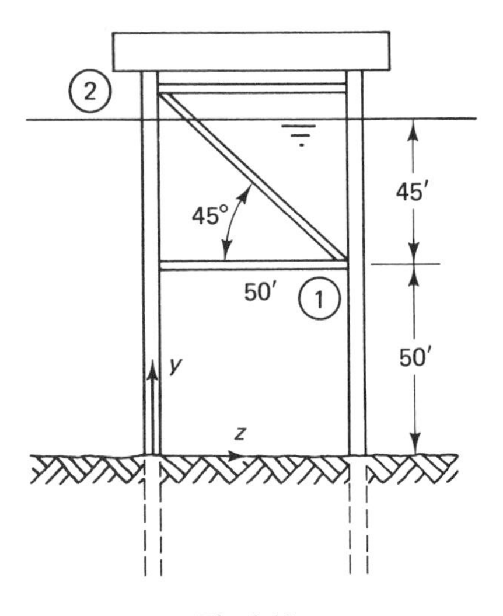

Fig. 3.19

The water level at the instant the above values apply is 95 ft above the seafloor. The diameter of the member is 2 ft.

From Fig. 3.19, we have the angle θ, ϕ defining the orientation of the member given as

$$\theta = 90^\circ, \qquad \phi = 135^\circ$$

Using equation (70), we then find

$$c_x = 0, \qquad c_y = -0.707, \qquad c_z = 0.707$$

The velocity and acceleration components normal to the member are given from equations (69) and (71) as

$$u_n = 14 \text{ ft/sec}, \qquad v_n = w_n = 2 \text{ ft/sec}$$
$$a_{nx} = 4 \text{ ft/sec}^2, \qquad a_{ny} = a_{nz} = -3 \text{ ft/sec}^2$$

The magnitude of the velocity, normal to the member, is determined from equation (68) as

$$v = 14.3 \text{ ft/sec}$$

and the force components f_x, f_y, f_z are determined from equation (72), with $C_D = 1$, $C_I = 2$, $\rho = 1.99$ slugs/ft^3, $D = 2$ ft, as

$$f_x = 448 \text{ lb/ft}, \qquad f_y = f_z = 19.4 \text{ lb/ft}$$

The length of the member in the water is $L = 45\sqrt{2} = 63.6$ ft, so that, finally, we have the total force components determined (because of the assumed uniform flow) from equation (74) as

$$F_x = 28.5 \text{ kips}, \qquad F_y = F_z = 1.23 \text{ kips}$$

It may be noted from Fig. 3.19 that we could equally well have chosen the the angles θ, ϕ defining the member orientation as $\theta = -90^\circ$, $\phi = 45^\circ$, rather than those chosen above, since the assumed positive direction along the member axis is arbitrary for these calculations. With this alternative choice, the signs of C_y and C_z will be reversed from those given above but the normal velocity and acceleration components—and hence the forces—will remain unchanged. #

3.6 MAXIMUM WAVE FORCE ON AN OFFSHORE STRUCTURE

The equations developed above for vertical piles and arbitrarily oriented cylinders may be applied to the individual members of an offshore structure in regular waves and the results combined to determine the maximum horizontal wave force on the structure. If the main piles of the structure are vertical, equation (46) or (57) may be used to calculate the total force on them, the first equation applying when Airy waves are assumed and the second when Stokes waves are assumed. For piles located at the (arbitrarily chosen) origin $x = 0$, these equations apply directly and provide the total force on the pile at any instant ωt. For piles located a distance x_0 from those at $x = 0$, the equations must, however, be modified in order to obtain the force at the same instant ωt. Since the wave-induced water motion depends only on the parameter $\omega t - kx$, the modification is seen to involve simply the replacement of ωt with $\omega t - kx_0$ in the equations, with k denoting, of course, the wavenumber.

The horizontal force on members other than the vertical piles must generally be calculated by numerical integration of the first of equations (72), taking into account the variation of the water motion over the member. When the main piles are inclined rather than vertical, this equation must also be used for calculating the force on them. The maximum horizontal force on the structure is found by considering all members struck by the waves and numerically examining the total force for various values of ωt. The details will be illustrated in Example 3.6-1.

Figure 3.20 shows the results of calculations of the type above for a small test structure experiencing regular waves of various frequencies (Rolfes, 1980). All four faces of the structure were similar, with the diagonal members on any one face framing into those on the adjacent faces. Note that the frequencies are expressed in units of Hz (1 Hz = 2π rad/sec). The wave heights assumed in the calculations were 1/20 of the corresponding wave lengths. The Stokes wave theory was assumed in all calculations. Also shown are measurements made when the test structure was placed in a wave tank and regular waves of the prescribed kind run against it. It can be seen that the agreement between theory and measurement is good in spite of the simple choice of values for the drag and inertia coefficients ($C_D = 1$, $C_I = 2$).

EXAMPLE 3.6-1. A simple offshore structure consists of four vertical piles braced with horizontal and diagonal members. Side and front views of the structure are shown in Fig. 3.21 on page 130. The corresponding side and front faces behind those shown are identical. For a wave of height 20 ft and length 300 ft, in water of depth 80 ft, estimate the total horizontal wave force on the structure. The piles have diameters of 4 ft and the horizontal and diagonal members have a diameter of 2 ft. Assume that $C_D = 1$, $C_I = 2$.

The maximum horizontal wave force on the structure is determined by trial calculations for various values of ωt. As an illustration, we first assume the value $\omega t = 6.0$. To simplify the calculations, we use Airy wave theory, although similar procedures apply using the more accurate Stokes theory. Coordinate axes are chosen as indicated in Fig. 3.21.

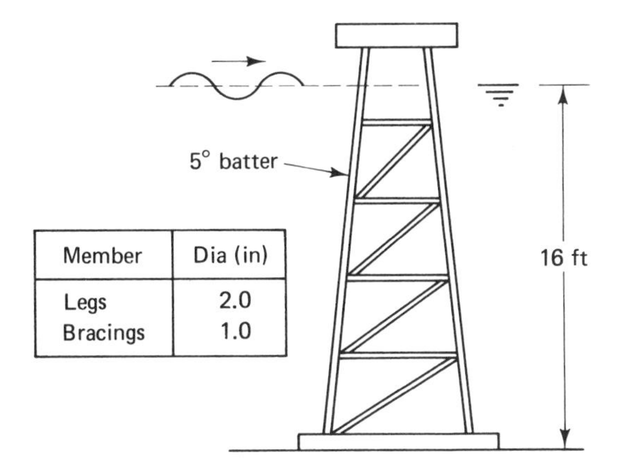

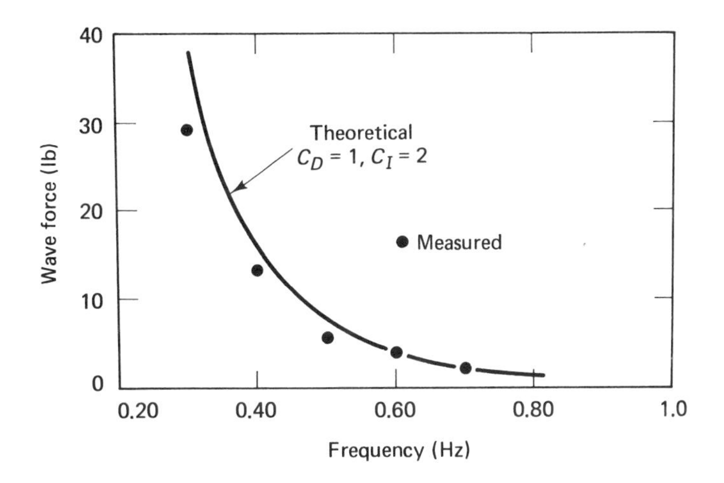

Fig. 3.20. Comparison of the theoretical and measured total wave force on a small offshore test structure.

Vertical Piles.

For the wave force on the main vertical piles, we use equations (46)–(48). These equations were developed for the case of a pile located at $x = 0$, and, hence, apply directly to the pile 1–3 for any instant ωt. For pile 4–6, located at $x = 50$, the equations must be modified in order to obtain the force at the same instant ωt. Since, as noted above, the wave-induced water motion depends only on the parameter $\omega t - kx$, the modification is seen to involve simply replacing ωt by $\omega t - 50k$ in equations (47) and (48). Thus, for pile 1–3, at $x = 0$,

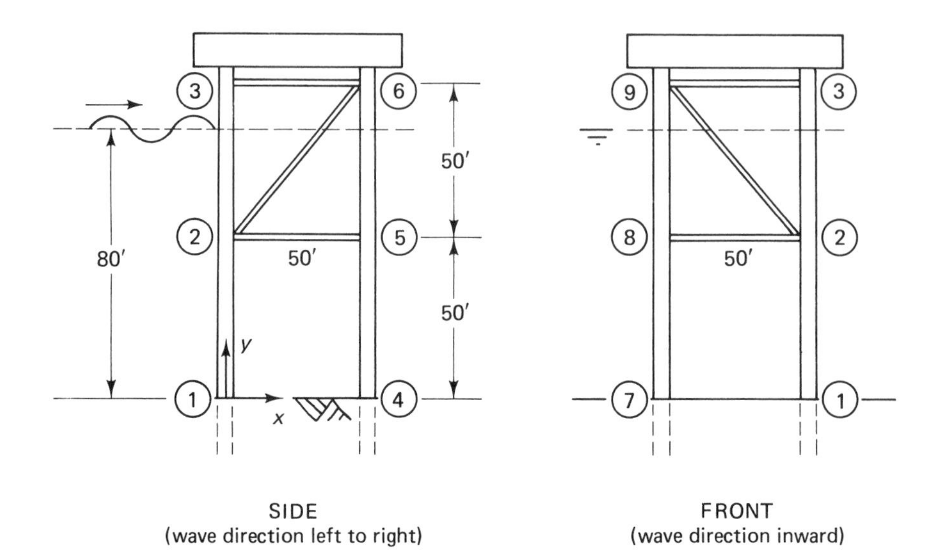

Fig. 3.21. Offshore structure: (a) side view (wave direction left to right); (b) front view (wave direction inward).

we have, on substituting numerical values $k = 0.0209$ ft, $\omega = 0.793$ rad/sec, $H = 20$ ft, $\rho = 1.99$ slugs/ft³, $h = 80$ ft, that

$$F_{1\text{-}3} = 450(\sinh 2ky + 2ky)|\cos \omega t| \cos \omega t - 5825 \sinh ky \sin \omega t$$

where

$$y = h + \eta = 80 + 10 \cos \omega t$$

And for pile 4–6, we have

$$F_{4\text{-}6} = 450(\sinh 2ky + 2ky)|\cos(\omega t - 50k)| \cos(\omega t - 50k) - 5825 \sinh ky \sin(\omega t - 50k)$$

where

$$y = h + \eta = 80 + 10 \cos(\omega t - 50k)$$

There are two piles at $x = 0$ and two at $x = 50$ in the full three-dimensional structure, so that the total force on the four piles F_p is

$$F_p = 2(F_{1\text{-}3} + F_{4\text{-}6})$$

Substituting $\omega t = 6.0$, we find

$$F_p = 2(15{,}580 + 15{,}870) = 62{,}900 \text{ lb}$$

Horizontal Front-face Members

The horizontal force per unit of length acting on a front-face horizontal member is determined from the first of equations (72), with $\theta = -90°$, $\phi = 90°$. We have

$$f_x = 1.99 \mathsf{v} u + 12.5 a_x$$
$$\mathsf{v}^2 = u^2 + v^2$$

For member 2–8, with $x = 0$, $y = 50$, the velocity and acceleration components are expressible from equations (7), (8), and (13) at $\omega t = 6.0$ as

$$u = 4.73 \text{ ft/sec}, \qquad v = 1.07 \text{ ft/sec}, \qquad a_x = 1.09 \text{ ft/sec}^2$$

These values apply along the entire 50-ft member length since the x-y coordinates do not change from one part of the member to another. Substituting, we find $f_x = 59.2$ lb/ft, which is constant along the member. The total horizontal force $F_{2\text{-}8}$ is accordingly determined as

$$F_{2\text{-}8} = (59.2)(50) = 2960 \text{ lb}$$

A similar calculation for the horizontal member on the other front face, say, member 5–11, gives (with $x = 50$, $y = 50$) $F_{5\text{-}11} = 2460$ lb.

Diagonal Front-face Members

The horizontal wave force f_x per unit of length on diagonal front-face members is determined from the first of equations (72), with $\theta = -90°$ $\phi = 45°$. We find

$$f_x = 1.99 \mathsf{v} u + 12.5 a_x$$
$$\mathsf{v}^2 = u^2 + \tfrac{1}{2} v^2$$

For member 2–9, we have $x = 0$ and y varying from 50 to the water surface. The elevation of the water surface on the member is determined from the equation

$$y = 80 + 10 \cos \omega t$$

which, for $\omega t = 6.0$, gives $y = 89.6$ ft. Since the member lies at an angle of 45° to the y-axis, that part of its length over which the force acts is $(89.6 - 50)/0.707 = 56.0$ ft. The wave force per unit of length varies along this member, because of the variation of the water motion with the y-coordinate and, hence, the total horizontal wave force must be determined by numerical integration. As an approximation, we divide the 56-ft length into two segments of 28.0-ft length and calculate the force f_x at the midlength of each segment. The total force is then calculated by assuming the f_x values constant over the respective segments. The calculations are summarized in Table 3.13 for $\omega t = 6.0$. The

Table 3.13

x (ft)	y (ft)	u (ft/sec)	v (ft/sec)	a_x (ft/sec²)	f_x (lb/ft)
0	59.9	5.60	1.38	−3.77	79.5
0	79.7	8.12	2.20	−5.99	156.8

total horizontal force $F_{2\text{-}9}$ is calculated as

$$F_{2\text{-}9} = (79.5 + 156.8)(28) = 6620 \text{ lb}$$

A similar calculation for the diagonal member, say 5–12, on the other front face gives $F_{5\text{-}12} = 3590$ lb.

Diagonal Side-face Members

The horizontal force per unit of length acting on a side-face diagonal is determined from the first of equations (72), with $\theta = 0$, $\phi = 45°$. We find

$$f_x = 0.995v(u - v) + 6.25(a_x - a_y)$$
$$v^2 = \tfrac{1}{2}(u - v)^2$$

For member 2–6, we have both x and y varying along the member to its intersection with the water surface.

The elevation of the water surface on the member is calculated in this case from the equations

$$y = 80 + 10 \cos(\omega t - kx)$$
$$y = 50 + x$$

The first of these gives the water-surface elevation for given values of t and x and the second is the geometric equation for member. Combining these equations, we have

$$x = 30 + 10 \cos(\omega t - kx)$$

For $\omega t = 6$, this equation gives (by iteration or trial-and-error solution) $x = 35.2$ ft. The elevation of the water surface on the member is thus $50 + 35.2 = 85.2$ ft and the length of that part of the member struck by the wave is $35.2/0.707 = 49.8$ ft. As in the previous calculation, the wave force per unit length varies along the member and we estimate the total force by dividing the member length struck by the wave into two segments each of length 24.9 ft. The wave forces f_x are calculated at the midlength of each segment and the total force calculated by assuming these values constant over the respective segments. The calculations are summarized in Table 3.14. The total force is thus calculated as

$$F_{2\text{-}6} = (40.2 + 53.4)24.9 = 2330 \text{ lb}$$

Table 3.14

x (ft)	y (ft)	u (ft/sec)	v (ft/sec)	a_x (ft/sec^2)	a_y (ft/sec^2)	f_x (lb/ft)
8.8	54.8	5.10	2.17	2.04	−3.41	40.2
26.4	76.4	5.31	5.42	4.66	−3.88	53.4

An identical calculation applies for the corresponding diagonal member on the other side face, say member 8–12, so that the total force on this member is $F_{2\text{-}8} = 2330$ lb.

Total Wave Force

The upper horizontal members of the structure are not struck by the wave and hence have no wave force acting on them. The lower horizontal members on the side face are parallel to the wave direction and similarly have no horizontal wave force acting on them. The total horizontal wave force acting on the structure at $\omega t = 6.0$ is thus obtained by adding all of the forces on the piles and members calculated above, giving a value of 83,200 lb.

To find the maximum total horizontal force on the structure, the process described above must be repeated for other values of ωt. In this way, the maximum total horizontal force is found to occur approximately at $\omega t = 5.8$ and to equal 83,600 lb.

3.7 JOINT LOADS FROM WAVE FORCES

The analysis of a framed offshore structure by the methods of Chapter 2 requires that the distributed wave loadings on the members be replaced by equivalent joint forces and moments (see Section 2.4). In determining such loadings, it is customary to select the instant in the wave cycle when the horizontal wave force on the structure is at its maximum value. The contributions to the equivalent loadings from each member are then calculated assuming an idealized distribution of the wave force over the member, or segments of it. The total joint loads are finally determined as the sum of the contributions from each member framing into the joint.

Equivalent joint loads for a member having a uniform load distribution have been given previously in Chapter 2. A generalization of this case is shown in Fig. 3.22 where equivalent joint loads are given for a member having a bilinear force distribution over a portion of its length. The inclusion of a segment of the member over which no forces act is appropriate for the upper members of an offshore structure not struck entirely by waves. The bilinear distribution over the loaded segment of the member also allows a better representation of the wave force distribution than a single linear distribution. A more

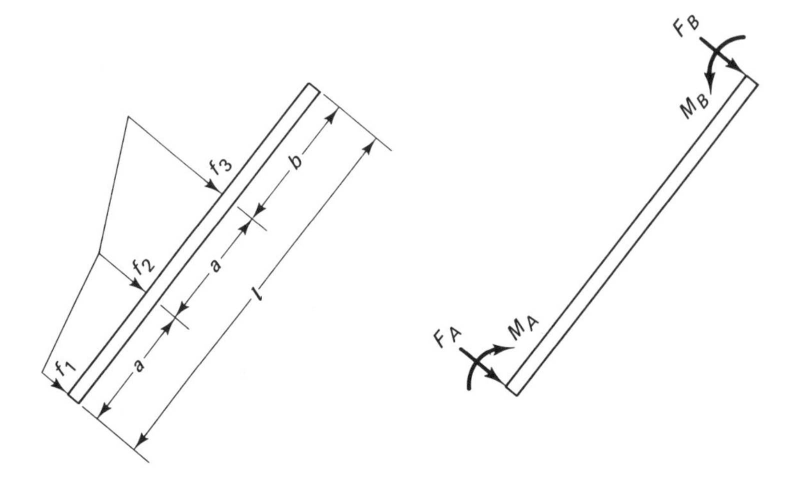

$$P_1 = f_1 + 2f_2 + f_3$$
$$P_2 = 5f_1 + 6f_2 + f_3$$
$$P_3 = 17f_1 + 14f_2 + f_3$$
$$P_4 = 49f_1 + 30f_2 + f_3$$

$$F_A = \frac{P_1 ab^2}{2l^3}(3l - 2b) + \frac{P_2 ba^2}{l^3}(l - b) + \frac{P_3 a^3}{4l^3}(l - 2b) - \frac{P_4 a^4}{10l^3}$$

$$M_A = \frac{P_1 ab^2}{2l^2}(l - b) + \frac{P_2 ba^2}{6l^2}(2l - 3b) + \frac{P_3 a^3}{12l^2}(l - 3b) - \frac{P_4 a^4}{20l^2}$$

$$F_B = \frac{P_1 a}{2} - F_A$$

$$M_B = \frac{P_1 ab}{2} + \frac{P_2 a^2}{6} - F_A l + M_A$$

Fig. 3.22. Equivalent end loads for an assumed bilinear force distribution.

refined description involving three or more linearly loaded segments may, of course, be developed and employed if the bilinear distribution is not considered an adequate representation of the actual load distribution.

EXAMPLE 3.7-1. Determine equivalent joint loads from wave forces for a side frame of the offshore structure considered in Example 3.6-1. The maximum horizontal wave force on the structure occurs at the instant $\omega t = 5.8$. The position of the wave on the structure at this instant is shown in Fig. 3.23(a).

We first consider the diagonal member 2–6. The water surface at the instant $\omega t = 5.8$ is at an elevation of 83.7 ft on this member. The idealized

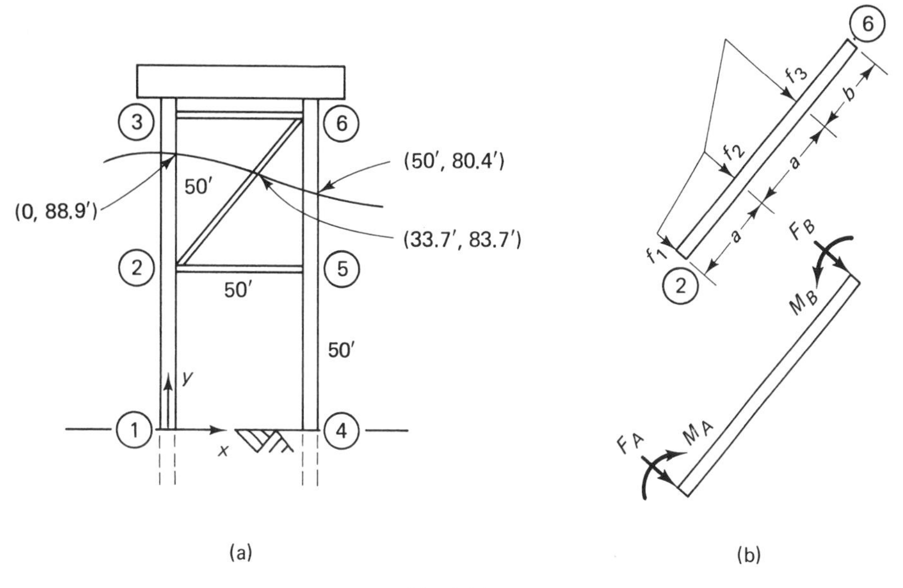

Fig. 3.23

loading is shown in Fig. 3.23(b) with a and b accordingly determined (since the member is inclined 45° to the vertical) as 23.8 ft and 23.0 ft, respectively. The normal forces f_1, f_2, f_3 are determined from equation (73) as

$$f_1 = 46.5 \text{ lb/ft} \quad (x = 0,\ y = 50 \text{ ft})$$
$$f_2 = 62.0 \text{ lb/ft} \quad (x = 16.9,\ y = 66.9 \text{ ft})$$
$$f_3 = 61.1 \text{ lb/ft} \quad (x = 33.7,\ y = 83.7 \text{ ft})$$

The corresponding equivalent joint loads are thus found from Fig. 3.22 as

$$F_A = 1.88 \text{ kips}, \qquad M_A = 22.0 \text{ ft-kips}$$
$$F_B = 0.87 \text{ kip}, \qquad M_B = 15.3 \text{ ft-kips}$$

Resolving the equivalent joint forces into x and y components and choosing positive forces as those along the coordinate axes and positive moments as clockwise in conformity with the convention used in the structural theory of Chapter 2, we thus have equivalent joint loadings at joints 2 and 6 from this member to be

$$F_{2x} = 1.32 \text{ kips}, \qquad F_{2y} = -1.32 \text{ kips}, \qquad M_2 = 22.0 \text{ ft-kips}$$
$$F_{5x} = 0.62 \text{ kip}, \qquad F_{5y} = -0.62 \text{ kip}, \qquad M_5 = -15.3 \text{ ft-kips}$$

Similar calculations apply to the remaining members, with results summarized below.

Member 1–2:

$$F_{1x} = 2.46 \text{ kips}, \qquad M_1 = 21.6 \text{ ft-kips}$$
$$F_{2x} = 3.26 \text{ kips}, \qquad M_2 = -24.9 \text{ ft-kips}$$

Member 2–3:

$$F_{2x} = 6.46 \text{ kips}, \qquad M_2 = 60.2 \text{ ft-kips}$$
$$F_{3x} = 5.52 \text{ kips}, \qquad M_3 = -60.2 \text{ ft-kips}$$

Member 4–5:

$$F_{4x} = 3.35 \text{ kips}, \qquad M_4 = 29.0 \text{ ft-kips}$$
$$F_{5x} = 4.08 \text{ kips}, \qquad M_5 = -32.0 \text{ ft-kips}$$

Member 5–6:

$$F_{5x} = 5.64 \text{ kips}, \qquad M_5 = 46.3 \text{ ft-kips}$$
$$F_{6x} = 2.31 \text{ kips}, \qquad M_6 = -29.7 \text{ ft-kips}$$

Member 2–5:

$$F_{2y} = -0.27 \text{ kip}, \qquad M_2 = 1.05 \text{ ft-kips}$$
$$F_{5y} = 0.29 \text{ kip}, \qquad M_5 = 1.25 \text{ ft-kips}$$

In addition to the foregoing joint forces, we must also consider contributions from members on front faces of the structure framing into the joints of the side face. Neglecting out-of-plane forces and moments, as is necessary in a two-dimensional analysis of the side frame, the front-face horizontal framing into joint 2 is found by numerical integration of equations (75) to have a total horizontal force of 3.18 kips and a total vertical force of −0.86 kip. Thus, assuming one-half of these forces to act at joint 2, we have the additional joint loadings

$$F_{2x} = 1.59 \text{ kips}, \qquad F_{2y} = -0.43 \text{ kip}$$

For the front-face diagonal framing into joint 2, we similarly have the joint loadings

$$F_{2x} = 3.25 \text{ kips}, \qquad F_{2y} = -0.21 \text{ kip}$$

The net contributions from the horizontal and diagonal front-face members framing into joint 5 are similarly calculated as

$$F_{5x} = 2.68 \text{ kips}, \qquad F_{5y} = 1.17 \text{ kips}$$

Finally, adding together all the results, above, we find the joint forces to be as summarized in Table 3.15. It may be noted that the front face diagonals

Table 3.15

Joint	F_x (kips)	F_y (kips)	M (ft-kips)
1	2.46	0	21.6
2	15.9	−2.23	58.4
3	5.52	0	−60.2
4	3.35	0	29.0
5	12.4	1.46	15.6
6	2.93	−0.62	−45.0

framing into joints 2 and 5 also have out-of-plane forces F_z given, respectively, by −0.420 and 0.954 kip. These forces are neglected in the two-dimensional representation of the joint loads given above. #

3.8 BUOYANT FORCES

Pressure loadings on a fully or partially submerged object arise from the weight of the water above it and from the movement of the water around it by wave action. Equation (15) gives, for example, the magnitude of the pressure as determined by the Airy theory. One effect of pressure on a submerged member of an offshore structure is to induce stresses in the member not accounted for in the structural analysis described in Chapter 2. These stresses are discussed in Chapter 4. Another effect of the pressure is to exert horizontal and vertical forces on the member. The forces arising from the pressure associated with wave action are included in the generalized Morison relation of equation (72). However, an additional *bouyant force* also arises from the hydrostatic pressure p given by

$$p = \gamma(h - y) \tag{76}$$

where γ denotes the specific weight of the water, h denotes the depth of water, and y denotes vertical distance from the seafloor. This force exists even when wave action is absent and must be accounted for separately.

To analyze this force, consider the arbitrarily shaped body shown in Fig. 3.24 and imagine a vertical volume element of it of end area dA and height $y_2 - y_1$. The pressure forces acting on the ends of this element are equal simply

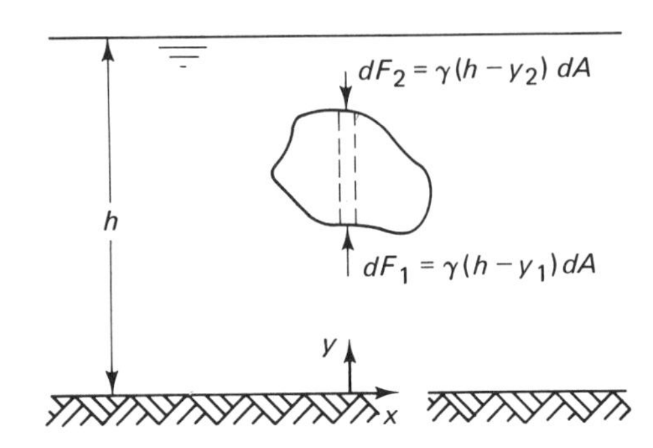

Fig. 3.24. Hydrostatic forces on a submerged body.

to the product of the local pressure and the area dA. Using equation (76), the net vertical force is therefore

$$dF = dF_1 - dF_2 = \gamma(y_2 - y_1)\, dA$$

and the toal upward force on the body is

$$F = \gamma \int_A (y_2 - y_1) dA = \gamma B \tag{77}$$

where B denotes the total volume of water displaced by the body. A similar calculation using horizontal volume elements also shows that the net horizontal force on the body is zero. Thus, the hydrostatic pressure is seen to cause the body to be buoyed up by a vertical force equal to the weight of water displaced by the body (Archimedes' principle). This result also remains valid when the body is only partially submerged.

In accounting for the buoyant force on an offshore structure, it is convenient to combine this with the weight of the structure and treat the two as an effective weight. In particular, if W denotes the weight of the structure in air, it is easily seen from the results above that its effective weight W' in water is given simply by

$$W' = W - \gamma B \tag{78}$$

where B denotes the total volume of water displaced by the structure.

This equation applies for the entire structure. Some care must, however, be exercised when applying it to the individual members, as in a detailed stress analysis of the structure. To illustrate, consider the vertical pile, consisting of members 1–2 and 2–3, shown in Fig. 3.25. The weights (in air) of the members below and above the seafloor are denoted, respectively, by w_1 and w_2. Now, the soil beneath the seafloor can generally be regarded as porous, with the pores

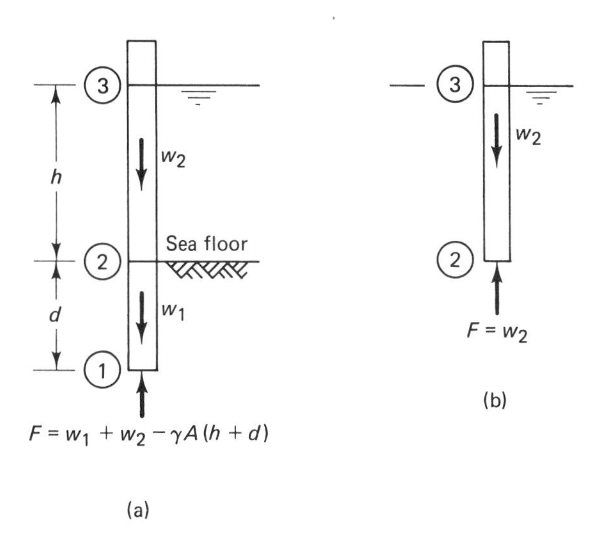

Fig. 3.25. Buoyant and weight forces on a pile.

filled with water at the same hydrostatic pressure that would exist if the soil were absent. The buoyant force acting at the base of the pile is accordingly determined as $\gamma A(h + d)$, where A is the base area of the pile over which the hydrostatic pressure acts. But this is simply equal to the weight of the water displaced by the entire pile, and hence its effective weight is its actual weight less the weight of water displaced. However, since the buoyant force acts only at the base of the pile, the effective weight of member 2–3 above the seafloor is seen to be its actual weight, as indicated in Fig. 3.25. This result is significant in determining joint loads on a structure arising from weight and buoyant forces.

EXAMPLE 3.8-1. Determine equivalent joint loads arising from weight and buoyant forces acting on a steel structure, having side face as shown in Fig. 3.26, for the instant when the wave force is at its maximum value ($\omega t = 5.8$). All four faces of the structure are similar, with diagonals on the front faces framing into joints 2 and 5. The position of the water surface on the structure at the instant $\omega t = 5.8$ is as indicated in Fig. 3.26. Vertical members have diameters of 4 ft and wall thicknesses of 1.5 in. Horizontal and diagonal members have diameters of 2 ft and wall thicknesses of 0.5 in.

We first consider the diagonal member 2–6. This member is assumed unflooded (water tight), so that its weight is calculated simply as the product of the specific weight of steel (0.484 kip/ft³) and the volume of steel in the member (18.1 ft³), that is, 8.76 kips. Since the member is inclined 45° to the vertical, this

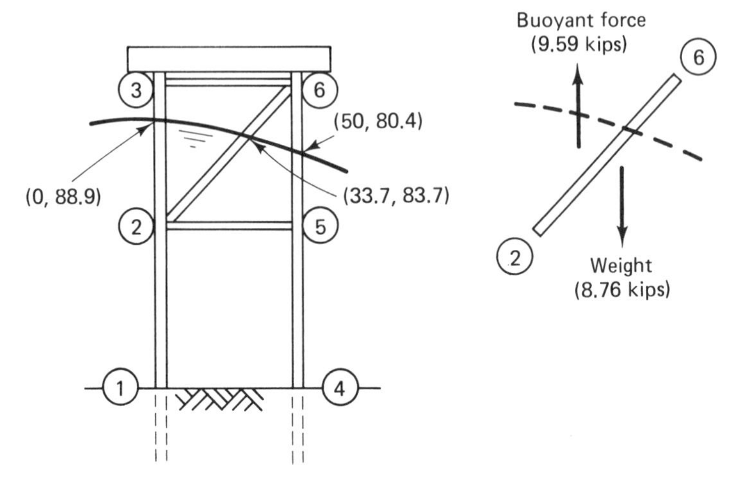

Fig. 3.26

downward vertical force has equal components of 6.20 kips parallel and normal to the member axis. Moreover, these are uniformly distributed along the entire length of the member so that equivalent end loads may be calculated according to the results given in Fig. 2.28. These are illustrated in Fig. 3.27(a).

To calculate the buoyant force on this member, we note that the water surface at the instant $\omega t = 5.8$ intersects the member a distance of 47.7 ft from end 2 (Fig. 3.26). Buoyant force thus acts only on this segment and is equal simply to the specific weight of water (64 lb/ft³) and the volume of water displaced (149.9 ft³), that is, 9.59 kips. This vertical force has components 6.78 kips parallel and normal to the member, uniformly distributed over the loaded segment. Equivalent end loads for the normal force may be calculated from the results given in Fig. 3.20 and those for the parallel force from the results given in Fig. 2.28. The results are shown in Fig. 3.27(b). On adding the end loads from the weight and buoyant forces, we have finally the end loads shown in Fig. 3.27(c), both for end forces normal and parallel to the member and when resolved into horizontal and vertical components. These end loads act at joints 2 and 6 of the side frame.

The joint loadings from the weight and buoyant forces acting on the remaining members of the side face may be calculated in a similar manner. Notice that the vertical members 1–2, 2–3, etc. have no buoyant forces acting on them so that joint loads from these members arise solely from the weight of steel in them (36.8 kips). This is true regardless of whether or not the members are flooded since any water inside will be supported by water in the members below the seafloor. For member 1–2, for example, vertical downward forces of 18.4 kips thus exist at joints 1 and 2 (see Fig. 2.28). Also, since the midlevel member 2–5 is completely submerged, its effective weight is uniformly distributed over the entire length of the member. Assuming the member unflooded, this

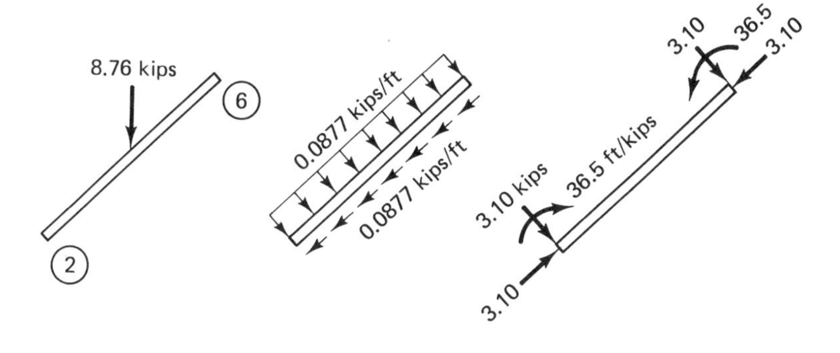

(a) Weight Loading

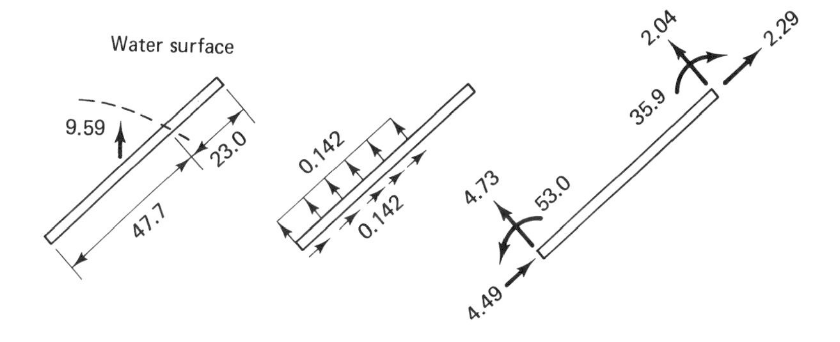

(b) Buoyant force

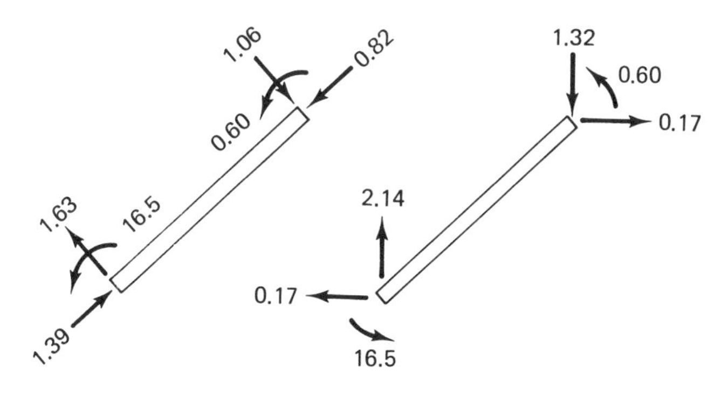

(c) Net End Loading

Fig. 3.27. (a) Weight loading; (b) buoyant force; (c) net end loading.

effective weight is calculated from equation (78) as 3.85 kips, acting upward, corresponding to a uniformly distributed upward loading of 0.077 kip/ft. Vertically upward forces of 1.92 kips thus exist at joints 2 and 5, together with moments of 16.0 ft-kips, that at joint 2 being counterclockwise and that at joint 5 being clockwise. In contrast, for the upper horizontal member 3–6, no buoyant force exists since it is not submerged and its uniformly distributed weight loading (0.124 kip/ft) gives rise to downward vertical forces at joints 3 and 6 of 3.10 kips, together with moments of 25.8 ft-kips, that at joint 3 being clockwise and that at joint 6 being counterclockwise.

Finally, joint forces from the effective weights of the members on the front faces of the structure framing into the joints of the side face are calculated assuming that their effective weights are carried equally by the two side faces. For example, the effective weight of member 2–9 is 2.29 kips, acting upward, so that it contributes an upward force of 1.15 kips at joint 2. Similarly, member 2–8 contributes a vertically upward force of 1.93 kips at joint 2 and member 3–9 contributes a vertically downward force of 3.10 kips at joint 3. The diagonal and horizontal members framing into joint 5 from its adjacent front face contribute a vertically downward force of 0.06 kip at joint 5 and the horizontal front-face member framing into joint 6 contributes a downward vertical force of 3.10 kips.

Combining all the results above, we find ultimately the joint loadings resulting from weight and buoyant forces to be as listed in Table 3.16.

Table 3.16

Joint	F_x (kips)	F_y (kips)	M (ft-kips)
1	0	−18.4	0
2	−0.17	−29.7	−32.5
3	0	−24.6	25.8
4	0	−18.4	0
5	0	−33.0	16.0
6	0.17	−25.9	−26.4

3.9 CURRENT LOADINGS

Currents refer to the relatively constant motion of water resulting from such sources as tidal action, wind drag, or river discharge. The most common currents considered in offshore structural analysis are tidal currents and wind-drift currents, the latter arising from the drag of local wind on the water surface.

Both of these curents are usually regarded as horizontal and varying with depth.

The magnitude and direction of the tidal current at the water surface is generally estimated from local field measurements, the direction of the current reversing with the rise and fall of the tide. In contrast, the wind-drift current at the water surface is normally calculated by assuming it to be about 1% of the overhead wind, 30 ft above the water. For engineering purposes, it is frequently assumed that the depth variation for tidal currents is governed by a one-seventh power law and that for wind-drift current is governed by a linear law (Olsen, 1976). These variations are illustrated in Fig. 3.28.

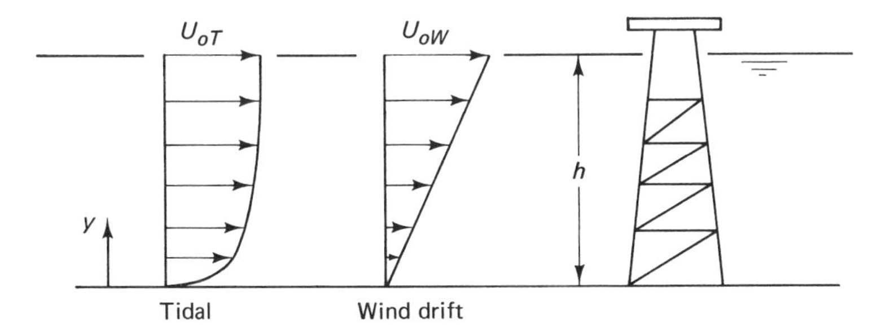

Depth variation: $U_T = U_{oT}(y/h)^{1/7}$ (tidal),
$U_W = U_{oW}(y/h)$ (wind drift)

Fig. 3.28. Assumed vertical distribution of tidal and wind-drift currents.

During storm conditions, currents exist simultaneously with the water motion arising from surface waves. The direction of the tidal current may, of course, be at angle to that of the direction of wave propagation. However, the wind-drift current is always assumed to be in the same direction as that of the wave movement.

Surface waves that exist simultaneously with currents are modified from those that would exist in the absence of any currents. The strength of the modification can generally be expected to vary with the ratio of maximum current to wave·speed. For design waves representative of extreme storm conditions, this ratio is usually sufficiently small to allow neglect of any modification caused by the current. Thus, for regular design waves and a horizontal current of arbitrary depth variation, the force exerted on an offshore structure is normally calculated by simply adding the horizontal water velocity caused by the waves to that component of the current velocity in the direction of wave propagation and using this resultant horizontal velocity together with the vertical water velocity and the two acceleration components in the manner outlined in Sections 3.6 and 3.7 for determining the maximum force and joint loadings.

3.10 *ADDITIONAL ENVIRONMENTAL LOADINGS*

Ice Loadings

Ice loadings can be important in certain offshore site locations, especially in polar regions, where thick sheets of ice can move with the tide and strike the legs of an offshore with significant force (Fig. 3.29).

Fig. 3.29. Ice loading on a structure.

The force F exerted by ice crushing against the leg of a structure is determined by the equation

$$F = Cf_c A \tag{79}$$

where f_c denotes the crushing strength of ice, C a force coefficient, and A the area struck by the ice. Typical values of C range between 0.3 and 0.7 and values of f_c range between 200 and 500 lb/in². In the absence of definite experimental information, a value of $Cf_c = 350$ lb/in² may be employed, representative of extreme conditions. For ice of thickness t striking a pile of diameter D, the area A may be taken as tD, again representative of extreme conditions. Since tidal action can cause significant variation in the still-water level at certain offshore sites, consideration must be given not only to the magnitude of the ice force but also to possible locations on the structure where this force can act.

Mud Loadings

Mud loadings can arise from mud slides in certain offshore sites, particularly in active delta regions, where soft soil is continually being deposited by river discharge (Fig. 3.30).

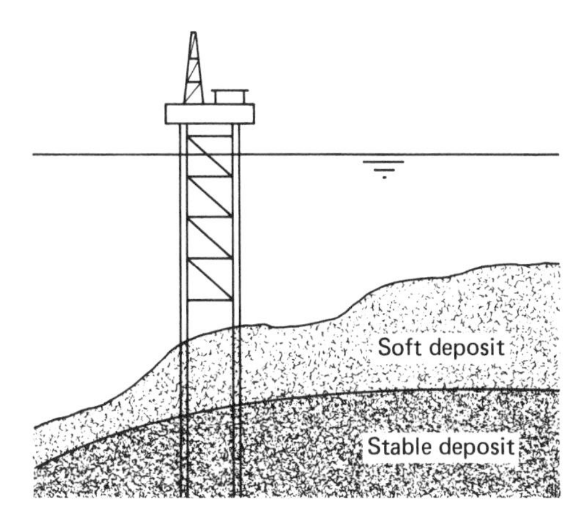

Fig. 3.30. Mud loading on a structure.

The force f acting on a unit length of pile embedded in sliding soil is expressible as

$$f = N\tau D \tag{80}$$

where N denotes a force coefficient, τ denotes the shearing strength of the soil, and D denotes the diameter of the pile. Typical values of N used in such calculations range from about 7 to 9. The shear strength of the soft deposit is customarily determined from laboratory tests of soil samples taken in the area of interest. Typical values are of the order of 100 to 200 lb/ft^2.

Because the soft deposit may extend appreciably below the seafloor, mudslide forces are customarily treated as distributed forces acting on the pile legs. This is in contrast with ice loadings, which may usually be regarded as concentrated forces because of the relatively small thickness of ice considered.

REFERENCES

API (1980). *Recommended Practice for Planning, Designing and Constructing Fixed Offshore Platforms*, American Petroleum Institute Publication RP-2A, Dallas, Tex.

ASCE (1961). Wind Forces on Structures, *Transactions, American Society of Civil Engineers*, Vol. 126, Part II, p. 1124.

Bretschneider, C. L. (1969). Overwater Wind and Wind Forces, *Handbook of Ocean and Underwater Engineering* (ed. J. J. Myers et al.), McGraw-Hill Book Company, New York, Chapter 12.

Chakrabarti, S. K., W. A. Tam, and A. L. Wolbert (1975). Wave Forces on a Randomly Oriented Tube, *Proceedings, Seventh Annual Offshore Technology Conference*, pp. 433–441.

Kinsman, B. (1965). *Wind Waves*, Prentice-Hall, Inc., Englewood Cliffs.

MacCamy, R. C., and R. W. Fuchs (1954). Wave Forces on Piles: A Diffraction Theory, *U.S. Army Corps of Engineers Technical Memorandum No. 69.*

McCormick, M. E. (1973). *Ocean Engineering Wave Mechanics*, John Wiley & Sons, Inc., New York.

Milne-Thomson, L. M. (1950). *Jacobian Elliptic Functions Tables*, Dover Publications, Inc., New York.

Mogridge, G. R., and W. W. Jamieson (1976). Wave Loads on Large Circular Cylinders: A Design Method, *National Research Council Report*, NRC No. 15827.

Morison, J. R., M. P. O'Brien, J. W. Johnson, and S. A. Schaaf (1950). Forces Exerted by Surface Waves in Piles, *Petroleum Transactions*, *American Institute of Mining Engineering*, Vol. 189, pp. 149–154.

Olsen, O. A. (1976). Wind, Wave and Current Forces on Offshore Structures, *The Technology of Offshore Drilling*, *Completion and Production*, The Petroleum Publishing Company, Tulsa, Okla., Chapter 3.

Rolfes, M. H. (1980). Wave Force and Structural Response, *Trident Scholar Project Report No. 108*, U. S. Naval Academy, Annapolis, Md.

Skjelbreia, L. (1959). Gravity Waves, Stokes Third Order Approximation, Tables of Functions, *Council on Wave Research*, University of California, Berkeley.

Skjelbreia, L., and J. A. Hendrickson (1961). Fifth Order Gravity Wave Theory, *Proceedings*, *Seventh Conference on Coastal Engineering*, pp. 184–196.

Thom, H. C.. S. (1968). New Distribution of Extreme Winds in the United States, *Journal of the Structural Division*, ASCE, Vol. 94, pp. 1787–1801.

Wade, B. G., and M. Dwyer (1976). On the Application of Morison's Equation to Fixed Offshore Platforms, *Proceedings*, *Eighth Annual Offshore Technology Conference*, pp. 1181–1186.

Weigel, R. L. (1960). A Presentation of Cnoidal Wave Theory for Practical Application, *Journal of Fluid Mechanics*, Vol. 7, pp. 273–286.

Weigel, R. L. (1964). *Oceanographical Engineering*, Prentice-Hall, Inc., Englewood Cliffs, N. J.

PROBLEMS

1. Consider the small light tower shown in Fig. P.1 and determine the total wind force F on the deck and lantern, assuming a wind speed (with gust factor) of 170 mph. Also locate its resultant distance b above the base of the deck.

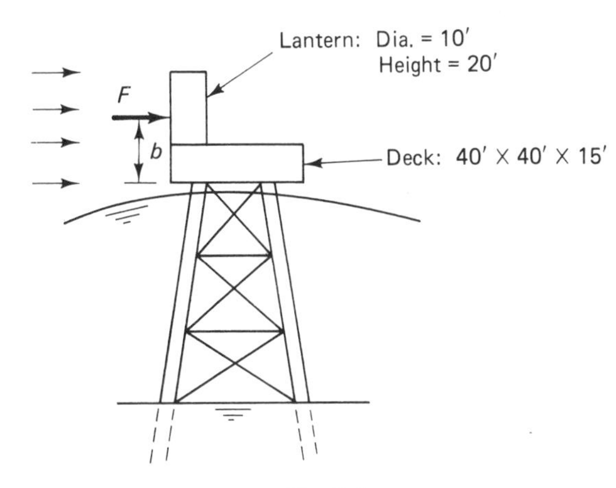

Fig. P.1

Ans. $F = 74.0$ kips, $b = 9.25$ ft.

2. Determine the period of an Airy wave of length 500 ft running in 100 ft. of water.
Ans. $T = 10.7$ sec.

3. Determine the length of an Airy wave of period 9 sec running in 80 ft of water.
Ans. $\lambda = 365$ ft.

4. For the wave of Problem 2, determine the maximum horizontal water velocity under the wave crest, assuming a wave height $H = 20$ ft.
Ans. $u = 7.70$ ft/sec.

5. Determine the wave speed c for the Airy wave of Problem 2.
Ans. 46.7 ft/sec.

6. Calculate the horizontal velocity and acceleration of the water at $x = 50$ ft, $y = 80$ ft, $t = 6$ sec, resulting from an Airy wave with surface deflection described by equation (6) with $H = 20$ ft, $k = 0.0180$ ft^{-1}, and $\omega =$ 0.732 rad/sec. The water depth is 90 ft.
Ans. $u = -6.31$ ft/sec, $a_x = 1.69$ ft/sec^2.

7. Repeat Problem 6 for the vertical velocity and acceleration.
Ans. $v = 6.01$ ft/sec, $a_y = -4.13$ ft/sec^2.

8. Determine the elevation of the crest of a Stokes wave of height $H = 60$ ft, length $\lambda = 800$ ft, running in water of depth 320 ft.
Ans. $\eta = 34.0$ ft.

9. Determine the period of the Stokes wave of Problem 8.
Ans. $T = 12.1$ sec.

10. For a Stokes wave, with surface deflection described by equation (18), determine the horizontal velocity and acceleration of water at $x = 0$,

$y = 50$ ft, $\omega t = 6.0$ resulting from a wave height of 35 ft and wave length of 375 ft in 75 ft of water.
Ans. $u = 9.31$ ft/sec, $a_x = 2.35$ ft/sec².

11. Repeat Problem 10 for the vertical velocity and acceleration.
Ans. $v = 2.23$ ft/sec, $a_y = -4.15$ ft/sec².

12. Determine the elevation of the crest of a cnoidal wave of height $H = 10$ ft, length $\lambda = 400$ ft running in water of depth 40 ft.
Ans. $\eta_c = 6.14$ ft.

13. Determine the period of the wave of Problem 12.
Ans. $T = 10.5$ sec.

14. Determine the maximum wave force exerted by Airy waves of height 20 ft and length 300 ft on a 4-ft-diameter pile, extending from the seafloor to above the maximum water elevation. The water depth is 80 ft. Assume that $C_D = 1$, $C_I = 2$.
Ans. 17.6 kips.

15. For a vertical pile of 8-ft diameter extending from the seafloor to above the maximum water elevation, determine the maximum horizontal force and associated base moment caused by a Stokes wave of height 70 ft and length 750 ft running in 150 ft of water. Assume that $C_D = 1$, $C_I = 2$.
Ans. $F = 475$ kips, $M = 16{,}280$ ft-kips.

16. Determine the horizontal and vertical components of the wave force per unit length acting at the midlength of member 1–2 of the side face shown in Fig. P.16, assuming wave-induced water motion described by

$$u = 13.7 \text{ ft/sec}, \qquad v = 4.21 \text{ ft/sec}$$

$$a_x = 4.01 \text{ ft/sec}^2, \qquad a_y = -6.76 \text{ ft/sec}^2$$

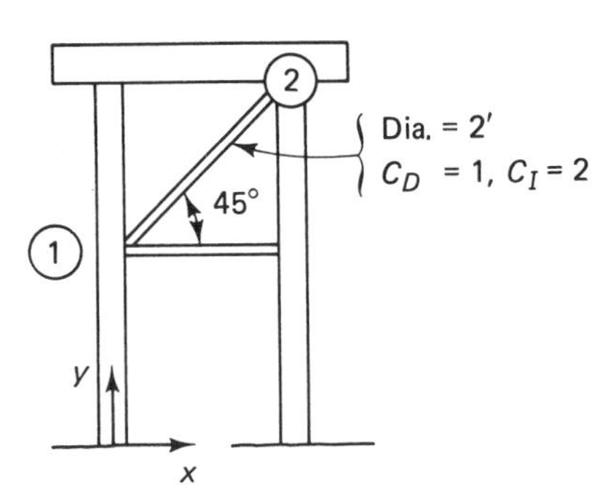

Fig. P.16

Ans. $f_x = -f_y = 131$ lb/ft.

17. For member 1–2 of Problem 16, recalculate the force components f_x, f_y assuming, in addition to the wave-induced water motion, a horizontal current of 2.2 ft/sec.
Ans. $f_x = -f_y = 164$ lb/ft.

18. For the four-legged structure shown in two dimentions in Fig. P.18, determine the maximum horizontal force, and associated time ωt, arising

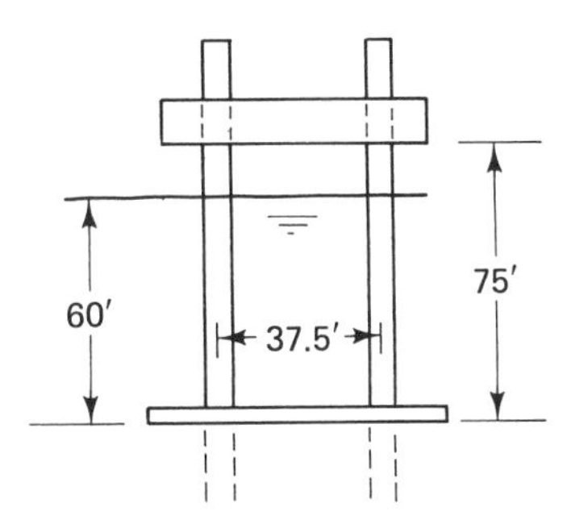

Fig. P.18

from Airy waves of height 15 ft and length 225 ft. The water depth is 60 ft. Assume that $C_D = 1$, $C_I = 2$. All four faces of the structure are identical. The outside diameter of each leg is 3 ft.
Ans. $F = 27.2$ kips, $\omega t = 5.8$.

19. Determine equivalent joint loads, for the vertical member 1–2–3 of Fig. P. 19, assuming an Airy wave of height 30 ft and length 450 ft in water

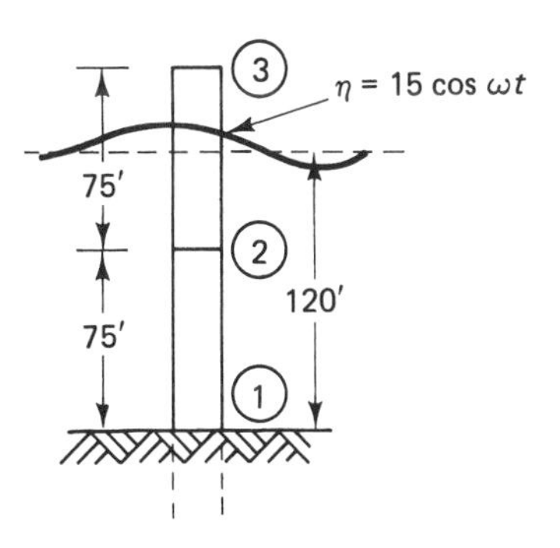

Fig. P.19

of depth 120 ft. The outside diameter of the member is 6 ft. Assume that $x = 0$, $t = 5.8$. Also assume that $C_D = 1$, $C_I = 2$.

Ans. $F_{1x} = 8.3$ kips, $M_1 = 109.4$ ft-kips
$F_{2x} = 32.8$ kips, $M_2 = 178.7$ ft-kips
$F_{3x} = 18.6$ kips, $M_3 = 304.8$ ft-kips

20. A horizontal steel bracing member of an offshore structure has an outside diameter of 3 ft and a wall thickness of 0.5 in. The member is completely submerged in salt water, with the inside unflooded. Determine the net vertical force (weight plus buoyant force) per unit length acting on the member. Assume that steel weighs 0.28 lb/in³.
Ans. 265 lb/ft (up).

21. The steel diagonal member of Fig. P.21 is completely submerged in salt water, with inside unflooded. Determine equivalent joint loads resulting from weight and buoyant forces.

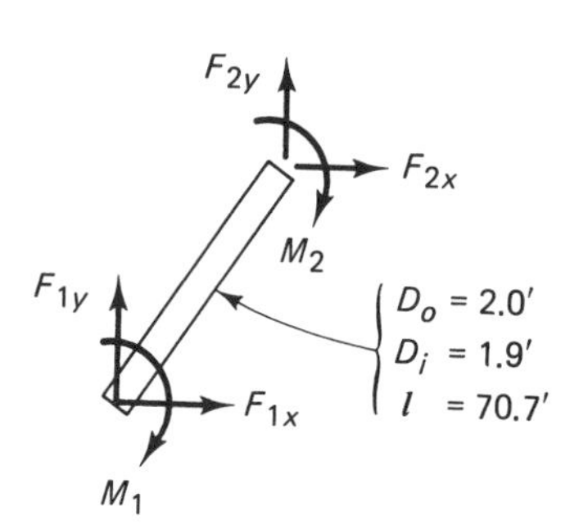

Fig. P.21

Ans. $F_{1x} = F_{2x} = 0$
$F_{1y} = F_{2y} = 1.87$ kips
$M_1 = -M_2 = 15.6$ ft-kips

22. An ice-resistant structure is supported by piles of 15 ft outside diameter. For ice sheets 3.5 ft thick, having a crushing strength of 300 lb/in², determine the maximum force exerted on each pile when struck by the ice.
Ans. 2268 kips.

23. A soft clay deposit 20 ft deep slides against six embedded piles of 4-ft outside diameter. Assuming that the shear strength of the clay is 200 lb/ft², determine the total force exerted on the piles. Use $N = 9$.
Ans. 216 kips.

24. For the steel structure, shown in two dimensions in Fig. P.24, determine the maximum total horizontal force and associated time ωt, assuming Airy waves of height 25 ft and length 350 ft. All four faces of the structure are identical. Vertical members have outside diameter of 4 ft and wall thickness of 1.5 in. Horizontal and diagonal members have outside diameter of 2 ft and wall thickness of 0.5 in. Assume that $C_D = 1$, $C_I = 2$.

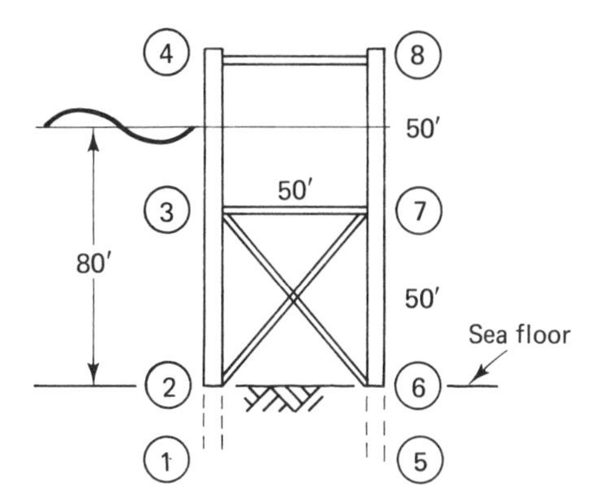

Fig. P.24

25. For the structure and wave of Problem 24, determine equivalent joint loads for the instant ωt when the horizontal wave force is maximum.

26. For the structure of Problem 24, determine equivalent joint loads from weight and buoyant forces. Assume horizontal and diagonal members unflooded.

Launching of a template substructure from a barge.
(Courtesy of McDermott Incorporated.)

4

STATIC METHODS OF ANALYSIS

CHAPTER 3 HAS BEEN concerned with engineering descriptions of the main environmental forces that can be exerted on an offshore structure. Maximum values of these forces at a particular offshore site may be combined with the structural methods of Chapter 2 to determine the adequacy of a proposed structure to withstand these loadings. Such a treatment neglects any inertia forces arising from the acceleration of the structure under the generally time-dependent loadings and is accordingly known as a *static* analysis. The present chapter describes details of this kind of analysis for both steel and concrete offshore structures. Consideration of pressure loading and joint failure are also included along with means for assessing the importance of *dynamic* effects.

4.1 DESIGN ENVIRONMENTAL CONDITIONS

The design of offshore structures is governed to a large degree by the severe environmental loadings that can be exerted on it. For most structures, these loadings arise from extreme storm conditions at the offshore site. Exceptions may exist in polar regions where large sheets of floating ice may strike the structure or in active seismic areas where earthquakes pose a severe hazard, but in most cases it is the storm conditions that dominate.

Design storm conditions are customarily chosen as extreme conditions having a recurrence interval of 100 years or so at the offshore site. The selection of a design wind can be made according to methods given in Chapter 3. In a static analysis, the associated surface waves are normally characterized by an extreme regular wave having specified height and length. The selection of these parameters is made most accurately by detailed study of the conditions in the area of interest (see, for example, Bea, 1974). Table 4.1 lists some typical maximum values for selected offshore areas of the United States.

Table 4.1 TYPICAL DESIGN STORM PARAMETERS FOR SELECTED U.S. WATERS

Location	Wave Height (ft)	Wave Length (ft)	Deck Clearance (ft)	Wind Speed (ft)
Gulf of Mexico	70	840	48	150
Cook Inlet (Alaska)	60	780	56	150
Santa Barbara Channel (California)	45	720	38	110

Source: After API (1980).

In addition to the selection of wind and wave conditions, it is also necessary to consider the rise in the water level in an area resulting from astronomical and storm tides, so that the height of a proposed structure can be chosen to minimize the possibility of storm waves striking the deck and its equipment. Some typical deck clearances are listed in Table 4.1 along with the wave data. For the Gulf of Mexico, these clearances refer to distance above the mean low water level of tidal variations and for Cook Inlet and the Santa Barbara Channel, where tidal variations are more complex, they refer to distance above the mean lower low water level.

Finally, currents must be assessed by study of the local conditions and these water velocities added to those caused by the wave motion. Of particular significance can be the wind-drift current associated with the design storm winds and caused by the drag of the wind on the water surface. Tidal and river-discharge currents can also be significant in certain areas.

4.2 FRAME ANALYSIS OF STEEL OFFSHORE STRUCTURES

Steel offshore structures are most commonly encountered in the form of template structures such as discussed in Chapter 1. A typical structure of this kind is illustrated in Fig. 4.1. It consists of a prefabricated steel framed substructure which is towed to the site, uprighted, and fixed by driving steel pipe piles through its hollow legs and into the seafloor, after which a prefabricated deck is attached.

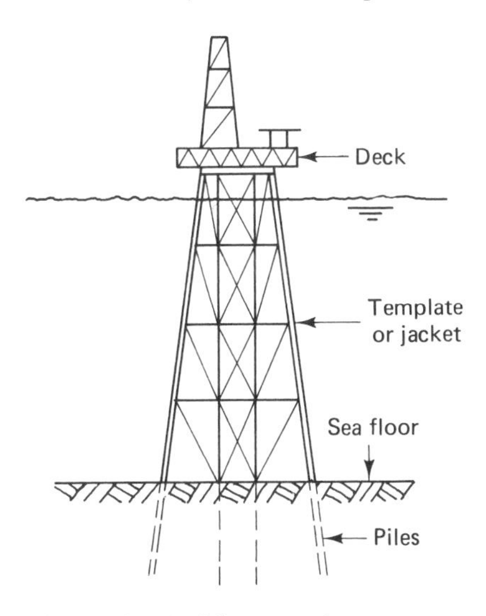

Fig. 4.1. Steel offshore template structure.

The substructure is normally constructed using cylindrical steel members both for main legs and the cross bracings. The template legs are usually vertical except for the outside legs and piles which are often battered (inclined) so as to provide a broader, more efficient, base for resisting horizontal loadings. These inclinations are, however, normally no more than 10° because of on-site installation problems. The support piles serve two purposes: They provide necessary vertical support for the deck loading and they secure the structure against horizontal wind and water loads. They are usually hollow circular cylinders with diameters ranging to 4 or more feet and wall thicknesses ranging to 1 or more inches. The depth of penetration of the piles into the ocean floor is typically required to be 200 ft or more in order to provide adequate resistance to loadings.

In analyzing a proposed template structure to determine its ability to withstand specified environmental conditions, maximum loadings from wind, waves, etc., may be estimated using methods outlined in Chapter 3 and these then combined with the operational loads to establish the maximum total loadings on the structure. Assuming that dynamic effects are negligible (see Section 4.10), these loadings may then be used with the structural methods of

Chapter 2 to find the maximum stresses existing in each of the various members. If the members are adequately sized, these stresses will all be within acceptable levels to prevent failure.

In performing this stress analysis, it is frequently sufficient to consider only the two cases where the direction of wave motion is along each of the principal horizontal axes of the structure and limit attention to a two-dimensional frame analysis, as indicated in Fig. 4.2. Out-of-plane members in each

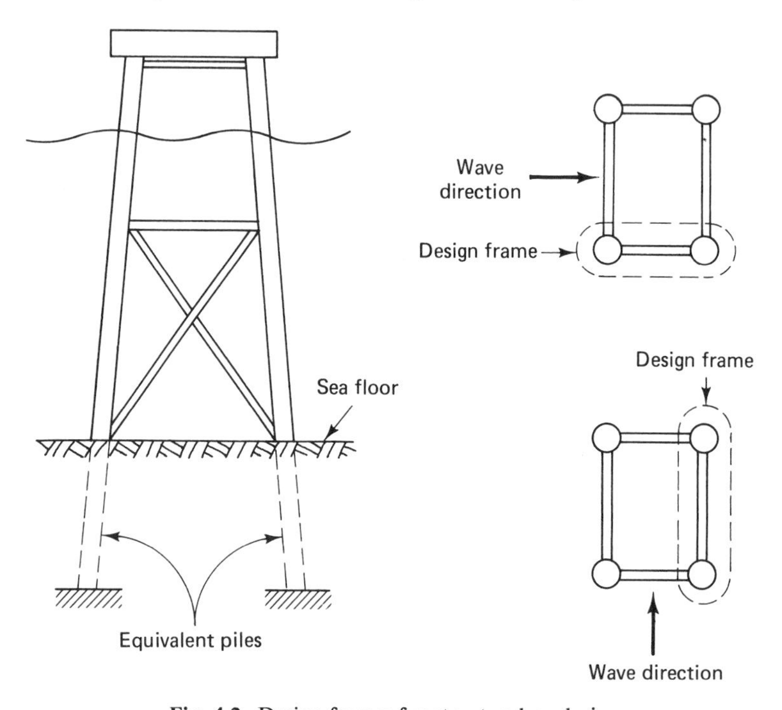

Fig. 4.2. Design frames for structural analysis.

case can be analyzed by considering the resultant distributed force acting on them and using the simplified assumption of pinned ends. Of course, when the geometry requires it, or when more accurate results are desired, a full three-dimensional structural analysis may be employed, with various wave directions assumed. Such analyses are readily carried out with fully automated digital-computer solutions for both the loadings and structural response.

The stress analysis described above generally requires that some consideration be given to the interaction of the structure with the support piles. This is especially important for soft-soil conditions, where large deflections and rotations of the piles and the connected structural elements can occur at the groundline. As illustrated in Fig. 4.2, this effect may be treated by assuming equivalent free-standing piles fixed at their base and having stiffness properties at the groundline approximating those of the actual embedded piles. Inclusion of these additional structural elements into the overall structure then allows direct analysis of the structure by the methods of Chapter 2 (see Example 2.4-3). Procedures for determining the stiffness properties of the equivalent piles are given in Chapter 5.

After determining that member sizes are adequate to withstand the environmental loadings expected on the structure at its offshore location, the structure must be analyzed to ensure that the member sizes are adequate to withstand the forces exerted during proposed transportation and installation procedures involved in placing the structure in its final operation position. A template structure is typically constructed on its side on land, then placed horizontally on a barge and towed to its offshore location and finally slid off the barge into the water and uprighted. Of the many loadings experienced during this process, two are especially important: the lifting of the structure onto the barge and the launching of the structure from the barge. For lifting loads, the structure may be analyzed when suspended in the proposed lifting manner [Fig. 4.3(a)] and, for launching, the structure may be analyzed under the support

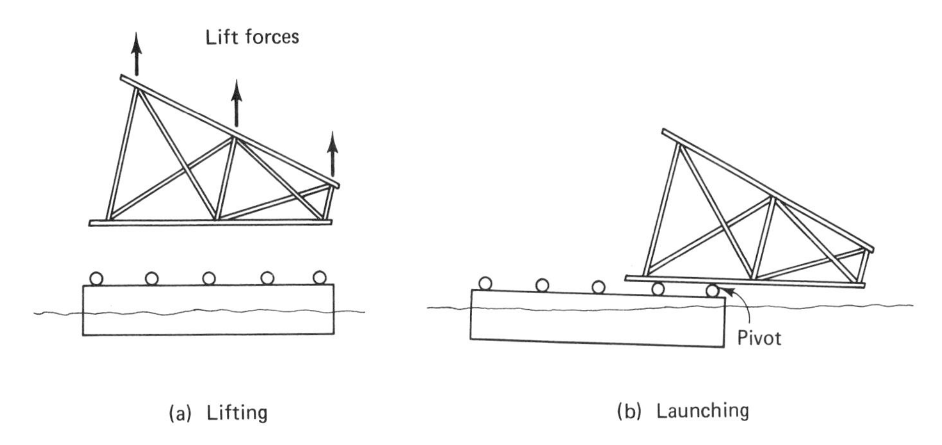

Fig. 4.3. Loading configurations during (a) lifting and (b) launching of a template substructure.

conditions existing as the structure is slid off the barge. A particularly severe case generally occurs when the structure is on the verge of tipping over into the water since the entire structure is then supported by a single pivot support [Fig. 4.3(b)]. In both lifting and launching, the applied structural loads arise, of course, from the weight of the structure, and the methods of Chapter 2 may be used to examine the resulting stresses in the individual members. In certain instances, inertia forces associated with the launching of the structure may also be important and these may be analyzed using methods described in Chapter 6.

If the installation stresses within certain members, or classes of members, of the substructure are found to be unacceptably large, their sizes must be increased and the structure reanalyzed to ensure that it can safely withstand the loadings involved. If changes are made, the structure must also then be reanalyzed to determine that it can still carry the environmental loads, since an increase in member dimensions causes, in turn, an increase in the environmental loads exerted on them.

The discussion above has been limited to the analysis of conventional steel template structures. Similar considerations will, however, apply for other types of steel offshore structures, such as ice-resistant and gravity structures described in Chapter 1. The foundational restraints on these structures will, of course, differ from the conventional template structure as will the loadings existing during transportation and installation to the offshore site, but the same general procedures of analysis will apply.

General guidance for the design of steel offshore structures can be found in the publication *Specifications for the Design, Fabrication and Erection of Structural Steel for Buildings* (AISC, 1969). Specific recommendations for the design of steel offshore structures can be found in the publication *Recommended Practice for Planning, Designing and Constructing Fixed Offshore Platforms* (API, 1980).

EXAMPLE 4.2-1. Consider a simple steel template platform with side face as shown in Fig. 4.4 and determine the stresses in member 3–8. The design wave has a height of 20 ft and a length of 300 ft and is assumed propagating parallel to the side face shown. The water depth under the storm conditions is 80 ft. All vertical members

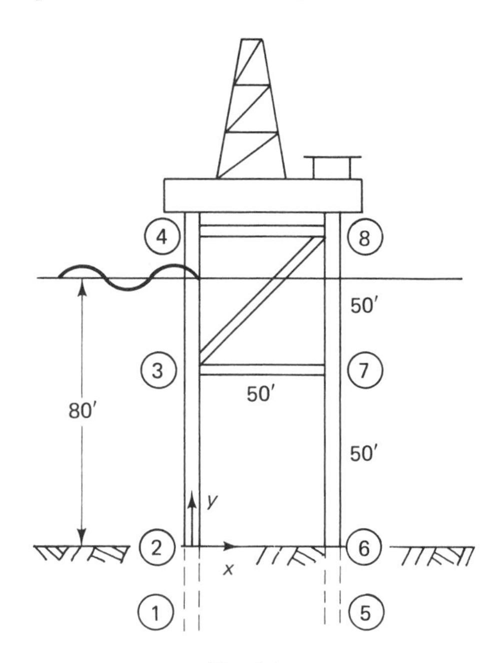

Fig. 4.4

have outside diameters of 4 ft and wall thicknesses of 1.5 in (including both the template leg and pipe pile driven through it). Horizontal and diagonal members have diameters of 2 ft and wall thicknesses of 0.5 in. All four faces of the structure are similar, with diagonal members from the two front faces framing into the corresponding joints 3 and 7. The deck weighs 500 kips and wind forces on the upper exposed part of the structure equal 100 kips.

As discussed above, the actual piles beneath the groundline are replaced by fictious free-standing ones, fixed at their base and having stiffness properties at the groundline approximating those of the actual piles. For the present problem these stiffness properties are assumed expressible, in units of ft and kips, as (see Example 5.6-2)

$$[K]_{1\text{-}2} = 10^4 \begin{bmatrix} 4.80 & 0 & -32.1 \\ 0 & 4.7 & 0 \\ -32.1 & 0 & 287 \end{bmatrix}, \qquad [K]_{5\text{-}6} = 10^4 \begin{bmatrix} 4.93 & 0 & -32.8 \\ 0 & 4.7 & 0 \\ -32.8 & 0 & 291 \end{bmatrix}$$

We note that the structure and design wave considered here is the same as that considered earlier in Examples 3.7-1 and 3.8-1, where joint loads due to the wave forces and buoyant and weight forces were calculated for the instant when the total wave force on the structure is at its maximum value. This is the normal assumption for joint load calculations; hence, these results may be used in the present analysis (with joint labelings 1, 2, 3 and 4, 5, 6 now replaced by 2, 3, 4 and 6, 7, 8, respectively). The joint loads due to the wave forces are shown in Fig. 4.5(a) and those due to weight and buoyant forces are shown in Fig. 4.5(b). Also shown are the particular loadings for member 3–8.

The deck weight and wind force on the structure also contribute loads to the upper joints 4 and 8 of the side frame. The deck weight of 500 kips and the wind force of 100 kips are both assumed carried equally by the two side frames of the structure. For the side face under consideration, we accordingly have the static results shown in Fig. 4.6, assuming that each leg carries an equal amount of the horizontal wind force and neglecting any concentrated moments existing at the deck-frame connections.

Collecting together all the results above, we find the total joint loads from wave action, from weight and buoyancy forces, and from the deck loading to be as summarized in Table 4.2.

Table 4.2 JOINT LOADS

Joint	F_x (kips)	F_y (kips)	M (ft-kips)
2	2.5	−18.4	21.6
3	15.7	−31.9	25.9
4	30.5	−134.6	−34.4
6	3.4	−18.4	29.0
7	12.4	−31.5	31.6
8	28.1	−166.5	−71.4

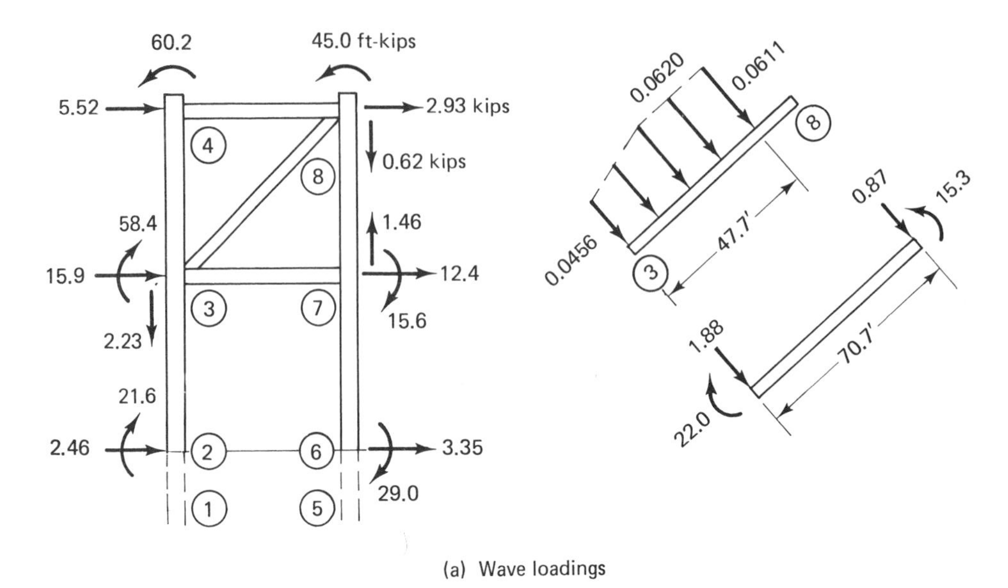

(a) Wave loadings

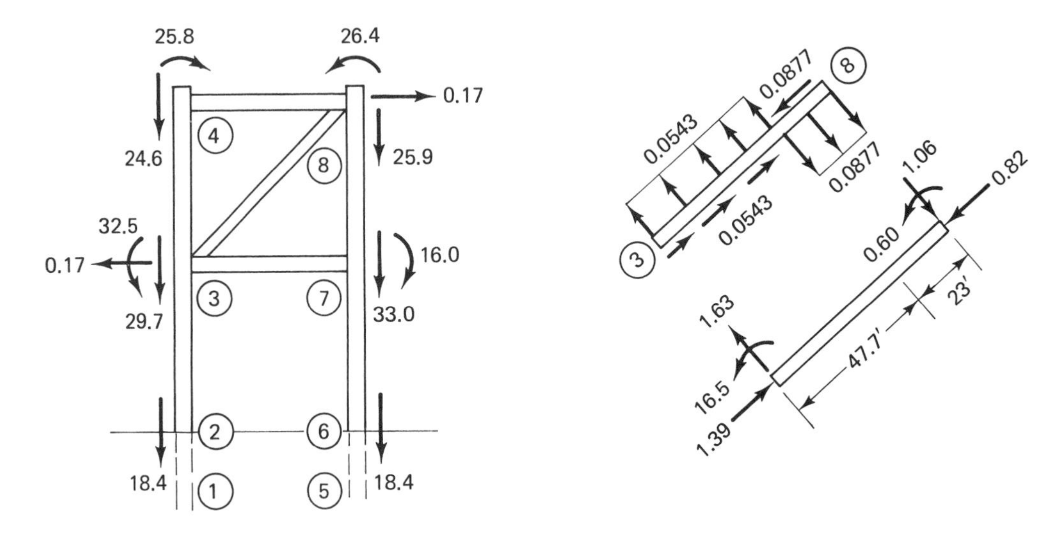

(b) Weight and buoyant forces

Fig. 4.5. (a) Wave loadings; (b) weight and buoyant forces.

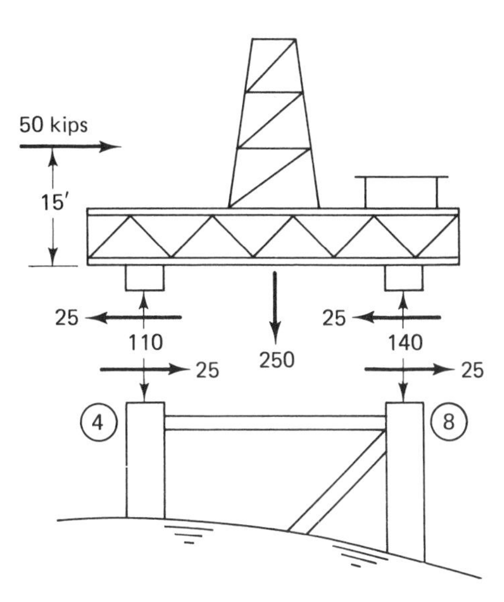

Fig. 4.6. Deck loadings.

To analyze the structure under these loads, we construct the reduced matrix stiffness equation, connecting known joint loads with unknown joint displacements, according to the methods of Chapter 2 (see, in particular, Example 2.4-3). Inverting this equation and solving for the displacements, we then find the results listed in Table 4.3, with U and V denoting, respectively, horizontal and vertical displacements and θ denoting rotation.

Table 4.3 JOINT DISPLACEMENTS

Joint	U (in)	V (in)	θ (rad)
2	0.169	−0.0214	1.95×10^{-3}
3	1.751	−0.0273	1.84×10^{-3}
4	2.033	−0.0397	-2.22×10^{-4}
6	0.164	−0.0808	1.91×10^{-3}
7	1.724	−0.1079	1.85×10^{-3}
8	2.005	−0.1321	-2.72×10^{-4}

Having the joint displacements, the internal force acting on the ends of the individual members can next be determined using the individual member stiffness matrices. In particular, for member 3–8, we have the total internal forces and moments $\{f^T\}$ acting on the ends of the member given in terms of the corresponding displacements and rotations $\{U\}$ by

$$\{f^T\} = [K]_{3\text{-}8}\{U\}$$

Carrying out the multiplication, we find

$$f_{3x}^T = -97.5 \text{ kips}, \qquad f_{3y}^T = -98.3 \text{ kips}, \qquad m_3^T = 37.6 \text{ ft-kips}$$
$$f_{8x}^T = 97.5 \text{ kips}, \qquad f_{8y}^T = 98.3 \text{ kips}, \qquad m_8^T = 5.92 \text{ ft-kips}$$

These total forces and moments are equal to the sum of the actual and equivalent forces and moments acting on the member ends. The equivalent values have been given in Fig. 4.5. Hence, resolving the total values into components parallel and normal to the member and subtracting the associated equivalent values, we may obtain the actual internal end forces and moments. These are shown in Fig. 4.7 together with the total distributed loading on the member resulting

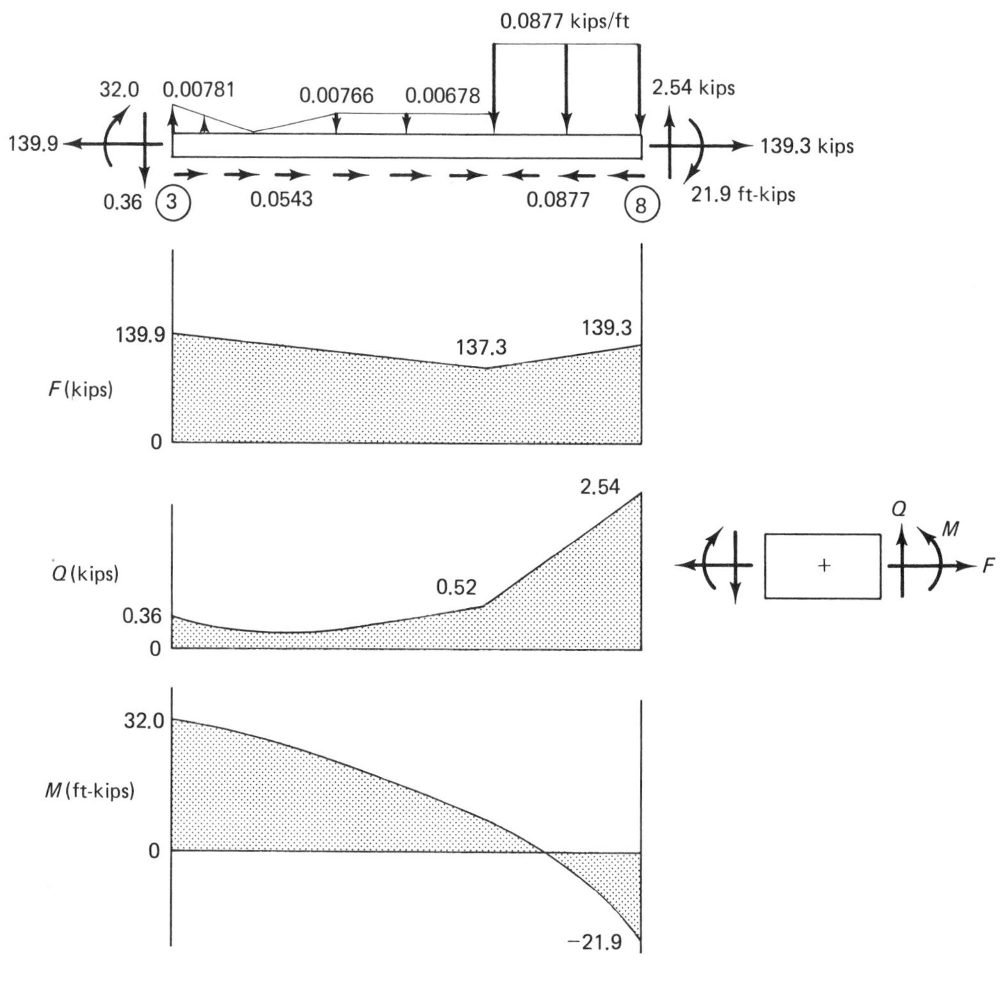

Fig. 4.7

from wave, weight, and buoyant forces. Also shown in Fig. 4.7 are sketches of the internal axial, shear, and moment variations along the member as determined from equilibrium considerations.

It can be seen that the axial force and bending moment have maximum values at end 3. The maximum axial and bending stresses in the member are accordingly determined as

$$\sigma_a = \frac{F}{A} = \frac{140}{0.256} = 547 \text{ kips/ft}^2 (3800 \text{ lb/in}^2)$$

$$\sigma_b = \pm \frac{MR}{I} = \pm \frac{(32)(1)}{0.123} = \pm 260 \text{ kips/ft}^2 (\pm 1800 \text{ lb/in}^2)$$

The maximum (tensile) stress in the member is thus 3800 + 1800 = 5600 lb/in². #

EXAMPLE 4.2-2. For the structure and loading of the preceding example, determine the maximum stress existing in the front-face horizontal member, say 3–11, framing into joint 3.

The total horizontal and vertical components of wave force exerted on this member have been given earlier in Example 3.7-1 and are expressible, respectively, as $F_x = 3.18$ kips and $F_y = -0.86$ kip. The net vertical force on the member arising from weight and buoyancy has also been determined in Example 3.8-1 and is expressible as $F_y = 3.85$ kips. Combining these, we have total components of $F_x = 3.18$ kips and $F_y = 2.99$ kips. The resultant total force F is thus

$$F = \sqrt{(3.18)^2 + (2.99)^2} = 4.37 \text{ kips}$$

Since the wave and buoyant forces act equally along the entire length (50 ft) of the member the distributed force f per unit of length is constant and given by

$$f = \frac{4.37}{50} = 0.0874 \text{ kip/ft}$$

Assuming the ends of the member to be simple supported (restrained against displacement, but free to rotate), the shear and moment diagram is easily constructed. The maximum moment M is found to occur at the midlength and to equal 27.3 ft-kips. Since the member has outside radius $R_o = 1$ ft and moment of inertia $I = 0.123$ ft⁴, the maximum bending stress is thus determined as

$$\sigma = \frac{MR_o}{I} = \frac{(27.3)(1)}{0.123} = 222 \text{ kips/ft}^2 \ (1540 \text{ lb/in}^2) \quad \#$$

4.3 BENDING-STRESS AMPLIFICATION

When members of an offshore structure are subjected to both bending and axial stresses, as is usually the case, the bending stress determined from matrix structural theory is subject to some error because of interaction with the axial stress. This is illustrated in Fig. 4.8, where it can be seen that the internal moment at any section of the member will depend in part on the axial loading and deflection of the member. This additional moment is not included in the matrix formulation described in Chapter 2 and hence it is necessary to provide a suitable correction.

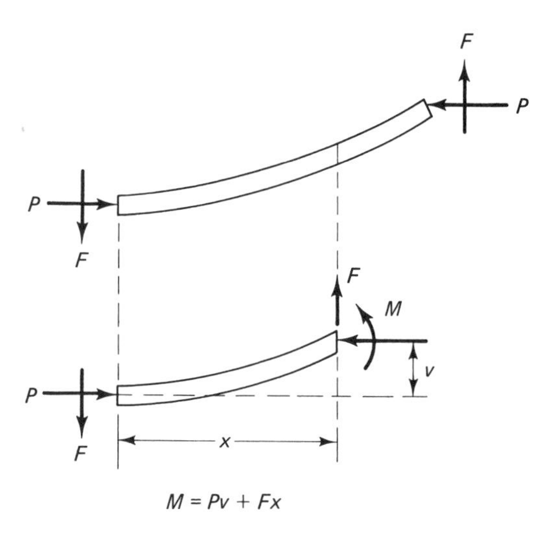

Fig. 4.8. Moment increase from axial loading.

With axial tensile stress, the effect is to reduce the bending stress from the value given by the theory, whereas with axial compressive stress, the effect is to increase it. The decrease in the bending stress resulting from axial tensile stress may safely be neglected, but the increase resulting from axial compressive stress clearly needs to be considered. From solid mechanics, this increase may be estimated using the amplification factor $\alpha \geq 1$ defined by

$$\alpha = \frac{C_m}{1 + \sigma_a/\sigma_e}, \qquad \sigma_a \leq 0 \tag{1}$$

where σ_a denotes the axial stress (negative in compression), σ_e the buckling stress of the member, and C_m a coefficient dependent on the loading and ranging from about 0.4 to 1.

The value of C_m equal to unity may always be chosen in the expression above for a conservative estimate. The evaluation of the buckling stress for members attached with rigid joints is more difficult. The general formula for the buckling stress of a member is expressible from solid mechanics as

$$\sigma_e = \frac{\pi^2 E}{(KL/r)^2} \tag{2}$$

where E denotes Young's modulus, L the length of the member, r the radius of gyration of the its cross section, and K an effective length factor dependent on the end conditions of the member. Typical values of K for various end conditions are shown in Fig. 4.9.

Case	(a)	(b)	(c)	(d)	(e)	(f)	(g)
Buckled shape of column is shown by dashed line. The ends are assumed free to experience relative vertical movement.							
Theoretical K value	0.5	0.7	1.0	1.0	2.0	2.0	∞

Fig. 4.9. Effective length factors.

For members whose ends are both restrained against any appreciable lateral movement by suitable bracing, as for example member 2–3 of the structure shown in Fig. 4.10(a), the value of K is seen from Fig. 4.9 to lie between

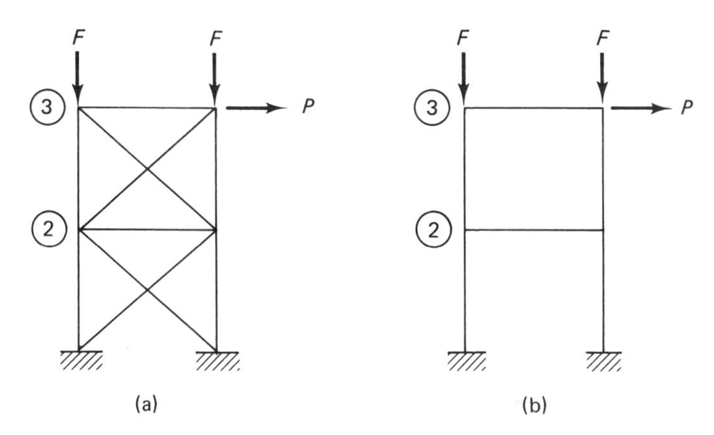

Fig. 4.10. Frames (a) restrained and (b) unrestrained against side sway.

0.5 (case a) and 1.0 (case d), depending on the degree of restraint existing. Thus, the value of K equal to unity may always be chosen for a conservative estimate.

If, however, both ends of the member are not restrained against appreciable lateral movement, as for example member 2–3 of the structure of Fig. 4.10(b), the value of K can be seen from Fig. 4.9 to range from unity (case c) to infinity (case g) depending on the amount of restraint existing. To estimate the value of K in this latter case, we may appeal directly to solid mechanics and determine the buckling stress with due consideration given to the elastic restraints existing at the ends of the member (see Gaylord and Gaylord, 1972). Alternatively, we may use an approximate method based on the nomograph shown in Fig. 4.11.

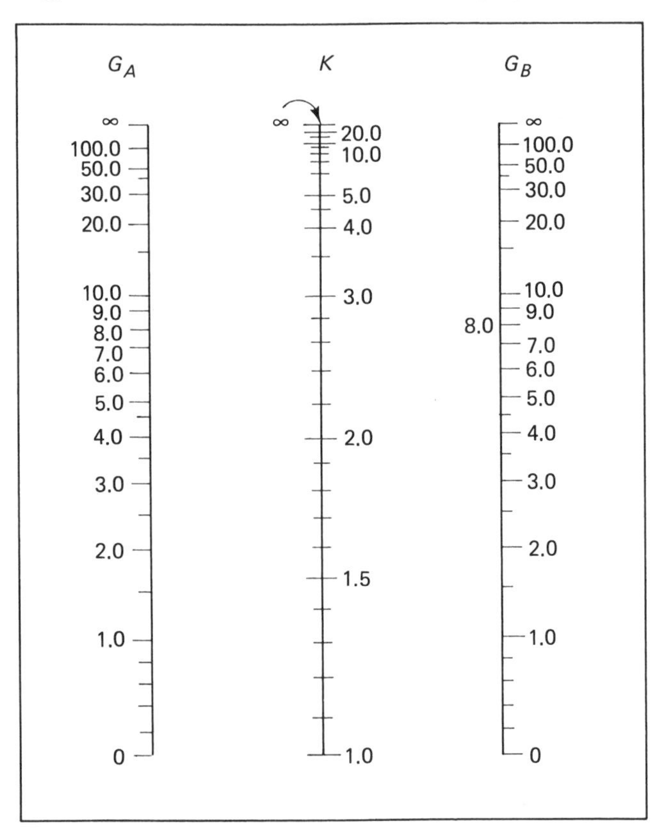

Fig. 4.11. Nomograph for determining the effective length factor.

The development of this chart is described in detail by McGuire (1968). It is based primarily on the assumptions that main compression members, such as member 2–3, in the structure of Fig. 4.10(b), all have the same cross sections and carry the same compressive loads and that all other members framing into the joints have identical cross sections. Under these assumptions, the value of K is then found to depend only on the values of

$$G = \frac{\sum I_c/L_c}{\sum I_b/L_b} \tag{3}$$

at each end joint of a compression member, where I_c and L_c denote the sectional moment of inertia and length of the main compression members framing into the joint and I_b and L_b denote corresponding quantities for the other members framing into the joint. With the corresponding values of G at each end, the appropriate value of K can be determined from the chart by simply constructing a straight line between the two values. For foundation restraint where the end of a member is fixed against rotation, the value of G is zero and where it is free to rotate the value is infinite.

Once the appropriate value of K is determined for a member subject to axial compressive stress and bending, the amplification factor α can be calculated and the bending stress in the member, as calculated from the matrix theory, then corrected by multiplying by this factor.

EXAMPLE 4.3-1. Consider the steel structure shown in Fig. 4.12 and determine corrected values of the bending stresses in members 1–2 and 2–3.

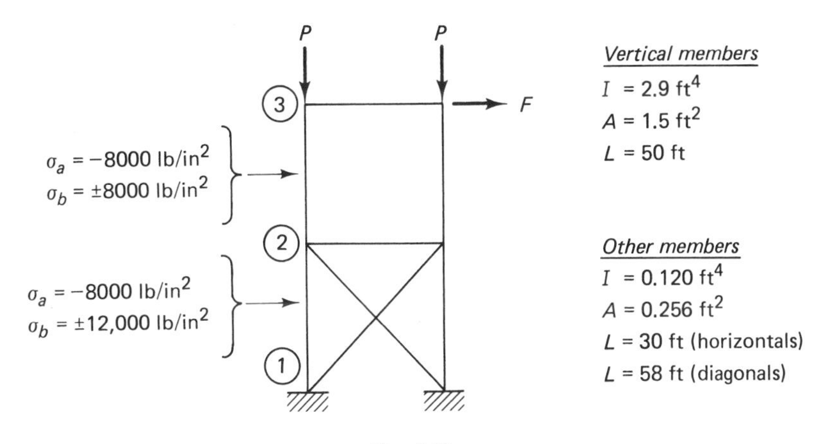

Fig. 4.12

Member 1–2 is in compression and is braced against lateral movement of its ends so that we may take $K = 1$ in equation (2). We then have for this member, with $E = 3 \times 10^7$ lb/in², $r = \sqrt{2.9/1.5} = 1.39$ ft,

$$\sigma_e = \frac{\pi^2(3 \times 10^7)}{(50/1.39)^2} = 2.29 \times 10^5 \text{ lb/in}^2$$

The axial stress is -8000 lb/in², so that $\sigma_a/\sigma_e = -0.035$. From equation (1), we then find the amplification factor α (choosing $C_m = 1$) to equal 1.04. The bending stresses from the theory are $\pm 12{,}000$ lb/in², so that the corrected values are

$$\sigma_b = (1.04)(\pm 12{,}000) = \pm 12{,}500 \text{ lb/in}^2$$

Member 2–3 is in compression and unbraced against lateral movement of its upper end. The value of K may accordingly be determined from Fig. 4.11. At joint 2, we have from equation (3)

$$G_2 = \frac{2.9/50 + 2.9/50}{0.120/30 + 0.120/58} = 19$$

and at joint 3,

$$G_3 = \frac{2.9/50}{0.120/30} = 15$$

Taking $G_A = G_2$ and $G_B = G_3$ in Fig. 4.11, we find $K = 3.8$. From equation (2) we then find

$$\sigma_e = \frac{\pi^2(3 \times 10^7)}{(190/1.39)^2} = 15{,}800 \text{ lb/in}^2$$

The ratio σ_a/σ_e is -0.506 and the amplification factor $\alpha = 2.0$. The corrected bending stresses are

$$\sigma_b = 2(\pm 8000) = \pm 16{,}000 \text{ lb/in}^2 \quad \#$$

4.4 PRESSURE-INDUCED STRESSES IN STEEL STRUCTURES

The effect of all-around pressure loading on submerged unflooded cylindrical members of offshore structures is to induce stresses in the members not accounted for in our previous discussion. The loading may arise simply from hydrostatic pressure in the water or from a combination of both hydrostatic pressure and that caused by wave action.

Free Cylindrical Member

We consider first the idealized case of an unrestrained cylindrical member subjected to a net external pressure $P = P_o - P_i$, where P_o and P_i denote, respectively, the external and internal pressure. The inside radius of the member is denoted by a, its wall thickness by t, and its length by L, as indicated in Fig. 4.13(a).

If, as shown in Fig. 4.13(b) we take a free-body section across a diameter of the tube, we have, on balancing forces in the vertical direction,

$$-2F_\theta - \int_0^L \int_0^\pi P(a + t) \sin\theta \, d\theta \, dz = 0$$

Assuming thin walls such that $t/a \ll 1$, this becomes

$$F_\theta = -LaP$$

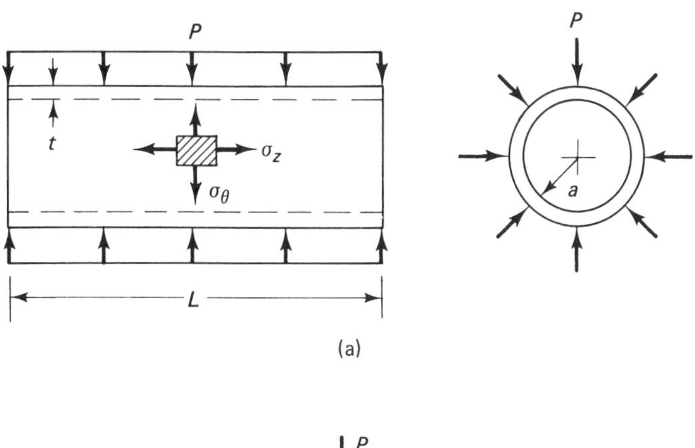

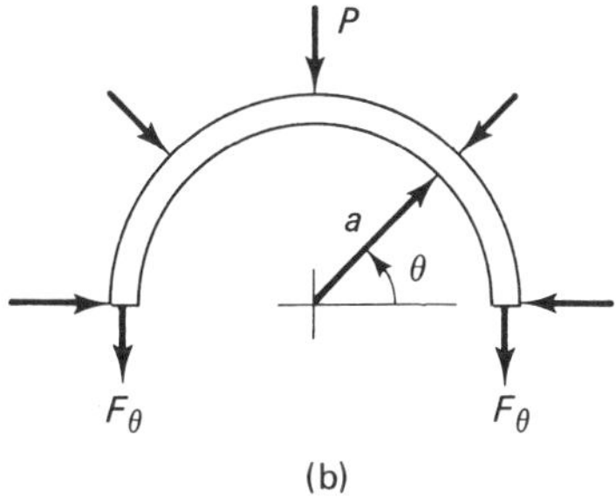

Fig. 4.13. Pressure loading on a long cylindrical member.

Dividing by the area Lt over which the tangential forces F_θ act, we thus have the tangential (or hoop) stress σ_θ given as

$$\sigma_\theta = -\frac{Pa}{t} \tag{4}$$

In addition to this stress, we also have a uniform axial stress σ_z caused by pressure loading on the ends of the shell. Since the axial force on the walls in this case is $P\pi a^2$ and the area of the walls carrying this load is $2\pi at$, we have

$$\sigma_z = -\frac{\pi a^2 P}{2\pi at} = -\frac{Pa}{2t} \tag{5}$$

For the stresses above, we have the following stress–strain relations from solid mechanics:

$$\begin{aligned} e_z &= \frac{1}{E}(\sigma_z - \nu\sigma_\theta) \\ e_\theta &= \frac{1}{E}(\sigma_\theta - \nu\sigma_z) \end{aligned} \tag{6}$$

where e_z is the strain, or change in length per unit initial length in the z-direction, e_θ is the change in length per unit initial length in the tangential direction, and E and ν denote Young's modulus and Poisson's ratio for the material. If ΔL denotes change in length in the z-direction and u denotes outward radial displacement of the shell, we have, in particular,

$$e_z = \frac{\Delta L}{L}$$

$$e_\theta = \frac{2\pi(a+u) - 2\pi a}{2\pi a} = \frac{u}{a} \tag{7}$$

Using equations (4) through (7), we accordingly find the following relations for the change in length and radial deflection of the cylinder:

$$\Delta L = -\frac{PaL}{Et}\left(\frac{1}{2} - \nu\right)$$

$$u = -\frac{Pa^2}{Et}\left(1 - \frac{\nu}{2}\right) \tag{8}$$

EXAMPLE 4.4-1. Determine the stresses existing in the unflooded member 1–2 of the offshore structure shown in Fig. 4.14. The member has an outside diameter of 2 ft and a wall thickness of 0.5 in.

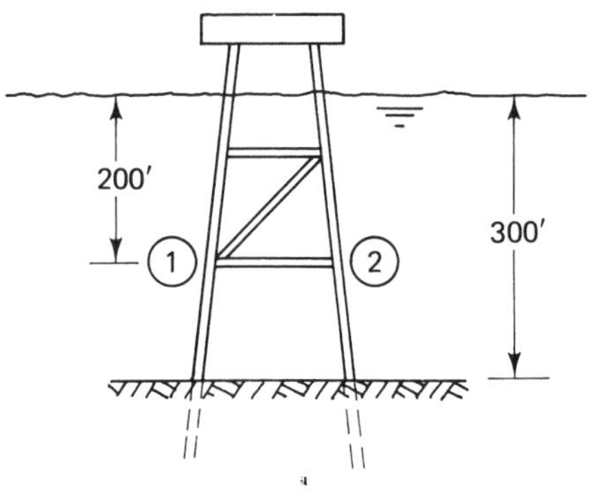

Fig. 4.14

The hydrostatic pressure P_o acting on the outside of the member is given by the equation

$$P_o = \gamma b + P_a$$

where γ denotes the specific weight of water, b denotes the depth of the member, and P_a denotes atmospheric pressure. The member is assumed unflooded with atmospheric pressure P_a inside. The net external pressure is thus

$$P = P_o - P_i = \gamma b = (64)(200) = 12{,}800 \text{ lb/ft}^2$$

Using equations (4) and (5) with $a = 12$ in, $t = 0.5$ in, and $P = 12{,}800/144 = 88.9$ lb/in², we find

$$\sigma_\theta = -\frac{(88.9)(12)}{0.5} = -2134 \text{ lb/in}^2$$

$$\sigma_z = -\frac{(88.9)(12)}{2(0.5)} = -1067 \text{ lb/in}^2 \quad \#$$

EXAMPLE 4.4-2. Reconsider Example 4.4-1 for the case of a design wave of height 70 ft and length 1000 ft.

Assuming Airy wave theory, the net external pressure P is given by equation (15) of Chapter 3 as

$$P = \frac{\gamma H}{2} \frac{\cosh ky}{\cosh kh} + \gamma(h - y)$$

where H denotes the wave height, k the wavenumber, h the water depth, y the distance from the seafloor, and γ the specific weight of the water. Substituting $k = 2\pi/1000$, $H = 70$, $y = 100$, $\gamma = 64$, $h = 300$, we find

$$P = 800 + 12{,}800 = 13{,}600 \text{ lb/ft}^2$$

From equations (4) and (5), we then have

$$\sigma_\theta = -2267 \text{ lb/in}^2, \qquad \sigma_z = -1133 \text{ lb/in}^2 \quad \#$$

Restrained Cylindrical Member

The theory discussed above assumes the radial deflection of the member to be constant along the entire length of the member. Restraints existing at the ends of the member prevents this assumption from being strictly satisfied. To include this effect, we need to consider bending of longitudinal strips of the member.

Consider, in particular, a thin-walled cylindrical member of radius a and wall thickness t and imagine a longitudinal strip AB taken from it of width $a\,d\theta$, as shown in Fig. 4.15. Treating this strip as a beam subjected to bending action, the longitudinal strain resulting from bending is given from beam theory as

$$e_z = -y \frac{d^2u}{dz^2} \tag{9}$$

where y denotes radial distance measured positive outward from the midsurface of the strip and u denotes the outward radial deflection of the strip at longitudinal distance z. If the strip is also subject to axial tension or compression loading, we have added to the bending strain the associated uniform axial

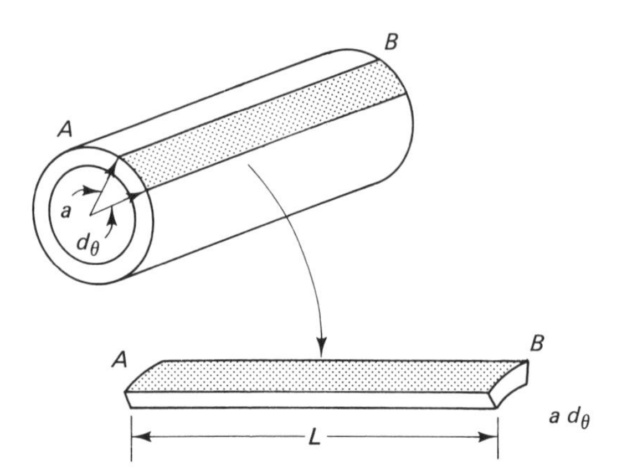

Fig. 4.15. Strip from a cylindrical member for bending analysis.

strain e_{z0} so that the complete strain is expressible as

$$e_z = -y\frac{d^2u}{dz^2} + e_{z0} \tag{10}$$

In addition to the longitudinal strain, we also have the tangential strain e_θ given by equation (7) as

$$e_\theta = \frac{u}{a} \tag{11}$$

Using the strain expressions above with the stress–strain relations of equation (6), we find the hoop stress σ_θ and longitudinal stress σ_z in the strip expressible as

$$\sigma_\theta = \frac{E}{1-\nu^2}\left(\frac{u}{a} - \nu y\frac{d^2u}{dz^2} + \nu e_{z0}\right) \tag{12}$$

$$\sigma_z = \frac{E}{1-\nu^2}\left(-y\frac{d^2u}{dz^2} + e_{z0} + \nu\frac{u}{a}\right) \tag{13}$$

It remains to determine the appropriate expression for the radial deflection u. To do this, we first determine the internal moment M per unit of circumference of the strip. Taking this to be positive counterclockwise on a right-hand face at any section of the strip (Fig. 4.16) we have

$$M = -\int_{-t/2}^{t/2} y\sigma_z\,dy \tag{14}$$

Substituting for σ_z from equation (13), we thus find

$$D\frac{d^2u}{dz^2} = M \tag{15}$$

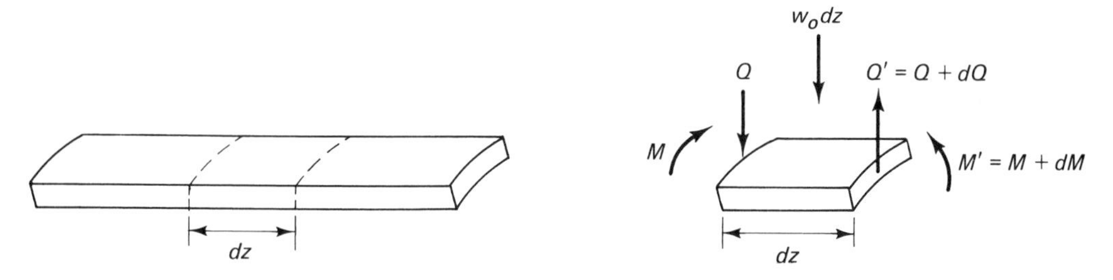

Fig. 4.16. Shear force and moment on a strip from a cylindrical member.

where D is defined by

$$D = \frac{Et^3}{12(1 - \nu^2)}$$

If w_0 denotes the net downward vertical loading on the strip, per unit of area, we also have from equilibrium considerations of the element shown in Fig. 4.16 that

$$\frac{dQ}{dz} = w_0, \qquad \frac{dM}{dz} = -Q \tag{16}$$

where Q denotes the shear force per unit of circumference, positive upward on a right-hand face as shown in Fig. 4.16. Combining these last two equations to eliminate the shear force Q and making use of equation (15), we obtain

$$D\frac{d^4u}{dz^4} = -w_0 \tag{17}$$

We have finally to relate the loading w_0 to the applied net external pressure loading P on the strip. If we look edge-on at the strip, we have the free-body view of Fig. 4.17. From statics, we easily find the net vertical distributed loading

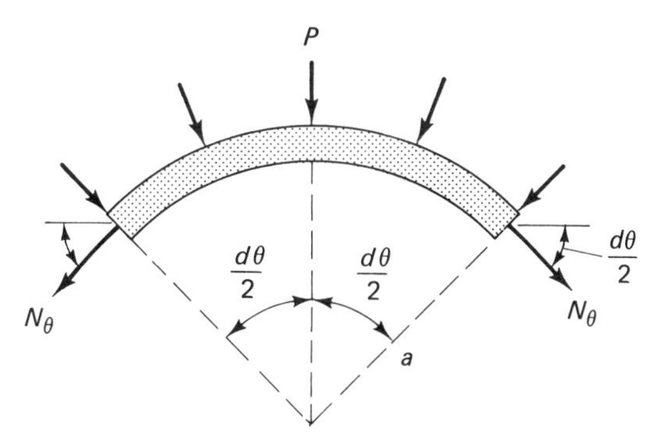

Fig. 4.17. Edge view of a strip, showing hoop forces and pressure loading.

w_0 expressible as

$$w_0 a \, d\theta = Pa \, d\theta + N_\theta \, d\theta$$

or

$$w_0 = P + \frac{1}{a} N_\theta \tag{18}$$

where N_θ is given by

$$N_\theta = \int_{-t/2}^{+t/2} \sigma_\theta \, dy$$

Substituting for σ_θ from equation (12), we find

$$N_\theta = \frac{Et}{1 - \nu^2}\left(\frac{u}{a} + \nu e_{z0}\right) \tag{19}$$

Also, if σ_{z0} denotes the applied axial stress, given by equation (5), we have on setting $y = 0$ in equation (13),

$$e_{z0} = \frac{1 - \nu^2}{E}\sigma_{z0} - \nu\frac{u}{a} \tag{20}$$

Thus, on combining equations (17)–(20), we have finally the governing equation for the radial deflection expressible as

$$D\frac{d^4 u}{dz^4} + \frac{Et}{a^2}u = -P' \tag{21}$$

where P' is given by

$$P' = P + \nu\frac{t}{a}\sigma_{z0}$$

The solution of equation (21) is expressible for constant P' as

$$u = e^{-\alpha z}(C_1 \cos \alpha z + C_2 \sin \alpha z) + e^{\alpha z}(C_3 \cos \alpha z + C_4 \sin \alpha z) - \frac{P' a^2}{Et} \tag{22}$$

where C_1, C_2, C_3, and C_4 denote constants and where α is given by

$$\alpha = \left(\frac{Et}{4a^2 D}\right)^{1/4} = \left[\frac{3(1 - \nu^2)}{a^2 t^2}\right]^{1/4} \tag{23}$$

The constants above are determined by the restraint conditions existing at the ends of the member. If the ends are assumed fixed against displacement and rotation, we have $u = du/dz = 0$ at each end.

A special case of considerable interest is that of a semi-infinite member built in at its end (Fig. 4.18). For finite deflections as $z \longrightarrow \infty$, we choose $C_3 =$

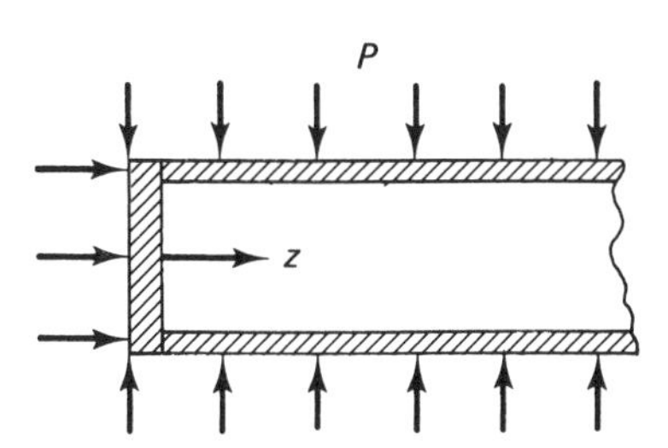

Fig. 4.18. End region of a cylindrical member.

$C_4 = 0$ in equation (22). Also, since

$$u = \frac{du}{dz} = 0 \qquad \text{at } z = 0$$

we have

$$C_1 = C_2 = \frac{P'a^2}{Et}$$

Thus, the radial deflection is given by

$$u = \frac{P'a^2}{Et}[1 - e^{-\alpha z}(\cos \alpha z + \sin \alpha z)] \tag{24}$$

where P' is given by

$$P' = P + \nu \frac{t}{a}\sigma_{z0} = P\left(1 - \frac{\nu}{2}\right)$$

For large values of αz, the radial deflection approaches that given by equation (8), thus indicating that the edge effect is a local one. If the member is sufficiently long (say, $L > 6/\alpha$), the solution above may therefore be used to describe the deflections in the vicinity of the ends when both ends are built in.

Of course, once the radial displacement is known, the stresses can be calculated using equations (12), (13), and (20). These may be expressed for the outside and inside surfaces ($y = \pm t/2$) as

$$\begin{aligned} \sigma_z &= -\frac{Pa}{2t} \pm \frac{3Pa(2 - \nu)\beta}{2t} e^{-\alpha z}(\cos \alpha z - \sin \alpha z) \\ \sigma_\theta &= -\frac{Pa}{t} + \frac{Pa(2 - \nu)}{2t} e^{-\alpha z}[(1 \pm 3\nu\beta) \cos \alpha z + (1 \mp 3\nu\beta) \sin \alpha z)] \end{aligned} \tag{25a}$$

where $\beta = [3(1 - \nu^2)]^{-1/2}$.

In addition to these normal stresses there also exists a shearing stress associated with the radial shear force Q, determined from equations (15), (16), and (24) as

$$Q = -D\frac{d^3u}{dz^3} = -\frac{P(2 - \nu)}{2\alpha} e^{-\alpha z} \cos az$$

If, as in beam theory, the shear stress is assumed to be distributed parabolically over the wall thickness, the maximum shear stress will occur on the midsurface and is given by $3Q/2t$. Thus, the radial shear stress τ is

$$\tau = -\frac{3}{4}\frac{P(2-\nu)}{\alpha t}e^{-\alpha z}\cos\alpha z \tag{25b}$$

A negative value of this shear stress (such as occurs at $z = 0$) implies that the stress acts outward on cylinder cross sections whose outward normals point in the direction of the build-in end (Fig. 4.16).

Inspection of the equations above shows that the extreme lognitudinal and shear stresses occur at the restrained end. The extreme longitudinal stresses occur on the inner and outer surfaces, where the shear stress is zero, and the extreme shear stress occurs on the midsurface, where the longitudinal stress is reduced and equal to that given by equation (5). The extreme longitudinal and shear stresses, together with the associated hoop stress, are expressible as

$$\begin{aligned}
\sigma_z &= -\frac{Pa}{2t}\left[1 \mp \frac{\sqrt{3}(2-\nu)}{(1-\nu^2)^{1/2}}\right] \quad \left(\text{on } y = \pm\frac{t}{2}\right) \\
\sigma_\theta &= \nu\sigma_z \\
\tau &= -\frac{3}{4}\frac{(2-\nu)P}{[3(1-\nu^2)]^{1/4}}\sqrt{\frac{a}{t}} \quad (\text{on } y = 0)
\end{aligned} \tag{26}$$

Choosing $\nu = 0.3$, as is typical for steel, we have at the outside surface

$$\sigma_z = 1.04\frac{Pa}{t}, \qquad \sigma_\theta = 0.313\frac{Pa}{t} \tag{26a}$$

and at the inside surface

$$\sigma_z = -2.04\frac{Pa}{t}, \qquad \sigma_\theta = -0.612\frac{Pa}{t} \tag{26b}$$

At distances removed from the end, we also have from equations (25a)

$$\sigma_z = -\frac{Pa}{2t}, \qquad \sigma_\theta = -\frac{Pa}{t}$$

in agreement with equations (4) and (5).

As in ordinary beam theory, the shear stress given by the last of equations (26) is generally small in comparison with the maximum longitudinal stress and can normally be neglected in design considerations of steel members.

EXAMPLE 4.4-3. Reconsider the stress analysis of Example 4.4-1 for the case where the ends of the member are assumed fixed against displacement and rotation.

The member has radius $a = 12$ in and wall thickness $t = 0.5$ in. The external pressure is 88.9 lb/in². Assuming a steel member with $E = 3 \times 10^7$ lb/

in² and $\nu = 0.3$, we have from equation (23)

$$\alpha^4 = \frac{3(1 - \nu^2)}{a^2 t^2} = 0.0758$$

so that $\alpha = 0.525$ in^{-1}. The solution of equation (24) applies for any member of length $6/\alpha > 11.4$, so that we may assume it appropriate here. Using equations (26) with $Pa/t = (88.9)(12/0.5) = 2130$ lb/in², we accordingly have for the outside surface at the restrained end

$$\sigma_z = (1.04)(2130) = 2215 \text{ lb/in}^2$$
$$\sigma_\theta = (0.313)(2130) = 667 \text{ lb/in}^2$$

and at the inside surface

$$\sigma_z = -(2.04)(2130) = -4350 \text{ lb/in}^2$$
$$\sigma_\theta = -(0.612)(2130) = -1300 \text{ lb/in}^2$$

At distances removed from the ends we also have

$$\sigma_z = -(0.5)(2130) = -1070 \text{ lb/in}^2$$
$$\sigma_\theta = -2130 \text{ lb/in}^2$$

The variation of σ_z and σ_θ along the member is described by equations (25a). This is shown in Fig. 4.19 for the inside surface of the member where the stresses are greatest. #

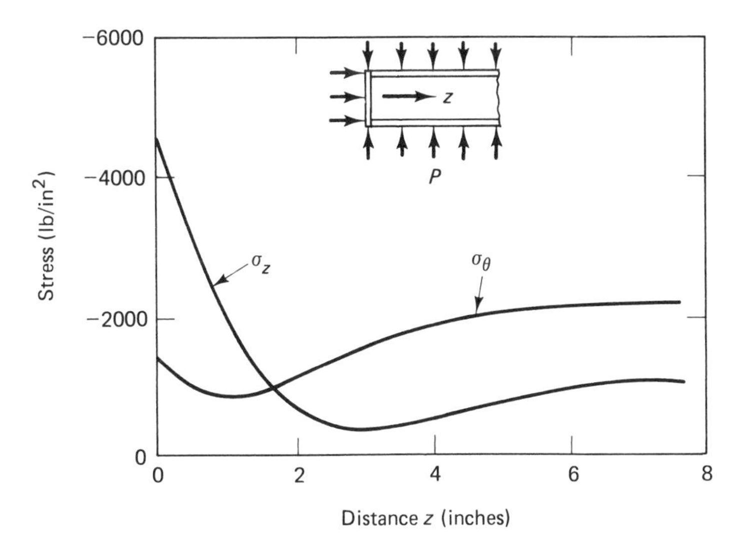

Fig. 4.19

4.5 DESIGN STRESS CRITERIA FOR STEEL MEMBERS

Ordinary structural steels are normally used in the construction of an offshore structure. Figure 4.20 shows a typical stress–strain curve for these steels. The yield stress is typically about 40 kips/in^2 or less and the maximum stress is about 60 kips/in^2.

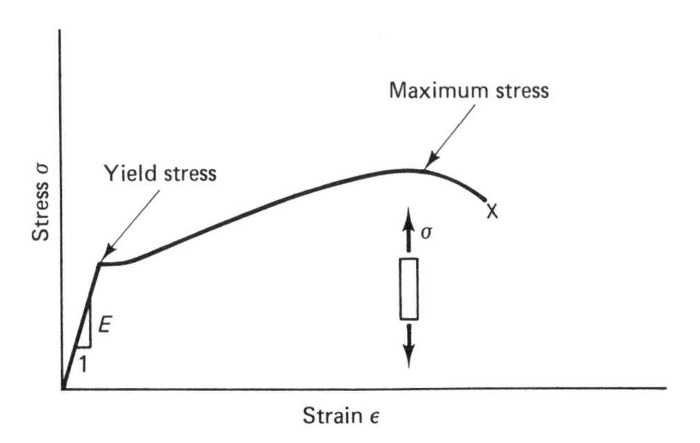

Fig. 4.20. Typical stress-strain curve of construction steel.

As long as the stress levels in a member of the structure never exceed the yield stress of the material, its behavior will be elastic, with strains disappearing with the disappearance of stress. A main objective in design is to choose the proper size of the members to ensure that this condition is always met under design-level loadings. In fact, appropriate factors of safety are employed to obtain an allowable stress (yield stress divided by the factor of safety) and the members are sized to ensure that this allowable stress is never exceeded under design loads.

Working levels for the longitudinal stress σ_z and the hoop stress σ_θ existing in members of an offshore structure may be determined using the maximum shear criterion for combined stresses from solid mechanics. More detailed criteria are given by the American Petroleum Institute (API, 1980). The longitudinal stress will, in general, consist of contributions from both pressure loading and frame response under storm conditions while the hoop stress will arise solely from pressure loading.

If we regard the total longitudinal stress as the sum of a uniform axial stress σ_a and a bending stress σ_b, we have, from the maximum shear criterion, the design condition for $\sigma_z > 0$, $\sigma_\theta < 0$ (or $\sigma_z < 0$, $\sigma_\theta > 0$) expressible as

$$\left|\frac{\sigma_a}{S_a} + \frac{\sigma_b}{S_b} - \frac{\sigma_\theta}{S_\theta}\right| \leq 1 \tag{27}$$

and for $\sigma_z < 0$, $\sigma_\theta < 0$, (or $\sigma_z > 0$, $\sigma_\theta > 0$) the conditions

$$\left|\frac{\sigma_a}{S_a} + \frac{\sigma_b}{S_b}\right| \leq 1, \qquad \left|\frac{\sigma_\theta}{S_\theta}\right| \leq 1 \tag{28}$$

where S_a, S_b, and S_θ denote corresponding allowable stress values for the axial, bending, and hoop stresses, if acting alone. These values are governed, in general, by the yield stress σ_Y of the material, by the local longitudinal compressive buckling stress σ_{zc} at which longitudinal wrinkling (crushing) of the member occurs, and by the hoop compressive buckling stress $\sigma_{\theta c}$ at which the member looses its circular shape and collapses.

The local longitudinal buckling stress may be expressed approximately as

$$\sigma_{zc} = 0.3E\frac{t}{a} \tag{29}$$

where E denotes Young's modulus of the material, t the wall thickness of the member, and a its radius. Similarly, for a member having no auxiliary circumferential reinforcement, the hoop buckling stress for regions away from the restrained ends is expressible approximately as

$$\sigma_{\theta c} = 0.22E\left(\frac{t}{a}\right)^2 \tag{30}$$

For positive (tensile) values of σ_z and σ_θ, buckling is not possible and the allowable stresses S_a, S_b, and S are governed solely by the yield stress σ_Y of the material. These are normally taken, with appropriate factors of safety, as

$$S_a = 0.6\sigma_Y, \qquad S_b = 0.67\sigma_Y, \qquad S_\theta = 0.5\sigma_Y \tag{31}$$

When σ_z or σ_θ is compressive, the corresponding allowable stress must take into account the possibility of buckling. Typical values are listed in Table 4.4. For $\sigma_{zc}/E \geq 0.010$ and $\sigma_{\theta c}/\sigma_Y \geq 4$, these are the same as those given by equation (31).

Table 4.4 ALLOWABLE STRESSES

Compressive Longitudinal Stress			Compressive Hoop Stress	
$\frac{\sigma_{zc}}{E}$	$\frac{S_a}{\sigma_Y}$	$\frac{S_b}{\sigma_Y}$	$\frac{\sigma_{\theta c}}{\sigma_Y}$	$\frac{S_\theta}{\sigma_Y}$
≥0.010	0.60	0.67	≥4.0	0.50
0.008	0.58	0.65	3.0	0.48
0.006	0.55	0.61	2.0	0.45
0.004	0.50	0.56	1.0	0.38
0.002	0.41	0.46	≤0.5	$\frac{1}{2}\frac{\sigma_{\theta c}}{\sigma_Y}$

The allowable stress values S_a, S_b, and S_θ given above refer to cases where stress contributions from design environmental loadings are not considered. When the longitudinal stresses from frame response under these loadings are included, together with additional hoop stress arising from wave-induced water pressure, it is normally considered satisfactory to increase the allowable stress values by one-third to account for the generally conservative description of the design conditions.

EXAMPLE 4.5-1. Member 1–2 of the offshore structure shown in Fig. 4.21 has a net external pressure acting at end 1 of 178 lb/in² (depth of 400 ft). The member is made of steel with wall thickness of 0.5 in and a radius of 12 in. Examine the stress levels assuming a yield stress of 36,000 lb/in².

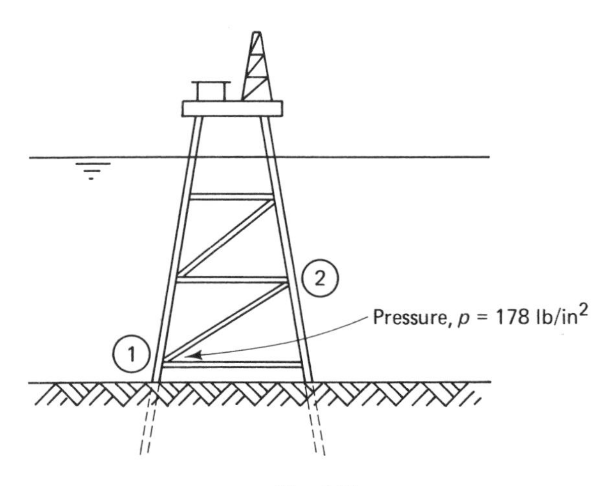

Fig. 4.21

From the analysis of Section 4.4, we find the following longitudinal and hoop stresses existing at the inside surface of end 1:

$$\sigma_z = -8730 \text{ lb/in}^2, \qquad \sigma_\theta = -2620 \text{ lb/in}^2$$

At the outside surface of the restrained end, we also find

$$\sigma_z = 4465 \text{ lb/in}^2, \qquad \sigma_\theta = 1340 \text{ lb/in}^2$$

and at regions away from the restrained ends, we find

$$\sigma_z = -2135 \text{ lb/in}^2, \qquad \sigma_\theta = -4270 \text{ lb/in}^2$$

Near the restrained end, we have from equations (29) that $\sigma_{zc}/E = 0.0125$, so that local longitudinal buckling is of no consequence when σ_z is compressive (Table 4.4). Hoop buckling is also of no consequence because of the radial restraint existing there. Thus, the allowable stresses are determined for this region from equation (31) as

$$S_a = 21{,}600 \text{ lb/in}^2, \qquad S_b = 24{,}000 \text{ lb/in}^2, \qquad S_\theta = 18{,}000 \text{ lb/in}^2$$

Now, the foregoing values of the longitudinal stress σ_z at the end of the member consist of a uniform axial stress σ_a and bending stress σ_b. The uniform axial stress is the same as the longitudinal stress existing at regions away from the end, so that we have at the inside surface

$$\sigma_a = -2135 \text{ lb/in}^2, \qquad \sigma_b = -6600 \text{ lb/in}^2$$

Since $\sigma_z = \sigma_a + \sigma_b < 0$ and $\sigma_\theta < 0$, the design conditions (28) apply and we have

$$\left|\frac{-2135}{21{,}600} - \frac{6600}{24{,}000}\right| = 0.37, \qquad \left|\frac{-2620}{18{,}000}\right| = 0.15$$

both of which are acceptable.

For the outside surface of the member at end 1, we similarly find

$$\sigma_a = -2135 \text{ lb/in}^2, \qquad \sigma_b = 6600 \text{ lb/in}^2$$

Since $\sigma_z = \sigma_a + \sigma_b > 0$ and $\sigma_\theta > 0$, condition (28) again applies and we have

$$\left|\frac{-2135}{21{,}600} + \frac{6600}{24{,}000}\right| = 0.18, \qquad \left|\frac{1340}{18{,}000}\right| = 0.07$$

which are acceptable.

Next consider the region away from the end. Here we have from equations (29) and (30) that $\sigma_{zc}/E = 0.0125$, as before, and $\sigma_{\theta c}/\sigma_Y = 0.32$. The corresponding allowable stresses S_a, S_b are accordingly the same as above, and the stress S_θ is found from Table 4.4 to equal 5760 lb/in². Thus, with $\sigma_a = -2135$ lb/in², $\sigma_b = 0$, $\sigma_\theta = -4270$ lb/in², we have from the relations (28)

$$\left|\frac{-2135}{21{,}600}\right| = 0.10, \qquad \left|\frac{-4270}{5760}\right| = 0.74$$

both of which are acceptable. #

EXAMPLE 4.5-2. Reconsider the preceding example for the case where end 1 of member 1–2 is subjected to a uniform axial stress of 10,000 lb/in² and bending stresses of 5000 lb/in² as a result of wind and wave action on the structure. Assume that the bending stresses have been corrected according to the methods of Section 4.3 and neglect any increase in the pressure arising from wave motion.

The axial and bending stress at end 1 resulting from the wind and wave action on the structure may be assumed to exist both at the end and at small distances from it where the pressure-induced stresses away from the restrained end apply. This is a reasonable assumption since the axial stress from wind and wave action is constant along the member and the change in the corresponding bending stress is generally small in comparison with that associated with the pressure-induced stresses.

We consider first the stresses existing at the restrained end 1. The allowable stresses are the same as in the preceding example. However, because we are now considering stress contributions from storm loadings, we may increase these values by one-third, so that we have

$$S_a = 28{,}800 \text{ lb/in}^2, \qquad S_b = 32{,}000 \text{ lb/in}^2, \qquad S_\theta = 24{,}000 \text{ lb/in}^2$$

At the inside surface, we have from the preceding example

$$\sigma_a = -2135 \text{ lb/in}^2, \qquad \sigma_b = -6600 \text{ lb/in}^2, \qquad \sigma_\theta = -2620 \text{ lb/in}^2$$

From the storm loading, we also have

$$\sigma_a = 10{,}000 \text{ lb/in}^2, \qquad \sigma_b = \pm 5000 \text{ lb/in}^2$$

Considering the tensile side of this latter bending stress, we thus have

$$\sigma_a = 7865 \text{ lb/in}^2, \qquad \sigma_b = -1600 \text{ lb/in}^2$$

together with the hoop stress given above. Since $\sigma_z = \sigma_a + \sigma_b > 0$ and $\sigma_\theta < 0$, relation (27) applies and we have

$$\left| \frac{7865}{28{,}800} - \frac{1600}{32{,}000} + \frac{2620}{24{,}000} \right| = 0.33$$

which is acceptable.

Considering next the compressive side of the storm-induced bending stress, we have

$$\sigma_a = 7865 \text{ lb/in}^2, \qquad \sigma_b = -11{,}600 \text{ lb/in}^2$$

since $\sigma_z = \sigma_a + \sigma_b < 0$ and $\sigma_\theta < 0$, relations (28) apply, giving

$$\left| \frac{7865}{28{,}800} - \frac{11{,}600}{32{,}000} \right| = 0.09, \qquad \left| \frac{2620}{24{,}000} \right| = 0.11$$

which are acceptable.

Similar considerations show also that the stress levels at the outside surface of end 1 are acceptable.

Finally, consider the stresses at distances away from the effect of the end restraint. From the preceding example, we have

$$\sigma_a = -2135 \text{ lb/in}^2, \qquad \sigma_b = 0, \qquad \sigma_\theta = -4270 \text{ lb/in}^2$$

as the pressure-induced stresses. Choosing the tensile side of the storm-induced bending stress, we thus have net axial and bending stresses

$$\sigma_a = 7865 \text{ lb/in}^2, \qquad \sigma_b = 5000 \text{ lb/in}^2$$

together with the value of σ_θ given above. Since $\sigma_z = \sigma_a + \sigma_b > 0$ and $\sigma_\theta < 0$, condition (27) applies. As in the preceding example, the allowable stresses S_a, S_b remain unchanged, but the allowable stress S_θ is reduced for this region. From the preceding example, this is equal to 5760 lb/in². Allowing for the one-third increase appropriate here, we have $S_\theta = 7680$ lb/in². Substituting, we find

$$\left| \frac{7865}{28{,}800} + \frac{5000}{32{,}000} + \frac{4270}{7680} \right| = 0.99$$

which is acceptable. A similar calculation for the compressive side of the storm-induced bending stress shows the resulting stresses also acceptable. #

4.6 CIRCUMFERENTIAL RING STIFFENERS

From the examples of Section 4.5, it can be seen that the critical element in the design of members subjected to a net external pressure is usually the hoop buckling stress. In cases where this is not sufficiently great compared with the pressure-induced hoop stress to ensure a safe design, the member may be reinforced radially with circular rings spaced along its length as shown in Fig. 4.22.

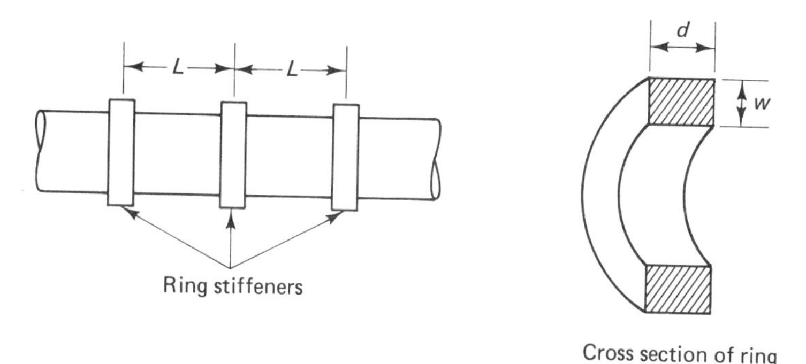

Fig. 4.22. Ring stiffeners for radial reinforcement.

The increased hoop buckling stress arising from these stiffening rings depends on the wall thickness t and radius a of the member and on the spacing L of the rings along the member. From solid mechanics, we have, in particular, that the hoop buckling stress $\sigma_{\theta c}$ depends only on the ratio t/a and the parameter β defined by

$$\beta = \frac{L}{a}\sqrt{\frac{a}{t}} \tag{32}$$

Figure 4.23 shows generally accepted values of $\sigma_{\theta c}$ for various values of t/a and β. Either this graph or more precise formulas summarized by the American Petroleum Institute (API, 1980) may be used for design purposes.

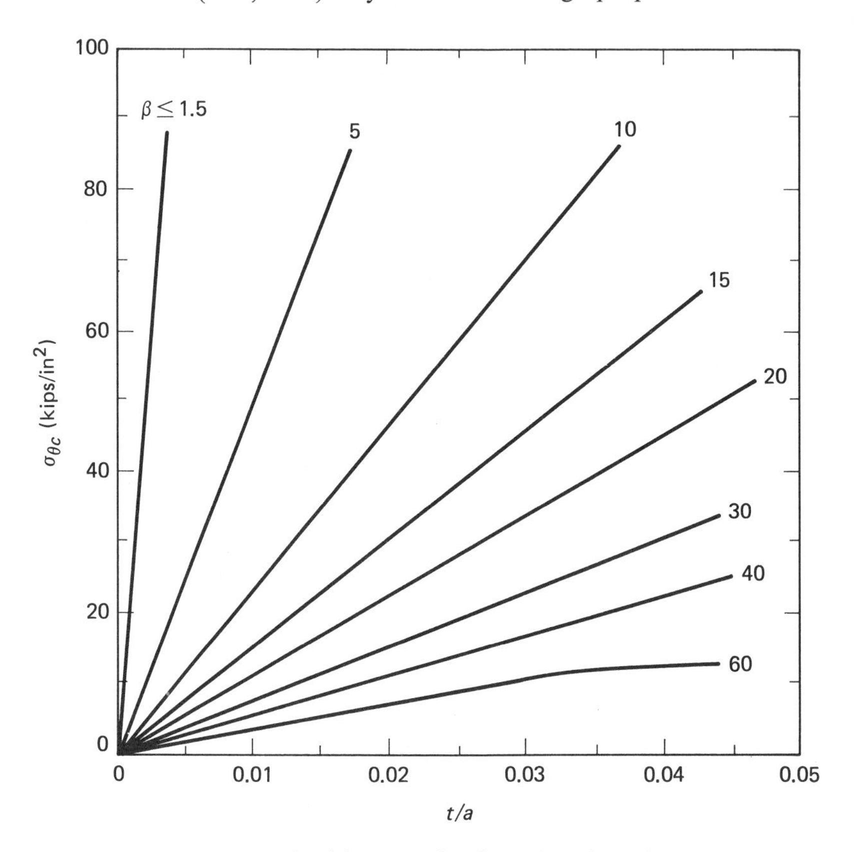

Fig. 4.23. Hoop buckling stress for ring-stiffened members.

The design of the ring may be based on its critical buckling load per unit of circumference q, which, from solid mechanics, is expressible as

$$q = \frac{3EI}{R^3} \tag{33}$$

where E denotes the Young's modulus of the ring material, R the radius of the centerline of the ring, and I the moment of inertia of its section. In terms of the width d and height w of the ring's cross section (Fig. 4.22), we have

$$R = a + \tfrac{1}{2}w, \qquad I = \tfrac{1}{12}\,dw^3 \tag{34}$$

If the pressure loading on the member between adjacent rings is assumed to be carried solely by the rings, we have $q = PL$, where P denotes the net external pressure loading. Taking P to be 1.5 times the critical buckling pressure of the shell, we thus have

$$P = 1.5\sigma_{\theta c}\frac{t}{a} \tag{35}$$

and equation (33) becomes expressible as

$$I = \frac{tLR^3}{2aE}\sigma_{\theta c} \tag{36}$$

which becomes the minimum acceptable value of I for design of the ring.

EXAMPLE 4.6-1. A steel member of an offshore structure, having a yield stress of 36,000 lb/in², a wall thickness of 0.5 in, and a radius of 12 in, is subjected to a net external pressure of 356 lb/in² (800 ft depth). Examine the hoop stress and design appropriate ring stiffeners if needed to prevent buckling.

The stresses away from the ends of the member are determined from equations (4) and (5) as

$$\sigma_\theta = -8530 \text{ lb/in}^2, \qquad \sigma_z = -4270 \text{ lb/in}^2$$

and the critical buckling stress is determined from equation (30) as

$$\sigma_{\theta c} = 11{,}460 \text{ lb/in}^2$$

We thus have $\sigma_{\theta c}/\sigma_Y = 0.318$, so that from Table 4.5 we have $\mathbf{S}_\theta = 5729$ lb/in². Using relations (28), we find

$$\left|\frac{\sigma_\theta}{S_\theta}\right| = \left|\frac{-8530}{5729}\right| = 1.5$$

which is unacceptable.

To increase the buckling stress, we use ring stiffeners and assume tentatively that they are spaced every 8 ft along the member. From equation (32) we find, with $t/a = 0.042$, that $\beta = 39$. From Fig. 4.23 we thus see that $\sigma_{\theta c} =$ 25 kips/in². With this value, we have $\sigma_{\theta c}/\sigma_Y = 0.69$. The allowable stress S_θ is determined from Table 4.4 by interpolation using the values of S_θ/σ_Y associated with σ_θ/σ_Y equal to 0.5 and 1.0. We find $S_\theta/\sigma_Y = 0.30$ and $S_\theta = 10{,}800$ lb/in².

Using relations (28), we find with this increased allowable stress that

$$\left|\frac{\sigma_\theta}{S_\theta}\right| = \left|\frac{-8530}{10{,}800}\right| = 0.79$$

which is acceptable.

For the design of the rings, we assume tentatively that the thickness w of the ring is 2.5 in and determine the required width d from equation (36). We have (assuming a steel ring)

$$I = \frac{(0.5)(96)(13.25)^3}{2(12)(3 \times 10^7)}(250{,}00) = 3.88 \text{ in}^4$$

From the definition of I given by equation (34), we find

$$d = \frac{12I}{W^3} = \frac{(12)(3.88)}{(2.50)^3} = 2.98 \text{ in.}$$

Thus, rings of 2.5 in. thickness and 3.0 in. width, with inside diameter of 2 ft, would be appropriate. #

4.7 ANALYSIS OF JOINTS

Punching Shear Failure

Joints of an offshore structure are formed at locations where cross members are welded onto the main legs of the structure. Figure 4.24(a) illustrates the general details of such a connection for the typical case where the cross members framing into the joint do not overlap. Under design wind and wave loadings, the forces in each cross member are seen to be transmitted directly to the wall of the leg. The possibility accordingly exists of a *punching shear failure* through the wall if its thickness is too small.

For an approximate analysis of the punching shear stress induced in a leg of the structure by a cross member, we may neglect the curvature of the leg and treat the problem as an inclined circular member framing into a flat plate [Fig. 4.24(b)]. In this case, the perimeter of the intersection will generally be elliptical in shape, with its major and minor axes equal, respectively, to $2R_b/\sin\theta$ and $2R_b$, where R_b denotes the radius of the cross member and θ its inclination, as defined in Fig. 4.24(a). Thus, if f_x denotes the normal force component exerted on the leg by the cross member, the punching shear τ_{pa} in the leg wall from this force is expressible as

$$\tau_{pa} = \frac{f_x}{tC} \tag{37}$$

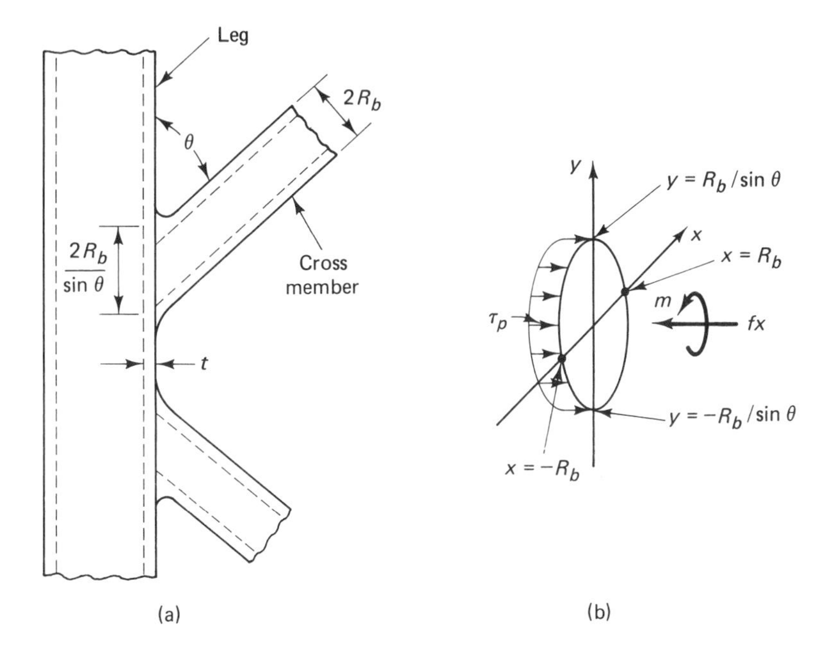

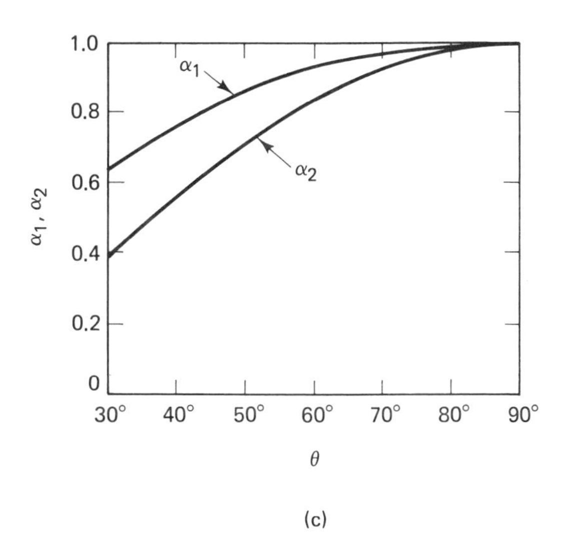

Fig. 4.24. Joint geometry and shear coefficients.

where t denotes the wall thickness of the leg and C denotes the circumference of the intersectional curve. Similarly, if m denotes the moment exerted on the leg by the cross member, the punching shear stress τ_{pm} resulting from this moment is expressible as

$$\tau_{pm} = = \frac{my}{tI_1}, \qquad I_1 = \int_C y^2\, ds \tag{38}$$

where y is measured from the center of the elliptical intersection curve [Fig. 4.24(b)]. The extreme values occur at $y = \pm R_b/\sin\theta$, so that on combining the two expressions above, the maximum punching (or pulling) shear stress τ_p may be expressed as

$$\tau_p = \alpha_1 \frac{|f_x|}{2\pi t R_b} + \alpha_2 \frac{|m|}{\pi t R_b^2} \tag{39}$$

where α_1 and α_2 denote dimensionless coefficients defined by

$$\alpha_1 = \frac{2\pi R_b}{C}, \qquad \alpha_2 = \frac{\pi R_b^3}{\sin\theta}\frac{1}{I_1} \tag{40}$$

These coefficients depend only on the inclination θ of the cross member and hence can be evaluated using the geometrical relations shown in Fig. 4.24(b). Numerical values are given in Fig. 4.24(c).

To ensure against punching shear failure of the joint, it is necessary that the shear stress τ_p be less than the shear yield stress of the material, with a suitable factor of safety. This may roughly be estimated as 0.4 times the tensile yield stress. More detailed criteria for selection of this value are given by the American Petroleum Institute (API, 1980), together with discussion of more complex overlaping joints.

EXAMPLE 4.7-1. Determine the minimum wall thickness of the vertical leg of Fig. 4.25(a) needed to prevent punching shear failure from the loads exerted by member 1–2. Assume an allowable shear yield stress of 14,400 lb/in².

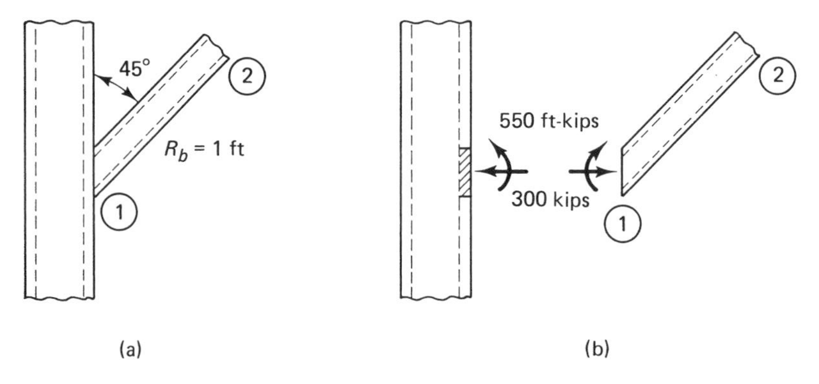

Fig. 4.25

The inclination angle θ equals 45° and we accordingly find from Fig. 4.24(c) that $\alpha_1 = 0.82$ and $\alpha_2 = 0.56$. Substituting values into equation (39), we thus have

$$14{,}400 = 0.82\frac{300 \times 10^6}{2\pi t(12)} + 0.56\frac{550 \times 10^6 \times 12}{\pi t(12)^2}$$

which gives $t = 0.79$ in as the required minimum wall thickness. #

Fatigue Failure

In addition to punching shear failure at a joint, there also exists the possibility of *fatigue failure* resulting from the growth of small flaws in the material under repeated cyclic loading from wave action. Such failures can occur even though the stress in the member never exceeds the yield stress of the material. The higher the cyclic stress, the lower will, however, be the number of cycles needed for failure. Because of the abrupt change in geometry at the end of a cross member framing into a joint, localized stress increases exist there so that fatigue failures, when they occur, can generally be expected at the ends of cross members or in the weld material of the joints.

The fatigue properties of a material are customarily represented by an *S-N curve*, showing the number of cycles N needed to cause failure at a given cyclic stress amplitude σ (or stress range 2σ). Figure 4.26 shows a typical curve

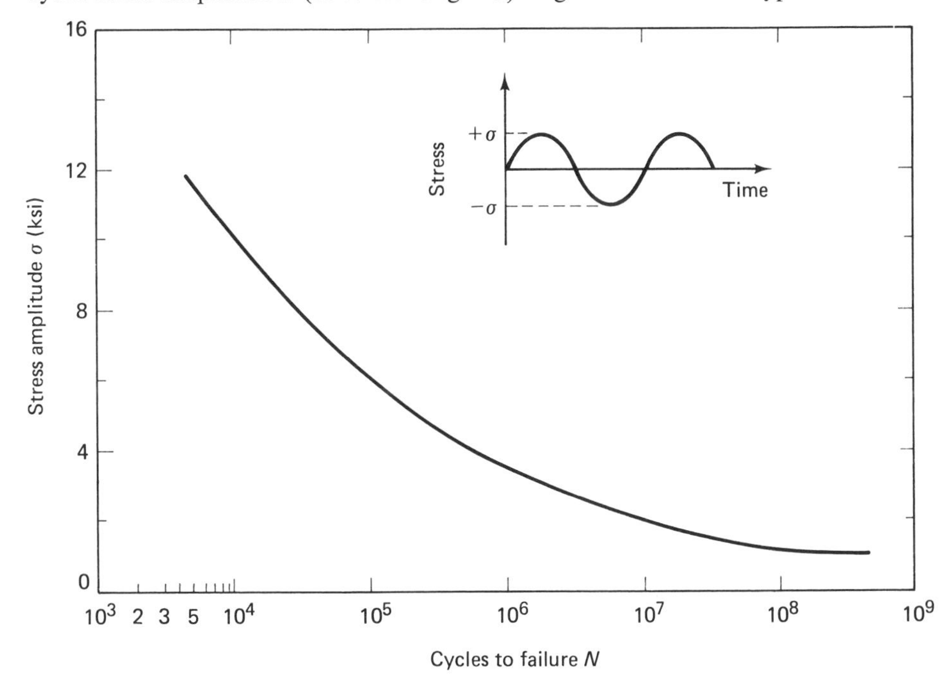

Fig. 4.26. *S-N* fatigue curve for structural steel in saltwater.

of this kind appropriate for steels commonly used in offshore structures. As indicated, for stress amplitudes less than about 1 ksi, no fatigue failure occurs, regardless of the number of cycles experienced by the material. This stress level is referred to as the *endurance limit*.

The data shown in Fig. 4.26 refer to cyclic loading at a fixed stress amplitude σ. With offshore structures, the members are subjected to variable stress amplitudes over the life of the structure as a result of the variable sea states experienced. In this case, each cyclic stress amplitude causes some fatigue damage to a member and fatigue failure is then ultimately determined by the cumlative damage from all the many stress amplitudes experienced. The simplest representation of the net cumulative damage needed for fatigue failure is expressible (the Palmgren–Miner rule) as

$$\frac{n_1}{N_1} + \frac{n_2}{N_2} + \cdots + \frac{n_m}{N_m} = 1 \tag{41}$$

where n_1 denotes the number of cycles at stress level σ_1, N_1 denotes the life of the member at stress level σ_1 (as determined from the *S-N* curve), n_2 denotes the number of cycles at stress level σ_2, etc.

In applying this relation to fatigue analysis of a member of an offshore structure, it is, of course, necessary to have a reasonable estimate of the various stress levels experienced by the member during its intended life, together with the number of cycles expected at each level. The formation of this estimate and the subsequent fatigue analysis is best carried out using statistical procedures as described, for example, by Maddox and Wildenstein (1975). A simplified analysis may, however, be made by considering various idealized discrete cyclic stress levels likely to be induced in the member by wave action during its intended life and estimating the period and percent of time of occurrence of each. The fatigue life of the member can then be estimated directly from equation (41). In particular, if T denotes the fatigue life and P_1, P_2, etc., denote fractions of time at which stress amplitudes σ_1, σ_2, etc., occur, the associated values of n_1, n_2, etc., in equation (41) are determined simply as

$$n_1 = \frac{P_1 T}{T_1}, \qquad n_2 = \frac{P_2 T}{T_2}, \qquad \text{etc.}$$

where T_1, T_2, etc., denote the periods of the cyclic stresses σ_1, σ_2, etc. Substituting these expressions into equation (41), we thus have

$$\frac{P_1 T}{N_1 T_1} + \frac{P_2 T}{N_2 T_2} + \cdots + \frac{P_m T}{N_m T_m} = 1 \tag{42}$$

from which the fatigue life T may be determined.

In carrying out the analysis, we may assume a distribution of wave heights and periods, together with the percent of time that they occur at a particular offshore site. For each, we may then calculate the maximum longitudinal stress (axial plus bending) at the ends of cross members of a structure to be located at the site using the methods of Section 4.2. These maximum stresses will equal the stress amplitudes in the cyclic wave loadings and the periods of the stress cycles will be the same as these. As noted above, the abrupt change in geometry at the joint causes local stress amplitudes there to exceed these calculated values. Because of their localized nature, such stress increases are not normally considered in proportioning the members by the methods of Section 4.2. However, they are important in fatigue analysis because they can accelerate the growth of small flaws in the affected regions. In fatigue analysis, the stress increases are therefore customarily accounted for with a *stress concentration factor* k_s such that if σ is the calculated stress amplitude, the actual value is $k_s\sigma$. Values of k_s used in engineering applications generally range from 2 to 3. In addition to this stress amplification, there can also exist a further amplification for certain wave periods from dynamic effects. This may be accounted for approximately using a dynamic amplification factor, as described later in Section 4.10. Once the best estimates of the stress amplitudes are determined, the values of N_1, N_2, etc., in equation (42) may be found from the *S-N* fatigue curve and the life of the member calculated by solution of this equation.

EXAMPLE 4.7-2. The heights (H) and periods (T) of waves at an offshore site are idealized into distinct levels and tabulated below, together with the fraction of time (P) that such waves can be expected. Stress amplitudes (σ) at a joint for a particular cross member of a proposed structure at the site are also shown in Table 4.5.

Table 4.5

H (ft)	T (sec)	P (%)	σ (kips/in²)
40–60	12	0.01	5.0
20–40	10	0.03	3.0
10–20	9	0.10	1.5
5–10	7	1.00	0.6
0–5	3	98.86	0.4

Determine the fatigue life T of the joint, assuming a stress concentration factor $k_s = 2$. Neglect any dynamic amplification of the stresses.

The actual stress associated with the nominal stress amplitude $\sigma_1 = 5$ kips/in² is $k_s\sigma_1 = 10$ kips/in². From the *S-N* curve of Fig. 4.26, the life N_1 of the joint at this stress level is found to equal 10^4 cycles. The corresponding fraction of time P_1 at which this stress level exists is 0.0001 and the period T_1 is 12 sec. Treating the other stress levels in a similar manner and substituting into equation (42), we have

$$\frac{0.0001T}{12 \times 10^4} + \frac{0.0003T}{10 \times 10^5} + \frac{0.001T}{18 \times 10^6} + \frac{0.01T}{35 \times 10^6} = 1$$

Notice that the number of cycles for failure at the lowest nominal stress level (0.4 kip/in²) is infinite since the actual stress level (0.8 kip/in²) is less than the endurance limit indicated in Fig. 4.26. The ratio P_5T/N_5T_5 for this stress level is accordingly zero and hence not included in the relation above. Solving for T, we find that $T = 6.78 \times 10^8$ sec or 21.5 years. #

4.8 ANALYSIS OF OFFSHORE CONCRETE PLATFORMS

Offshore concrete platforms were discussed in Chapter 1. These structures take the form of gravity structures in that they rely on their own weight to provide resistance to overturning from lateral loads and are supported by large foundational elements which bear directly on the seafloor. Figure 4.27 shows a typical

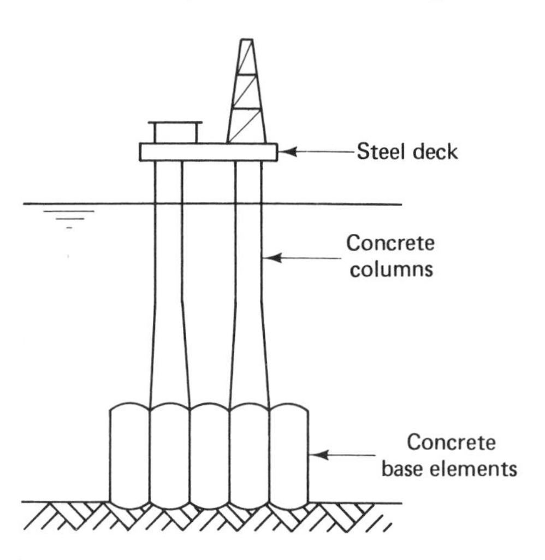

Fig. 4.27. Typical concrete gravity platform.

example of this kind of structure. As illustrated, these structures generally consist of a cellular base and one or more unbraced columns supporting a deck structure. The base and columns are generally made of concrete and the deck is made of steel. The main purpose of the base is to provide support and stability to the structure. These structures are constructed for the most part in a shallow harbor area in their upright position. The deck is next attached and the structure then towed semisubmerged to its offshore site, after which the base is flooded and the structure sunk to the seafloor.

Apart from installation requirements, the design and analysis of gravity structures generally parallels that described earlier for steel structures. Design wind and water conditions are specified and preliminary estimates of member sizes of a chosen design form then investigated, and improved if necessary, to ensure adequate strength under the loadings imposed. The structural methods of Chapter 2 are available, as in the case of steel structures, for determining the internal forces and moments within the individual members. The use of concrete introduces, however, additional design considerations not involved with steel structures.

As a construction material, concrete shows good strength under compressive loadings (typically 3000 to 6000 lb/in^2) but almost no strength in tension (less than 15% of the compressive strength). Thus, with concrete members subjected to tension either directly or through bending action, it is necessary to provide some additional load-carrying capacity for the tensile forces.

One way to do this is to cast the concrete around steel reinforcing bars aligned in the direction of the anticipated tensile forces. When the concrete hardens and the member is later subjected to loading, the resulting tensile stresses will then be carried by the steel reinforcing bars. One disadvantage of this *reinforced concrete* for offshore structures is that the concrete around the reinforcement cracks when it is subjected to tensile stress, thus exposing parts of the steel reinforcing bars directly to the corrosive action of the sea water.

An alternative method of providing tensile strength to a concrete member is to leave small ducts along the member as it is being cast, after which high-strength steel wires, strands, or bars (collectively called *tendons*) are threaded through the ducts, anchored at one end, and placed in tension and anchored to the concrete at the other end. The tensile force in the steel is thus transferred to the concrete as a compressive stress. When the member is then subjected to a tensile stress in the direction of the prestress, the net effect, if the prestress is sufficient, is simply to unload the concrete from its initial compressive state. Under these circumstances, the concrete itself experiences no tensile stress, and hence does not crack under loading.

The design of prestressed concrete members is carried out using elastic beam theory, with sufficient prestress being applied to prevent tensile cracking. Unlike steel, however, concrete does not have a well defined yield stress below which the stress–strain behavior is linear. Instead, stress–strain curves for concrete in compressions show a gradually changing slope that depends not only on the stress level but also on the strength of the particular concrete considered. Figure 4.28(a) shows typical stress–strain curves for concrete of two different strengths σ_c'. The slope of the initial, nearly straight, portion of the stress–strain curve is used to determine the elastic modulus E_c of the concrete and is described by the empirical equation

$$E_c = 33\gamma_c^{3/2}\sqrt{\sigma_c'} \qquad \text{lb/in}^2$$

where γ_c denotes the specific weight of the concrete in lb/ft^3 (typically about 145 lb/ft^3) and σ_c' is expressed in units of lb/in^2. Elastic calculations based on

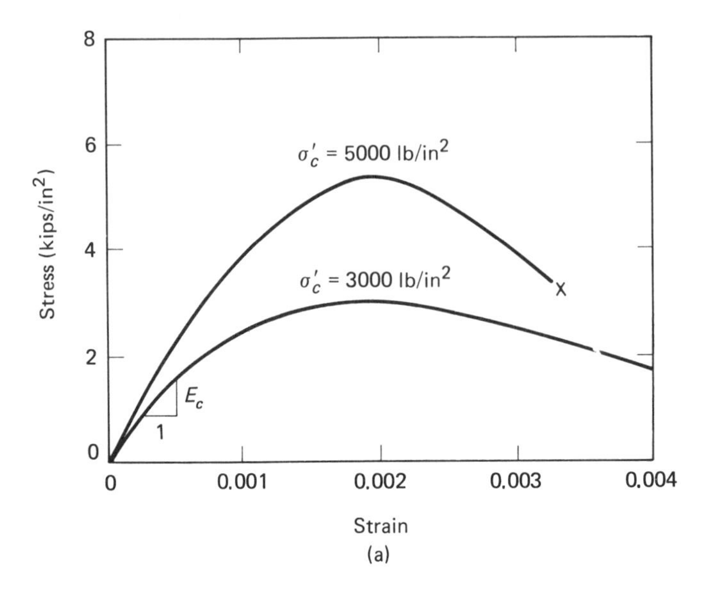

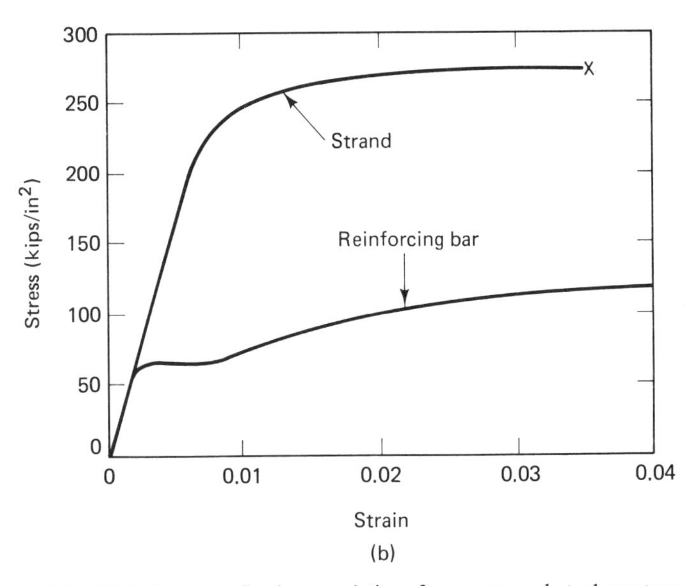

Fig. 4.28. Stress-strain characteristics of concrete and steel prestress strands and reinforcing bars: (a) typical stress-strain curves of concrete in compression; (b) typical stress-strain curves of steel prestress strands and reinforcing bars.

this modulus are generally limited to stress levels no greater than 45% of the concrete strength.

Even within this stress limit, uncertainties can, however, arise because of creep of the concrete, stress relaxation of the prestress tendons, anchor slip, etc. For this reason, and in contrast with the design of steel members, an overall factor of safety of the member is generally established on the basis of the ultimate load that the member can withstand without failure, rather than on the stress levels existing under design loads.

In a typical offshore structure such as shown in Fig. 4.27, prestress is generally used to prevent cracking of the concrete columns and base elements under design loads. However, with prestress alone considered in their design, the ultimate capacity of the members may not differ from the design loading by the desired overall factor of safety (typically in the range 1.5 to 2 as determined from a combination of load increase and capacity reduction). In this case, ordinary reinforcing bars are then added to the design to increase the ultimate capacity to the desired level. Typical stress–strain curves for prestress strands and ordinary reinforcing bars are shown in Fig. 4.28(b).

Recommendations for the design of general concrete structures can be found in the publication *Building Code Requirements for Reinforced Concrete* (ACI, 1977). Specific reference to offshore structures can be found in the publication *Guide for the Design and Construction of Fixed Offshore Concrete Structures* (ACI, 1978). General principals of concrete design are discussed in detail by Winter and Nilson (1973) and Lin (1963), among others.

To illustrate the main principles involved in prestress design, we consider a concrete member having a single prestress tendon along its centroidal axis as shown in Fig. 4.29. The member ends are subjected to axial force F, shear force

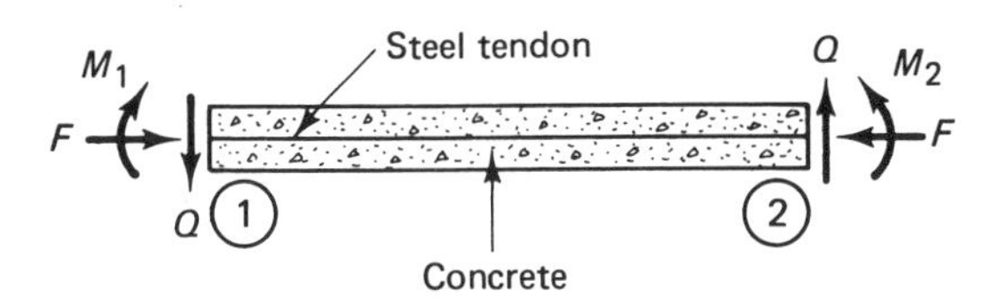

Fig. 4.29. Prestressed concrete member.

Q and bending moments M_1, M_2 about the centroid of the section. The area of the tendon is assumed sufficiently small that the stresses in the concrete can be calculated using the gross dimensions of the section.

If a denotes the distance from the centroidal axis to the outer surface, the extreme bending stresses σ_{cb} are expressible as

$$\sigma_{cb} = \pm \frac{Ma}{I} \tag{43}$$

where M denotes the maximum moment and I denotes the moment of inertia of the section. The axial stress is similarly calculated as

$$\sigma_{ca} = -\frac{F}{A} \tag{44}$$

where A denotes the total area of the section. The tensile force in the tendon is transferred to the concrete as compression so that the prestress in the concrete σ_{cp} is calculated as

$$\sigma_{cp} = -\frac{\sigma_s A_s}{A} \tag{45}$$

where σ_s denotes the tensile prestress in the tendon and A_s denotes its area.

Combining the stresses above, we have the net stress σ_c in the concrete expressible as

$$\sigma_c = \pm\frac{Ma}{I} - \frac{F}{A} - \frac{\sigma_s A_s}{A} \tag{46}$$

The condition that the tensile bending stress in the concrete is zero is given by

$$0 = +\frac{Ma}{I} - \frac{F}{A} - \frac{\sigma_s A_s}{A} \tag{47}$$

from which the area A_s of the tendon can be determined when the prestress σ_s is specified.

With this value, the maximum compressive stress in the concrete is then given by

$$\sigma_c = -\frac{Ma}{I} - \frac{F}{A} - \frac{\sigma_s A_s}{A} \tag{48}$$

Because elastic beam theory has been used to calculate the bending stress, this compressive stress is generally restricted to a value of no more than 45% of the compressive strength of the concrete. For specified sectional dimensions of the member, this condition accordingly limits the magnitudes of the moment and axial force that can be carried by the member.

The shear stress in the concrete may also be calculated using elastic beam theory and this combined with the axial stress to determine the principal tensile stress. This stress will act at an angle to the tensile bending stress and is sometimes referred to as diagonal tension. If τ denotes the shear stress at a point in the member and σ the corresponding longitudinal stress, the principal tensile stress σ_1 is determined from the Mohr circle analysis of Fig. 4.30 as

$$\sigma_1 = \frac{\sigma}{2} + \sqrt{\left(\frac{\sigma}{2}\right)^2 + \tau^2} \tag{49}$$

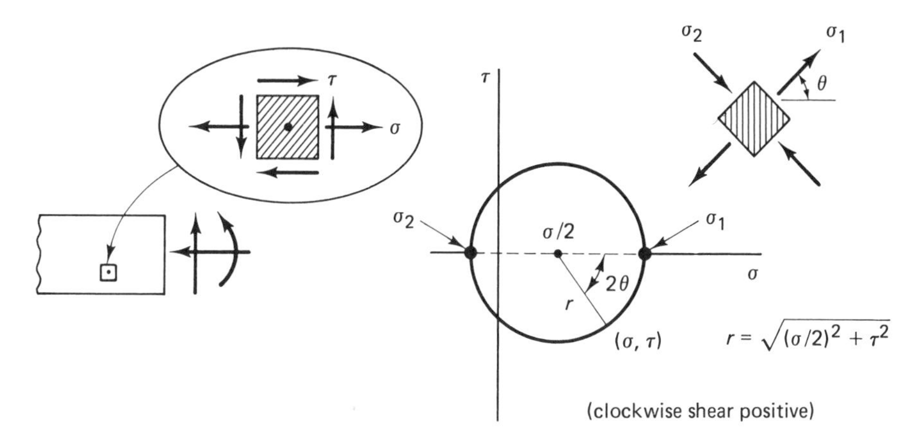

Fig. 4.30. Mohr's circle and principal stresses in a concrete member.

and the angle θ that it makes with the member axis is given by

$$\tan 2\theta = \frac{2\tau}{\sigma} \tag{50}$$

The critical fracture stress for this stress, in units of lb/in², is normally assumed to equal about $4\sqrt{\sigma_{c'}}$, where $\sigma_{c'}$ denotes the compressive strength of the concrete in units of lb/in². The condition for avoiding shear cracking is thus $\sigma_1 \leq 4\sqrt{\sigma'_c}$, with σ_1 being the greatest value of principal tensile stress existing in the member.

For relatively long members where bending stresses dominate, cracking from tensile bending stresses can generally be expected to occur before cracking from diagonal tension. To estimate the moment M_c at which this cracking would start, we may use equation (46) with the tensile stress in the concrete set equal to the critical cracking stress σ''_c. This critical stress is normally estimated (in lb/in²) as $7.5\sqrt{\sigma'_c}$, where σ'_c denotes the compressive strength of the concrete in units of lb/in². Thus, we have

$$\sigma''_c = \frac{M_c a}{I} - \frac{F}{A} = \frac{\sigma_s A_s}{A} \tag{51}$$

from which the cracking moment may be determined.

The results above have all been based on the assumption that the concrete could be treated as a perfectly elastic material. As noted earlier, because of uncertainties regarding creep, stress relaxation of the prestress tendons, etc., the results from this *working stress theory* are subjected to some error. Consistent with good engineering practice in concrete design, an overall factor of safety is accordingly determined by estimating the maximum moment that the member can withstand without failure. This analysis is carried out using *ultimate stress theory*.

To calculate this moment, we consider the limiting condition where the concrete in compression is on the verge of crushing and where that in tension is assumed fully cracked and incapable of carrying any tensile loading. The compressive strain at which crushing failure of the concrete occurs is typically chosen as 0.003.

The variation of the strain over the cross section of the column is assumed linear, as shown in Fig. 4.31(a), and the total strain variation is described by the equation

$$\epsilon = -\frac{0.003}{e}\xi \tag{52}$$

where ξ denotes distance from the neutral axis, measured positive upward, and e denotes the distance from the outer fiber to the neutral axis.

Because the concrete stress–strain curve will no longer be linear at this strain level, the compressive stress variation in the concrete will be as indicated in Fig. 4.31(b). As noted above, the tensile stress in the concrete is assumed zero; that is, at this level of strain, the concrete is assumed to have cracked and is no longer capable of carrying any tensile force.

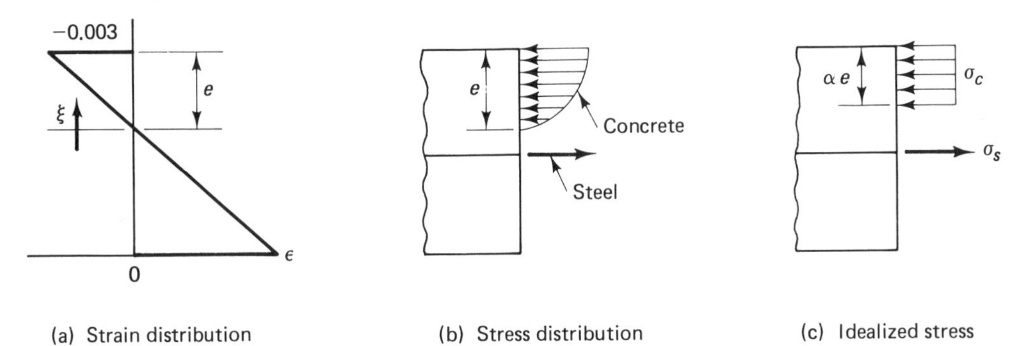

Fig. 4.31. Actual and idealized stress distribution for ultimate moment calculation: (a) strain distribution; (b) stress distribution; (c) idealized stress distribution.

To simplify calculations, it is customary to approximate the compressive stress distribution in the concrete by a uniformly distributed stress distribution as shown in Fig. 4.31(c), with the uniform stress σ_c given in terms of the concrete compressive strength σ_c' by $\sigma_c = 0.85\sigma_c'$ and the depth factor α chosen as 0.85 for $\sigma_c' \leq 4000$ lb/in^2 and decreased by 0.05 for every 1000 lb/in^2 increase in σ_c' over 4000 lb/in^2.

From equilibrium, the sum of the compressive force in the concrete and the tensile force in the steel tendon must equal the axial compressive load F, that is,

$$\sigma_s A_s - \sigma_c A_c = -F \tag{53}$$

where A_c denotes the area over which σ_c acts. This equation may be used to locate the neutral (bending) axis by solving for the distance e. The stress σ_s is determined from the stress–strain curve of the steel tendon for a strain equal to the sum of the initial prestrain and that given by equation (52). Since this strain depends on the distance e, solution of equation (53) for its value must be carried out by a trial-and-error calculation.

Having the value of e, the ultimate moment M_u, about the centroid of the section, is then determined as the sum of the moments from the compressive and tensile forces. In the present case, the tendon lies along the centroidal axis of the beam so that the moment arm of the tensile force is zero. The ultimate moment is accordingly determined from the resultant compressive force $\sigma_c A_c$ as

$$M_u = \sigma_c A_c \bar{x} \tag{54}$$

where $\bar{x}$ denotes the distance from the centroid of the section to the centroid of the area A_c.

If the ultimate moment does not differ from the design moment by a suitable factor of safety, it can be increased by including ordinary (nonprestressed) reinforcing bars in the member parallel to the member axis. If, for example, a single reinforcing bar of area A_r is located a distance j below the centroidal axis of the member, the stress σ_{sr} in this bar, as determined from the strain given by equation (52) and the stress--strain curve of the bar, gives rise to an additional force $\sigma_{sr}A_r$ that must be included in equation (53) and an additional moment $j\sigma_{sr}A_r$ that enters into equation (54).

In addition to the possible need for ordinary reinforcing bars to increase the ultimate moment of the member, it may also be necessary to include such bars for reinforcement where cracking under combined shear and longitudinal stress could occur under the ultimate load. The reason is that the required ultimate moment may not be realizable because of the inability of the cracked member to carry the associated shear force. In such a case, reinforcing bars, called *stirrups*, are placed perpendicular to the axis of the member to provide the necessary shear resistance.

As an illustration, consider the cracked portion of a member with stirrups as shown in Fig. 4.32. If Q_u denotes the shear force associated with the required

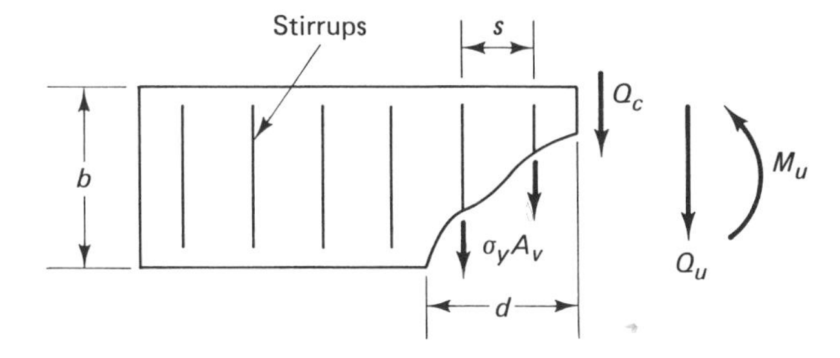

Fig. 4.32. Reinforcement for shear in a cracked section.

ultimate moment M_u and Q_c denotes that carried by the uncracked concrete, the excess shear force $Q_u - Q_c$ must be balanced by forces in the stirrups. With A_v denoting the total sectional area of stirrups at any longitudinal location, the maximum force developed there is approximately $\sigma_Y A_v$, where σ_Y denotes the yield stress of the stirrups. Finally, if d denotes the projection of the crack length on the member axis, and s the spacing of the stirrups, the number of stirrup locations is d/s, so that

$$\sigma_Y A_v \frac{d}{s} = Q_u - Q_c \tag{55}$$

The shear force Q_c is normally estimated as the lesser of the shear forces needed to initiate diagonal tension cracks in the central part of the member or that needed to cause spreading of the bending-induced tensile cracks at the outer surface. The former occurs when the diagonal tension (in lb/in²) equals about $4\sqrt{\sigma_c'}$ and the latter when the longitudinal tension (in lb/in²) equals about $10\sqrt{\sigma_c'}$, where σ_c' denotes the compressive strength of the concrete in units of lb/in². The projected length d of the crack depends on the average inclination of the crack and the distance it spreads into the member. For engineering purposes it is usually assumed to be about 0.8 times the depth b of the member. Having values of Q_c and d, equation (55) can then be solved for the spacing s of stirrups of specified strength and sectional area. As a general rule, this spacing should not be greater than about $d/2$, so that all potential cracks are intersected. Also when $Q_c \geq Q_u$, no shear reinforcement is predicted by equation (55). However, even when this condition obtains, most design recommendations include provisions for a minimum amount of shear reinforcement (see ACI, 1977).

EXAMPLE 4.8-1. Consider a proposed concrete offshore platform consisting of a massive base and single cylindrical column supporting an operating deck and equipment, as shown in Fig. 4.33. The column has an outside diameter of 15 ft and inside diameter of 11 ft. The deck and equipment weigh 5500 kips and the column weighs 948 kips. Assuming the resultant wind and wave forces on the column and deck to equal 400 kips and to act 62 ft above the base of the column, determine the net area of prestress tendons needed to prevent tensile stress in the concrete if the tendons are to be spaced symmetrically around the column section and to have an effective prestress of 150 kips/in². Also calculate the diagonal tension associated with the maximum shear stress and the cracking moment, assuming a concrete strength σ_c' of 5000 lb/in².

The internal forces and moments are maximum at the base of the column and are indicated in Fig. 4.33. The moment of inertia I and net area A of the column are calculated as

$$I = \frac{\pi}{64}(15^4 - 11^4) = 1766 \text{ ft}^4$$

$$A = \frac{\pi}{4}(15^2 - 11^2) = 81.7 \text{ ft}^2$$

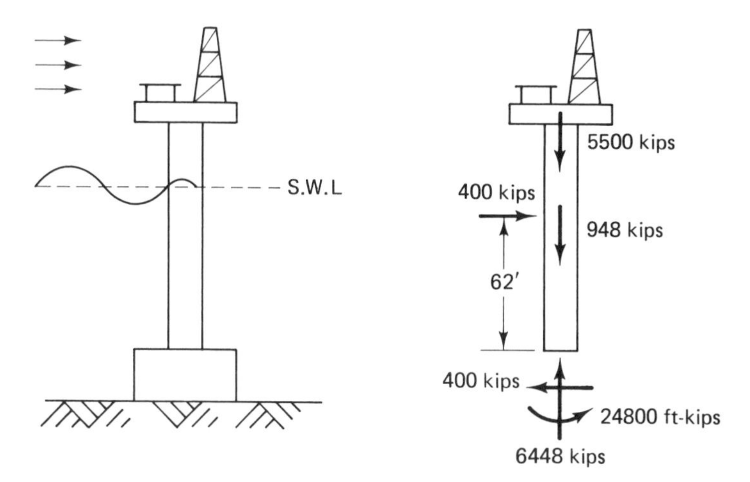

Fig. 4.33

The prestress is applied uniformly around the column so that its resultant force acts along the centroidal axis of the column. Equation (47) accordingly applies with $a = R_o = 7.5$ ft, $M = 24{,}800$ ft-kips, $F = 6558$ kips, so that

$$0 = \frac{(24{,}800)(7.5)}{1766} - \frac{6448}{81.7} - \frac{\sigma_s A_s}{81.7}$$

The stress in the tendons is 150 kips/in². Solving the equation above for A_s, we thus find the net area A_s of the tendons to equal 14.4 in². The associated maximum compressive stress in the concrete is

$$\sigma_c = -\frac{(24{,}800)(7.5)}{1766} - \frac{6448}{81.7} - \frac{(150)(14.4)}{81.7} = -211 \text{ kips/ft}^2$$

or 1465 lb/in², compression This may be compared with the maximum allowable value of $0.45\sigma_c' = 0.45(5000) = 2250$ lb/in².

From solid mechanics, the maximum shear stress τ_m in the column is known to occur on the axis of bending and to be given by the formula

$$\tau_m = \frac{Q_m S}{I(2t)}$$

where Q_m denotes the maximum shear force, I the moment of inertia of the cross section, t the wall thickness, and S the first moment of the area of the cross section above the bending axis, as given in terms of the outside radius $R_o = 7.5$ ft and the inside radius $R_i = 5.5$ ft by

$$S = \tfrac{2}{3}(R_0^3 - R_i^3) = \tfrac{2}{3}(7.5^3 - 5.5^3) = 170.3 \text{ ft}^3$$

The maximum shear force in the column equals 400 kips. The maximum shear stress is thus determined as

$$\tau_m = \frac{(400)(170.3)}{(1766)(4)} = 9.64 \text{ kips/ft}^2 \ (67.0 \text{ lb/in}^2)$$

The longitudinal stress σ on the axis of bending is calculated

$$\sigma = -\frac{6448}{81.7} - \frac{(150)(14.4)}{81.7} = -105 \text{ kips/ft}^2 \ (-732 \text{ lb/in}^2)$$

The maximum (principal) tensile stress resulting from the shear stress is thus determined from equation (51) as

$$\sigma_1 = -\frac{732}{2} + \sqrt{\left(\frac{732}{2}\right)^2 + (67)^2} = 6.1 \text{ lb/in}^2$$

This may be compared with the critical fracture value of $4\sqrt{\sigma_c'} = 4\sqrt{5000} =$ 283 lb/in².

The cracking moment M_c associated with bending tensile stress is calculated from equation (51) with $\sigma_c'' = 7.5\sqrt{\sigma_c'} = 7.5\sqrt{5000} = 530 \text{ lb/in}^2 =$ 76.3 kips/ft². Thus, we have

$$76.3 = \frac{M_c(7.5)}{1766} - \frac{6448}{81.7} - \frac{(150)(14.4)}{81.7}$$

from which we find $M_c = 42{,}780$ ft-kips. This is about 1.7 times the applied moment. #

EXAMPLE 4.8-2. Determine the ultimate moment capacity of the column of the preceding example, assuming the prestress tendons to be grouped at quarter points around the section at radius 6.5 ft as indicated in Fig. 4.34.

The assumed strain distribution corresponding to crushing failure of the concrete ($\epsilon = -0.003$) is as indicated in Fig. 4.34(a). The equation for this distribution is

$$\epsilon = -\frac{0.003}{e}\xi$$

where ξ denotes distance from the bending axis and e denotes the distance from the outside surface to the bending axis. The compressive force F_c in the concrete must combine with the net tensile force F_s in the steel to give the total compressive force F on the section. Thus, analogous to equation (53), we require that

$$F_s - F_c = -F = -6448 \text{ kips}$$

This equation is used to solve for the distance e by trial and error.

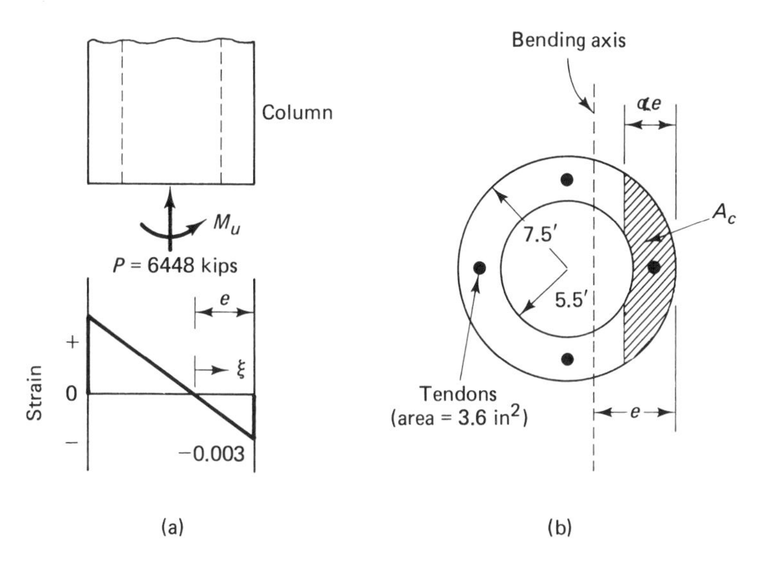

Fig. 4.34

As an approximation, the stress in the steel is assumed to vary linearly with the strain for strain values less than its yield strain and to equal the yield stress for greater strains. The elastic modulus of the tendons is assumed to equal 27×10^3 kips/in² and its yield stress to equal 212 kips/in². The yield strain is thus $212/27 \times 10^3 = 7.85 \times 10^{-3}$. The initial tensile prestress in the steel is 150 kips/in², so that the initial tensile strain is $150/27 \times 10^3 = 5.56 \times 10^{-3}$. The compressive stress distribution in the concrete is approximated by the uniform distribution shown in Fig. 4.30.

We first arbitrarily assume a value of $e = 3$ ft. For the steel tendons at position 1, we have $\xi = 6.5 - (7.5 - 3) = 2$ ft. The strain from the assumed bending distribution is

$$\epsilon = -\frac{0.003}{3}(2) = -0.002$$

Adding this to the initial prestrain of 5.56×10^{-3}, we find a net strain of 3.56×10^{-3} and a total stress of $(27 \times 10^3)(3.56 \times 10^{-3}) = 96.1$ kips/in². Similar consideration of the remaining tendons at positions 2, 3, and 4 shows that the net strain in these exceeds the yield strain, but not the breaking strain (typically about 0.030 or more). The stress in these tendons is thus equal to the yield stress. Since the area of each set of tendons is 3.6 in², the total force in the steel is thus calculated as

$$F_s = (96.1)(3.6) + 3(212)(3.6) = 2636 \text{ kips}$$

Assuming concrete of strength $\sigma_c' = 5$ kips/in², the compressive force in the concrete is determined, by the product $\sigma_c A_c$, where $\sigma_c = 0.85\sigma_c' = 0.85(5) =$

4.25 kips/in² and A_c is the shaded area of Fig. 4.34(b), given in terms of the inside radius $R_i = 5.5$ ft and outside radius $R_o = 7.5$ ft by

$$A_c = A_o, \qquad \alpha e \leq R_o - R_i$$
$$A_c = A_o - A_i, \qquad \alpha e > R_o - R_i$$

with A_o and A_i defined, respectively, for $R = R_o$ and $R = R_i$ by

$$A = \tfrac{1}{2}R^2(2\theta - \sin 2\theta)$$
$$\cos\theta = \frac{R_o - \alpha e}{R}$$

Using $e = 3$, $\alpha = 0.8$, we have $A_o = 18.25$ ft², $A_i = 1.11$ ft², so that $A =$ 17.14 ft². The compressive force in the concrete is thus calculated as

$$F_c = (4.25)(17.14 \times 144) = 10{,}490 \text{ kips}$$

The total force on the section is calculated as $F_s - F_c = -7854$ kips, which overestimates the actual compressive force of -6448 kips. Repeating the calculations above for an assumed value of e of 2.5 ft, we find $F_s - F_c = -5913$ kips, which underestimates the actual compressive force. By linear interpolation, we accordingly choose $e = 2.6$ ft, which gives $F_s = 2650$ kips and $F_c =$ -9009 kips. These combine to give $F_s - F_c = 6360$ kips, which is within 2% of the actual value.

With this value of e, we can now calculate the associated ultimate moment M_u about the centroid of the section. The force in the tendons at positions 2 and 4 do not contribute since their moment arms are zero. The force in tendons at position 1 is 346 kips and the moment arm is 6.5 ft. The force in tendons at position 3 is 763 kips and the moment arm is 6.5 ft. The net moment from the steel is $-(346)(6.5) + (763)(6.5) = 2710$ ft-kips. The compressive force in the concrete is 9009 kips and the centroid of the area A_c over which it acts is calculated from the equation

$$\bar{x} = x_o, \qquad \alpha e \leq R_o - R_i$$
$$\bar{x} = \frac{A_o x_o - A_i x_i}{A_i}, \qquad \alpha e > R_o - R_i$$

where x_o and x_i are determined, respectively, for R_o and R_i by

$$x = \frac{2}{3}R\frac{\sin^3\theta}{\theta - \cos\theta\sin\theta}$$

with A_o, A_i, and θ as defined earlier. For $e = 2.60$ ft, we find $A_o = 14.82$ ft², $A_i = 0.1$ ft², $x_o = 6.26$ ft, $x_i = 5.54$ ft, and $\bar{x} = 6.27$ ft. The moment from the concrete is thus (9009)(6.27) = 56,490 ft-kips. The ultimate moment is calculated as the sum of the contributions from the steel and concrete and is

$$M_u = 2710 + 56{,}490 = 59{,}200 \text{ ft-kips}$$

For a design moment of 24,800 ft-kips, as used in the preceding example, and an overall factor of safety of, say, 1.8, the required ultimate moment is only 44,640 ft-kips, so that the ultimate moment capacity of the column is more than adequate.

To examine the necessity for shear reinforcement at the required ultimate moment of 44,640 ft-kips, we may use equation (55). Since the maximum moment (at the base of the column) arises from a resultant force acting 62 ft above the column base (see Fig. 4.32), the shear force Q_u associated with the required ultimate moment is 44,640/62 = 720 kips. The shear force Q_c carried by the uncracked concrete in the section at this ultimate shear level is calculated as the lesser of that required to initiate diagonal-tension cracks at the centroid of the section or cause tensile bending cracks to spread inward toward the centroidal axis.

For diagonal-tension cracks, we assume the principal tensile stress (in lb/in²) at the centroid to be related to the concrete strength $\sigma_c' = 5000$ lb/in² by $\sigma_1 = 4\sqrt{\sigma_c'} = 283$ lb/in². From equation (49), we have (with $\sigma = -732$ lb/in², as determined in the preceding example)

$$283 = -\frac{732}{2} + \sqrt{\left(\frac{732}{2}\right)^2 + \tau^2}$$

Solving, we find $\tau = 536$ lb/in², or 77.2 kips/ft². The associated shear force Q at the centroid of the section is determined from (see shear–stress calculation of the preceding example)

$$\tau = \frac{QS}{I(2t)} = \frac{Q(170.3)}{(1766)(4)} = 77.2$$

so that $Q = 3202$ kips.

For spreading of tensile bending cracks, we assume a critical tensile stress of $10\sqrt{\sigma_c'} = 707$ lb/in², or 102 kips/ft², and use equation (46) in the form

$$102 = \frac{M_c'(7.5)}{1766} - \frac{6448}{81.7} - \frac{(150)(14.4)}{81.7}$$

where M_c' denotes the bending moment associated with the crack spreading. Solving this equation, we find $M_c' = 48{,}800$ ft-kips. For a resultant force 62 ft above the base, the associated shear Q is 48,800/62 = 787 kips.

Since $787 < 3202$, we choose $Q_c = 787$ kips in equation (55). Moreover, since this value is greater than the shear force $Q_u = 720$ kips, we see that no shear reinforcement is required by this equation. #

EXAMPLE 4.8-3. Suppose in Example 4.8-1 that the base of the structure consists of part of the main column and four additional concrete cylindrical members located around the column, as indicated in Fig. 4.35. For the total loading shown, determine the prestress requirements of the base, assuming that the outer members have an outside radius of 7.5 ft and a wall thickness of 1 ft. Also estimate the cracking moment and the maximum diagonal tension in the base.

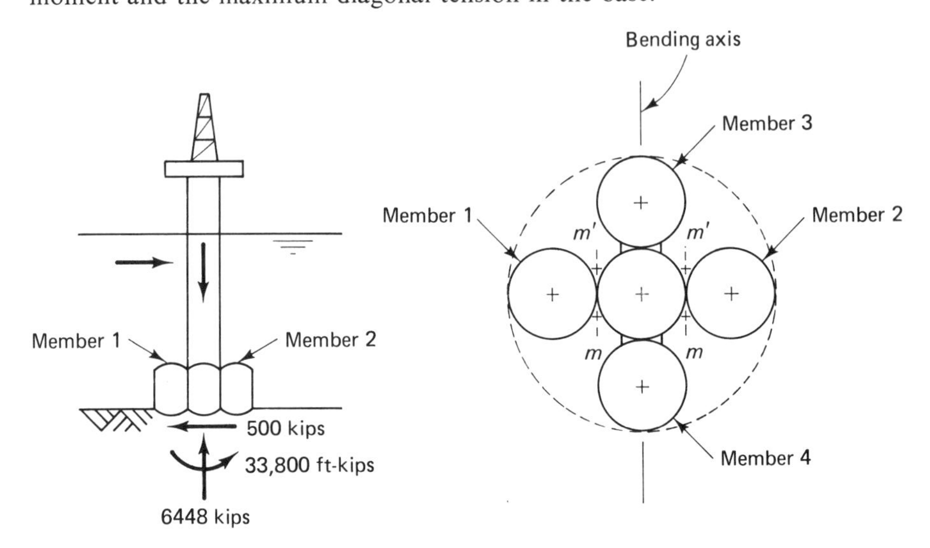

Fig. 4.35

The entire base is treated as a section of beam. The wind and wave forces are assumed acting in the direction from member 1 to member 2 and the bending axis is as indicated in Fig. 4.35. With this loading direction, the extreme bending tensile stress will occur at the outside surface of member 1 and the extreme compressive stress will occur at the outside surface of member 2. These extreme stresses are given by

$$\sigma_{cb} = \pm \frac{MR'}{I'}$$

where R' denotes distance from the neutral axis to the outer fibers of these members, M denotes the maximum moment from the applied loads, and I' denotes the moment of inertia of the entire base about the bending axis. From the data given, we have $R' = 22.5$ ft and $M = 33800$ ft-kips. The moment of inertia of the entire base is calculated from the values of each member. The moment of inertia of the sectional area of the main column about an axis through its centroid and parallel to the neutral axis of the entire section is 1766 ft^4 and that of each of the four surrounding members about corresponding individual axes as 1083 ft^4. To determine the combined moment of inertia of all these members about the neutral axis of the entire section, we use the parallel-axis theorem. Since the centroids of the main column and members 3 and 4 lie on the neutral axis, their contribution is simply $1766 + 2(1083) = 3932\ \text{ft}^4$. The remaining members 1 and 2 have their centroids located a distance of 15 ft from the neutral axis. With the sectional areas of these members equal to $44\ \text{ft}^2$, their individual contributions are $1083 + (15^2)(44) = 10{,}983\ \text{ft}^4$. Thus I' is finally determined as

$$I' = 3932 + 2(10{,}983) = 25{,}900\ \text{ft}^4$$

From these results, we accordingly determine the extreme bending stresses in members 1 and 2 as

$$\sigma_{cb} = \pm\frac{(33{,}800)(22.5)}{25{,}900} = \pm 29.4\ \text{kips/ft}^2\ (\pm 204\ \text{lb/in}^2)$$

To eliminate the tensile part of the bending stress, we must apply a prestress of $\sigma_{cp} = 29.4\ \text{kips/ft}^2$ (neglecting any compressive stress caused by the weight of the members or by hydrostatic pressure) to the concrete. Since the wind and wave forces may act in any direction, we apply this prestress to all four of the outer cylinders. If the prestress in the tendons is $150\ \text{kips/in}^2$, the net area of all the tendons in a single cylinder, assumed symmetrically spaced around its section, is

$$A_s = \frac{(29.4)(44)}{150} = 8.62\ \text{in}^2$$

With this prestress and the assumed loading direction from member 1 to member 2, the maximum tensile stress (in member 1) will be reduced to zero and the maximum compressive stress (in member 2) will be $-408\ \text{lb/in}^2$.

The cracking moment M_c is calculated by setting the maximum tensile stress in the base equal to $7.5\sqrt{\sigma_c'} = 530\ \text{lb/in}^2$ ($76.4\ \text{kips/ft}^2$) as in Example 4.8-1. Considering bending stress and axial prestress, we have

$$76.4 = \frac{M_c(22.5)}{25{,}900} - \frac{(8.62)(150)}{44}$$

which gives $M_c = 121{,}800$ ft-kips. This is 3.6 times the applied moment.

The shear stress τ in the base is given by the equation

$$\tau = \frac{QS'}{I't'}$$

where Q denotes the shear force, S' denotes the first moment of the area, about the bending axis, that lies away from the place where the shear stress is calculated, I' is as defined previously, and t' denotes the width of the concrete, parallel to the bending axis, at the place where the shear stress is calculated.

Inspection of the base geometry of Fig. 4.35 shows that the maximum shear stress τ_m will occur at the intersection of members 1 and 2 with the main column, if the fillet width (dimension along line m–m' in Fig. 4.33) is limited, Assuming this width to be 1 ft, we have, with $S' = 44 \times 15 = 660$ ft³, $Q = 500$ kips, $I' = 25{,}900$ ft⁴, $t' = 1$ ft,

$$\tau_m = \frac{(500)(660)}{(25{,}900)(1)} = 12.7 \text{ kips/ft}^2 \text{ (88.5 lb/in}^2\text{)}$$

We assume axial compressive prestress of 204 lb/in² to be applied in the fillets, as in the cylindrical members. The tensile bending stress in the fillet between member 1 and the main cylinder is determined from the bending-stress equation above (with $R' = 7.5$ ft) as 68 lb/in². The net stress σ in the fillet is thus $-204 + 68 = -136$ lb/in². The diagonal tension from the shear and compressive stresses is accordingly determined from equation (49) as

$$\sigma_1 = -\frac{136}{2} + \sqrt{\left(\frac{136}{2}\right)^2 + (88.5)^2} = 43.6 \text{ lb/in}^2$$

If, with a factor of safety of 1.8, we consider ultimate loads on the base of $M_u = 1.8 \times 33{,}800 = 60{,}840$ ft-kips and $Q_u = 1.8 \times 500 = 900$ kips, we may also calculate the associated diagonal tension in the same manner as above since the cracking moment is not exceeded. Thus, we find $\tau_m = 159$ lb/in² and $\sigma_1 = 105$ lb/in². This is less than the diagonal cracking tension ($4\sqrt{\sigma_c'} = 283$ lb/in²) by a factor of about 2.7, so that diagonal cracking of the base at the ultimate loads is not expected. #

4.9 PRESSURE-INDUCED STRESSES IN CONCRETE STRUCTURES

The stresses induced in submerged unflooded cylindrical members of an offshore concrete structure can be estimated using procedures similar to those given in Section 4.4 for steel structures.

As noted there, the effect of external pressure loading is to produce both longitudinal and hoop stresses in the member. At distances removed from the ends, these are described, respectively, by

$$\sigma_z = -\frac{Pa}{2t}, \qquad \sigma_\theta = -\frac{Pa}{t} \tag{56}$$

where P denotes the net external pressure, a the radius of the member, and t its wall thickness, assumed small in comparison with the radius. The longitudinal stress can be combined with that caused by other loadings to obtain the total longitudinal stress in the member acting together with the pressure-induced hoop stress. Because of the relatively low stress levels permitted in concrete members, circumferential buckling under the compressive hoop stress is generally not a consideration in failure. Moreover, experiments on concrete under biaxial stress states show only slight interaction between the two stress components insofar as failure is concerned, so that it is adequate in design to ensure that the maximum compressive values of the total longitudinal and hoop stresses are each less than the maximum allowable compressive stress, typically chosen as 45% of the compressive strength of the concrete. Of course, tensile values of the total longitudinal stress must be limited, as discussed in Section 4.8, because of the low tensile strength of concrete.

Effect of End Restraint

The effect of end restraint on concrete cylindrical members under external pressure can be significant both because of the increased longitudinal stress and the associated shear stress existing there. If flat end plates are used, the solution given in Section 4.4 for the extreme stresses can be used. Using equation (26) and assuming a typical Poisson's ratio for concrete $\nu = 0.15$, we have at the outside surface at the restrained end

$$\sigma_z = 1.12\frac{Pa}{t}, \qquad \sigma_\theta = 0.17\frac{Pa}{t} \tag{57a}$$

and at the inside surface

$$\sigma_z = -2.12\frac{Pa}{t}, \qquad \sigma_\theta = -0.32\frac{Pa}{t} \tag{57b}$$

The radial shear stress τ is zero on the inner and outer surfaces and maximum on the midsurface where the stresses are expressible as

$$\tau = -1.06P\sqrt{\frac{a}{t}}, \qquad \sigma_z = -\frac{Pa}{2t}, \qquad \sigma_\theta = -0.08\frac{Pa}{t} \tag{57c}$$

As noted earlier, the shear stress acts outward on cylinder cross sections whose outward normals point in the direction of the restrained end.

EXAMPLE 4.9-1. Consider the concrete cylindrical base element $ABCD$ of the offshore gravity structure shown in Fig. 4.36 and suppose, as a result of wind and wave forces F, that the maximum bending stress at side A is 300 lb/in², that the bending stress on side B is 100 lb/in², and that the associated shear stress on the midsurface at side B is 60 lb/in². Determine the additional longitudinal and shear stresses arising from a net external pressure of 80 lb/in² and estimate the amount of

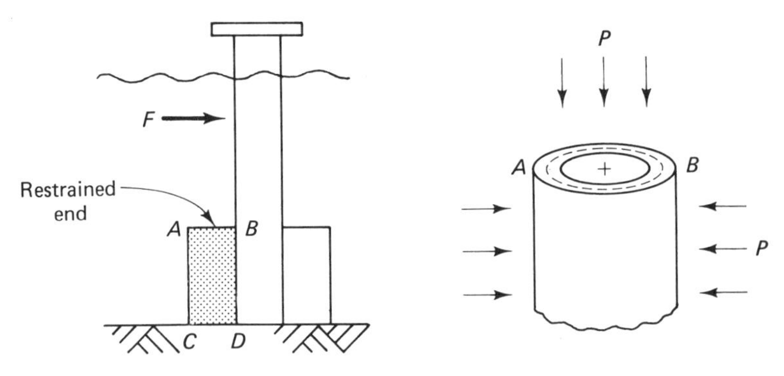

Fig. 4.36

prestress needed. The outside radius of the element is 10 ft and its wall thickness is 1 ft.

From equations (57a) and (57b), we have at the outside surface at the restrained end

$$\sigma_z = 896 \text{ lb/in}^2, \qquad \sigma_\theta = 136 \text{ lb/in}^2$$

and at the inside surface

$$\sigma_z = -1700 \text{ lb/in}^2, \qquad \sigma_\theta = -256 \text{ lb/in}^2$$

At the side A of the element, the bending stress from wind and wave action is 300 lb/in², so that, assuming that this does not change appreciably over the thickness, the extreme longitudinal stresses in the cylinder are

$$\sigma_z = 1200 \text{ lb/in}^2 \qquad \text{and} \qquad \sigma_z = -1400 \text{ lb/in}^2$$

To eliminate the tensile stress, we thus need a longitudinal prestress $\sigma_p = -1200$ lb/in², leading to resultant stresses of

$$\sigma_z = 0, \qquad \sigma_z = -2600 \text{ lb/in}^2$$

Note, however, that if the direction of the wind and wave forces were reversed, the associated bending stress would be -300 lb/in², so that the net stresses are then

$$\sigma_z = -600 \text{ lb/in}^2 \qquad \text{and} \qquad -3188 \text{ lb/in}^2$$

With a maximum allowable compressive stress in the concrete limited to 45% of its compressive strength, we thus see that a concrete with compressive strength equal to about 7000 lb/in² would be required.

Next let us examine the net shear stress. Form the bending action caused by the wind and wave loading, we have a maximum shear stress on the mid-surface at side B of 60 lb/in². The horizontal component of this shear stress acts in the same direction as the wind and wave forces (Fig. 4.36). For the external pressure loading, we also find from equations (57) a shear stress of magnitude $\tau_m = 268$ lb/in², the horizontal component of which acts radially outward. Thus, on side B the two shearing stresses act in the same direction and are additive to give a net stress $\tau = 328$ lb/in². For a prestress of -1200 lb/in² and a bending stress from wind and wave action of 100 lb/in² at side B, we have at the midsurface [where the pressure induced maximum shear stress exists and where, from equation (57c), the corresponding longitudinal stress is -400 lb/in²] a net compressive stress $\sigma = -1500$ lb/in². The diagonal tension is thus

$$\sigma_1 = \frac{\sigma}{2} + \sqrt{\left(\frac{\sigma}{2}\right)^2 + \tau^2} = 68.5 \text{ lb/in}^2$$

For a concrete of strength $\sigma_c' = 7000$ lb/in², the diagonal-tensile strength is approximately $4\sqrt{\sigma_c'} = 330$ lb/in², so that cracking from diagonal tension is not expected. The same is also found to be true if the direction of the wind and wave forces are reversed, leading to a shear stress of $268 - 60 = 208$ lb/in² and net compressive stress of -1700 lb/in².

Note finally that the magnitude of the maximum hoop compressive stress σ_θ in the cylinder, as determined from the second of equations (56), is 800 lb/in², which is well within an allowable stress of $(0.45)(7000) = 3150$ lb/in². Also, the maximum hoop tensile stress at the outside surface on the restrained end is 136 lb/in², which is well within the cracking tensile stress, estimated as $7.5\sqrt{\sigma_c'} = 628$ lb/in². #

Spherical End Caps

From Example 4.9-1, it can be seen that the effect of full restraint on the pressure-induced stresses is appreciable and leads to the requirements of large prestress and high-strength concrete for given cylinder dimensions. These requirements can, however, be much reduced if spherical (or near spherical) end caps are used in place of flat plates. The reason is simply that the restraint on the ends is relaxed by the flexible nature of the spherical caps.

To analyze the stresses existing under these circumstances, we consider the cylindrical element shown in Fig. 4.37(a) and note that the radial deflection of the cylinder, were it not restrained by the end caps, is [from equation (8)]

$$u = -\frac{Pa^2}{2Et}(2 - \nu) \tag{58}$$

By similar considerations, it can also be found that the radial deflection of the spherical cap, were it not restrained by the cylinder, is

$$u = -\frac{Pa^2}{2Et}(1 - \nu) \tag{59}$$

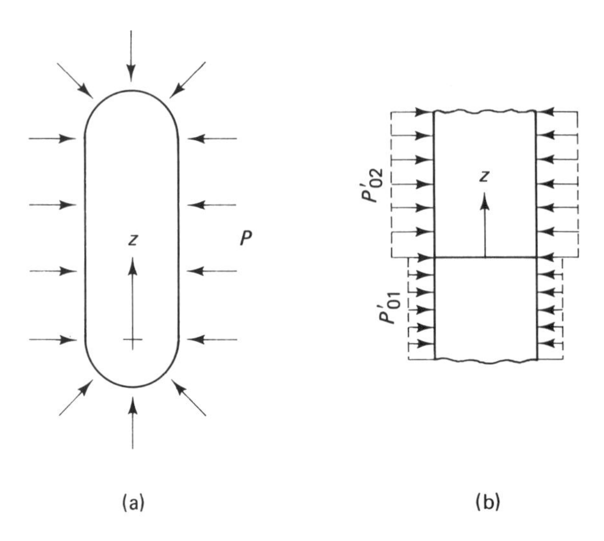

Fig. 4.37. Cylindrical member with spherical end caps.

Here, a denotes the radius of the cylinder and sphere, t their wall thicknesses, and E and ν denote Young's modulus and Poisson's ratio of the material.

Because the effect of the restraint is only local (exponential decay from cylinder–sphere connection joint), we may consider the spherical surface to be cylindrical in the vicinity of the joint ($z = 0$). In this case, we then consider two cylinders joined together, as shown in Fig. 4.37(b), with the upper cylinder being the actual cylinder and the lower being representative of the spherical surface in the vicinity of the joint. The actual cylinder has effective external pressure P'_{01} and the representative cylinder has effective external pressure P'_{02}, so chosen as to give the radial deflections of equations (58) and (59) in the absence of restraint.

From equation (21) we thus have for the radial deflection u_1 of the actual cylinder

$$D\frac{d^4u_1}{dz^4} + \frac{Et}{a^2}u_1 = -P\left(1 - \frac{\nu}{2}\right) \tag{60}$$

and for the representative cylinder, we have the radial deflection u_2 governed by

$$D\frac{d^4u_2}{dz^4} + \frac{Et}{a^2}u_2 = -\frac{P}{2}(1 - \nu) \tag{61}$$

The boundary conditions are that, at $z = 0$,

$$u_1 = u_2, \qquad \frac{du_1}{dz} = \frac{du_2}{dz}$$

$$M_1 = M_2, \qquad Q_1 = Q_2$$

where M and Q denote the internal moment and shear force given by equations (15) and (16).

The solutions of equations (60) and (61) consistent with the boundary conditions above are expressible as

$$u_1 = -\frac{Pa^2}{2Et}[(2 - \nu) - \tfrac{1}{2}e^{-\alpha z} \cos \alpha z] \tag{62}$$

$$u_2 = -\frac{Pa^2}{2Et}[(1 - \nu) + \tfrac{1}{2}e^{-\alpha z} \cos \alpha z] \tag{63}$$

where α is as defined earlier by equation (23).

With these solutions, the stresses may then be determined in a manner similar to that used for fully restrained ends.

In the cylinder, where stress conditions are most severe, the longitudinal stress σ_z and hoop stress σ_θ are found expressible on the outside and inside surfaces ($\pm t/2$) as

$$\begin{aligned} \sigma_z &= -\frac{Pa}{t} \pm \frac{3Pa}{4t}(\beta e^{-\alpha z} \sin \alpha z) \\ \sigma_\theta &= -\frac{Pa}{t} + \frac{Pa}{4t}e^{-\alpha z}(\cos \alpha z \pm 3\nu\beta \sin \alpha z) \end{aligned} \tag{64a}$$

with $\beta = [3(1 - \nu^2)]^{-1/2}$.

The radial shear stress τ on the midsurface is similarly expressible as

$$\tau = -\frac{3P}{16\alpha t}e^{-\alpha z}(\cos \alpha z - \sin \alpha z) \tag{64b}$$

The extreme longitudinal stresses σ_z occur in the cylinder a distance $\alpha z = 0.8$ from the joint. This stress and the associated hoop stress σ_θ are expressible (assuming that $\nu = 0.15$) at the outside surface as

$$\sigma_z = -0.64\frac{Pa}{t}, \qquad \sigma_\theta = -1.01\frac{Pa}{t} \tag{65a}$$

and at the inside surface as

$$\sigma_z = -0.36\frac{Pa}{t}, \qquad \sigma_\theta = -0.84\frac{Pa}{t} \tag{65b}$$

Thus, in contrast with flat end plates, no tensile stresses are developed because of end restraint and the governing compressive stresses are those on the outside surface. The radial shear stress τ is zero on the inner and outer surfaces and is greatest at the midsurface. It has a maximum value at the joint, where its value and that of the associated stresses are expressible (for $\nu = 0.15$) as

$$\tau = -0.14P\sqrt{\frac{a}{t}}, \qquad \sigma_z = -\frac{Pa}{2t}, \qquad \sigma_\theta = -\frac{3}{4}\frac{Pa}{t} \tag{65c}$$

The results above apply for spherical end caps, but the same procedures can be employed to determine the stresses in ellipsoidal caps of circular section having radius a and maximum perpendicular distance (height) b from the joint. Equations (65a), which give the greatest compressive stresses, are then replaced by

$$\sigma_z = -\left(1 + 0.28\frac{a^2}{b^2}\right)\frac{Pa}{2t}$$
$$\sigma_\theta = -\left(1 + 0.01\frac{a^2}{b^2}\right)\frac{Pa}{t} \tag{66a}$$

and equations (65b) for the inside surface are replaced by

$$\sigma_z = -\left(1 - 0.28\frac{a^2}{b^2}\right)\frac{Pa}{2t}$$
$$\sigma_\theta = -\left(1 - 0.16\frac{a^2}{b^2}\right)\frac{Pa}{t} \tag{66b}$$

Note that these stresses can become positive (tensile) if the ratio a/b is sufficiently large. The expressions for the radial shear stress and associated longitudinal and hoop stresses given by equations (65c) are replaced by

$$\tau = -\left(0.14\frac{a^2}{b^2}\right)P\sqrt{\frac{a}{t}}$$
$$\sigma_z = -\frac{Pa}{2t} \tag{66c}$$
$$\sigma_\theta = -\left(1 - 0.25\frac{a^2}{b^2}\right)\frac{Pa}{t}$$

The shear stresses given by equations (65c) and (66c) act in the same manner as for flat end plates, that is, radially outward in cylinder cross sections whose outward normals point in the direction of the joint.

EXAMPLE 4.9-2. Repeat Example 4.9-1 for the case of ellipsoidal end caps having ratio of height to radius $b/a = \frac{2}{3}$.

The longitudinal stresses given by equations (66*a*) and (66b) are

$$\sigma_z = -652 \text{ lb/in}^2, \qquad \sigma_z = -148 \text{ lb/in}^2$$

For an additional bending stress of 300 lb/in² from wind and wave forces, the total extreme stresses are

$$\sigma_z = -352 \text{ lb/in}^2, \qquad \sigma_z = 152 \text{ lb/in}^2$$

Thus a longitudinal prestress of -152 lb/in² is needed to eliminate the tensile stress, leading to resultant extreme stresses of

$$\sigma_z = -504 \text{ lb/in}^2 \qquad \text{and} \qquad \sigma_z = 0$$

If the direction of the wind and wave forces are reversed, the extreme stresses are then

$$\sigma_z = -1104 \text{ lb/in}^2 \qquad \text{and} \qquad \sigma_z = -600 \text{ lb/in}^2$$

The maximum shear stress, at the joint between the cylinder and cap, is determined from equations (66c) as 79.7 lb/in². Adding to this a shear stress of 60 lb/in² from wind and wave action, we have a net shear stress of $\tau = 140$ lb/in². The net compressive stress consists of the prestress of -152 lb/in², a stress of 100 lb/in² from wind and wave action, and a pressure-induced longitudinal stress of -400 lb/in², giving $\sigma = -452$ lb/in². The diagonal tension is thus

$$\sigma_1 = \frac{\sigma}{2} + \sqrt{\left(\frac{\sigma}{2}\right)^2 + \tau^2} = 39.8 \text{ lb/in}^2 \quad \#$$

A comparison of these results with the corresponding ones of Example 4.9-1 shows that the stresses and prestress requirements have been much reduced by the use of ellipsoidal end caps instead of flat plates.

4.10 EXAMINATION FOR DYNAMIC EFFECTS

As noted at the beginning of this chapter, the analysis of a proposed offshore structure using the structural methods of Chapter 2 and joint loadings based on extreme environmental conditions necessarily neglects any dynamic effects associated with the wave-induced periodic motion of the structure. Such a static analysis can, therefore, only be applied when the dynamic loadings are small in comparison with the maximum static loadings.

To illustrate the dynamic response of a structure to wave loadings and develop a simple formula for checking for dynamic effects, we consider here an approximate analysis of a typical platform such as shown in Fig. 4.38. Regular

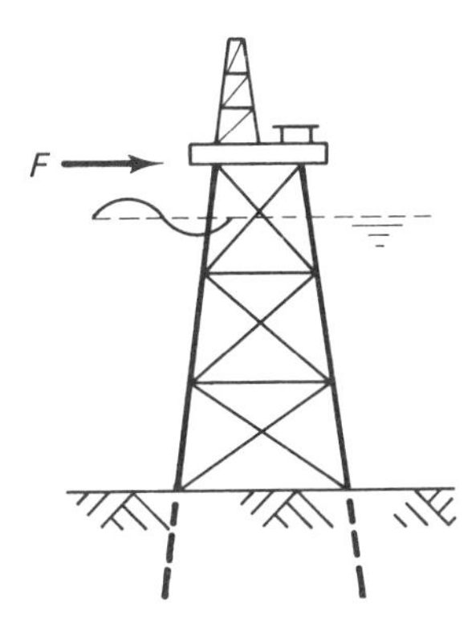

Fig. 4.38. Offshore structure with an equivalent wave force acting at deck level.

sinusoidal water waves are assumed and the forces on the structure represented approximately by a single concentrated force F acting at the top of the structure and of the form

$$F = F_0 \sin \omega t \tag{67}$$

where ω denotes the frequency of the wave, t denotes the time, and F_0 denotes the amplitude of the idealized wave force, chosen so as to give the same static deck deflection as that found from the actual distributed wave force acting on the structure.

As a further approximation, we assume one-half the mass of the support structure to be lumped into the deck mass to give an effective deck mass M given by

$$M = M_D + \tfrac{1}{2}M_S \tag{68}$$

where M_D denotes the deck mass and M_S denotes the total virtual mass of the support structure (actual mass plus apparent added mass resulting from its motion in water (see Section 3.4). With this simplification, the support structure itself may be regarded as massless and its response calculated using the equilibrium methods of Chapter 2. Because there remains only one movable mass (the effective mass at the top of the structure) and only one direction of sensible motion (the horizontal direction), the analysis for dynamic response in this case is known as a *single-degree-of-freedom dynamic analysis.*

Now, the total horizontal force F^T acting at the top of the structure can be regarded as the sum of the applied force F, the inertia force $-M\ddot{U}$, and a resistive damping force represented approximately by $-C\dot{U}$, where C denotes a constant damping coefficient, so that we have the total force acting at the top of the structure given by

$$F^T = F - M\ddot{U} - C\dot{U} \tag{69}$$

where overhead dots denote time derivatives.

From the static methods of Chapter 2, we may also relate the total force F^T to the horizontal displacement U at the top of the structure (equal) to the deck displacement by the equation

$$F^T = KU \tag{70}$$

where K denotes the stiffness of the structure.

Thus, on combining these last two equations and using the expression for the wave force given by equation (67), we have the governing differential equation for the deck motion expressible as

$$M\ddot{U} + C\dot{U} + KU = F_0 \sin \omega t \tag{71}$$

Defining the parameters ζ and Ω by the equations

$$2\zeta = \frac{C}{M}, \qquad \Omega^2 = \frac{K}{M} \tag{72}$$

and assuming light damping such that $\zeta^2 \ll \Omega^2$, the solution of equation (71) may be written as

$$U = e^{-\zeta t}(C_1 \cos \Omega t + C_2 \sin \Omega t) + \frac{F_0/M^*}{\sqrt{(\Omega^2 - \omega^2)^2 + (2\zeta\omega)^2}} \sin(\omega t - \phi) \tag{73}$$

where the phase angle ϕ is determined by

$$\tan \phi = \frac{2\zeta\omega}{\Omega^2 - \omega^2}, \qquad 0 \leq \phi \leq \pi \tag{74}$$

and where C_1 and C_2 denote constants which are expressible for initial conditions $U = \dot{U} = 0$ as

$$C_1 = -\frac{F_0/M}{\sqrt{(\Omega^2 - \omega^2)^2 + (2\zeta\omega)^2}} \cos \phi \tag{75}$$

$$C_2 = -\frac{F_0/M}{\sqrt{(\Omega^2 - \omega^2)^2 + (2\zeta\omega)^2}} \left(\frac{\omega}{\Omega} \sin \phi + \frac{\zeta}{\Omega} \cos \phi\right) \tag{76}$$

Inspection shows that the first, or *transient*, part of the solution above is ultimately damped out with the passage of time and that the solution approaches that described by the second, or *steady-state*, part. Thus, after a sufficiently long time, we have the deck displacement U described by a simple harmonic motion of frequency ω and phase angle ϕ whose amplitude may be written in terms of the maximum static deflection $U_s = F_0/K$ as

$$\frac{U_0}{U_s} = \frac{1}{\sqrt{\left[1 - \left(\frac{\omega}{\Omega}\right)^2\right]^2 + \left(\frac{2\zeta}{\Omega}\frac{\omega}{\Omega}\right)^2}} \tag{77}$$

Figure 4.39 shows the variation of U_0/U_s with ω/Ω for various degrees of damping. It will be seen that at the *resonant frequency* $\omega = \Omega$, the amplitude of the vibrations experiences a maximum, or near maximum, value which becomes increasingly larger as the degree of damping decreases. The value of Ω, given from equation (72) by

$$\Omega = \sqrt{\frac{K}{M}} \tag{78}$$

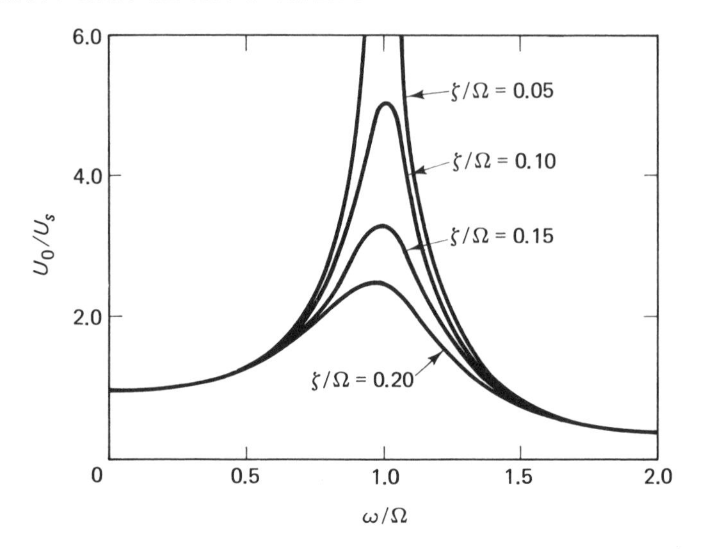

Fig. 4.39. Dynamic amplification of deck deflection.

is thus of major importance in the structural response. It is known as the *fundamental* or *natural flexural frequency* of the idealized structure and is seen to be dependent only on the structural properties.

Damping values ζ/Ω for offshore structures typically range from about 0.05 to 0.10, that is, from about 5 to 10% of *critical damping*, where critical damping is defined by $\zeta/\Omega = 1$. From Fig. 4.39 it can thus be seen that the deck deflection (and hence the stresses in the support structure) can be many times that determined statically ($\omega = 0$) when the wave-loading frequency is near the natural flexural frequency of the structure. This is known as *dynamic amplification*. For the design of this structure to be based on static methods of analysis and a regular design wave, it is thus necessary that the frequency of the design wave be sufficiently small in comparison with the natural flexural frequency of the structure to ensure that the *dynamic-amplification factor* U_0/U_s given by equation (77) does not differ appreciably from unity say, no more than 5 or 10%.

EXAMPLE 4.10-1. Consider the steel offshore structure with side face as shown in Fig. 4.40 and determine if a static analysis is appropriate for a design wave having height of 40 ft and period of 9 sec. All four sides of the structure are identical. Vertical members have outside diameter of 4 ft and wall thickness of 1.5 in. Horizontal and diagonal members have outside diameter of 2 ft and wall thickness of 0.5 in. The equivalent pile stiffness shown is that associated with the static analysis.

The appropriate relations are given by equations (77) and (78). We first determine the stiffness K of the structure relating horizontal force and displacement at the deck level. Considering the side frame and pile stiffness shown,

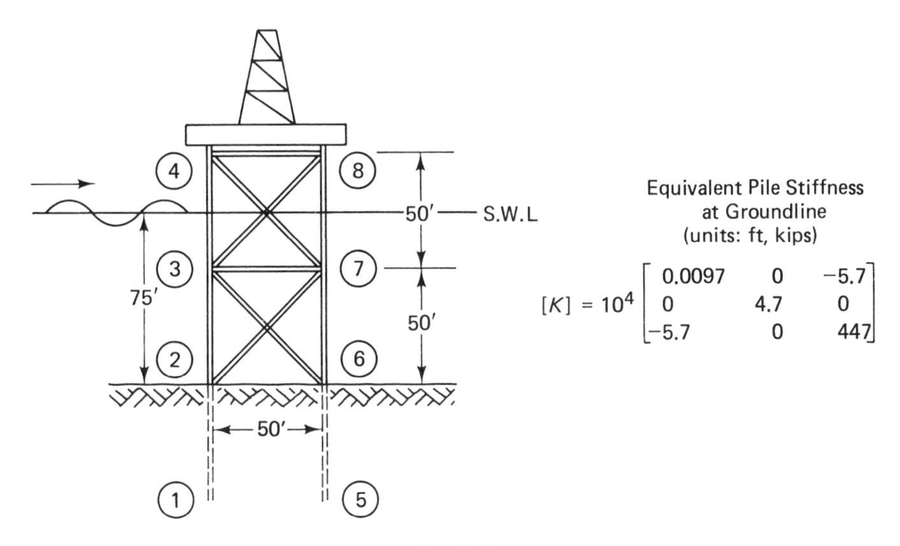

Fig. 4.40

and assuming 100-kip horizontal forces to act at joints 4 and 8, we find, by the matrix methods of Chapter 2, the resulting horizontal displacement to be equal to 0.378 ft. The stiffness for this frame is thus $2 \times 10^5/0.378 = 5.29 \times 10^5$ lb/ft. For the remaining side frame of the three-dimensional structure, we also have this same stiffness. Hence, the total stiffness K of the structure is $2(5.29 \times 10^5)$, i.e.,

$$K = 1.06 \times 10^6 \text{ lb/ft}$$

Next, we calculate the mass of the deck and support structure. The deck weight is equal to 500,000 lb. The mass in units of slugs is the weight in pounds divided by the acceleration of gravity (32.2 ft/sec²). Thus, $M_D = 500{,}000/32.2 = 15{,}530$ slugs. The weight of the support structure in air is easily determined from the sum of the individual member weights as 485 kips. This corresponds to a mass of 15,060 slugs. Assuming further that the vertical members of the structure are filled with water to the still-water level, we find this mass to equal 6585 slugs. The total actual mass of the support structure is thus determined as $15{,}060 + 6585 = 21{,}645$ slugs.

The total added mass of the structure resulting from its motion in water may be calculated from the sum of the added masses of the individual members. From equation (66) of Chapter 3, this mass per unit length for a vertical cylinder of outside diameter D is expressible in terms of the water density ρ and inertia coefficient C_I as $\rho(C_I - 1)\pi D^2/4$. Choosing $C_I = 2$ and multiplying by the length L of the member, its added mass is thus expressible as $\rho\pi D^2 L/4$, which is simply equal to the mass of water displaced by the member. For the two vertical members shown in Fig. 4.40, we have

$$2(1.99)\frac{(\pi)(4^2)}{4}(75) = 3750 \text{ slugs}$$

For the two lower diagonals, we have, on treating these as effective vertical cylinders of length 50 ft,

$$2(1.99)\frac{(\pi)(2^2)}{4}(50) = 625 \text{ slugs}$$

For the two upper diagonal members, we similarly have, on treating these as effective vertical cylinders of length 25 ft,

$$2(1.99)\frac{(\pi)(2^2)}{4}(25) = 313 \text{ slugs}$$

Thus on the side face shown, the added mass is 3750 + 625 + 313 = 4688 slugs. Doubling this to account for the added mass of the other side frame, we have 9376 slugs. In addition, from the front face we have for the lower diagonals

$$2(1.99)\frac{(\pi)(2^2)}{4}(70.7) = 884 \text{ slugs}$$

and for the upper diagonals

$$2(1.99)\frac{(\pi)(2^2)}{4}(35.4) = 443 \text{ slugs}$$

For the horizontal bracing at midheight, we also have

$$(1.99)\frac{(\pi)(2^2)}{4}(50) = 313 \text{ slugs}$$

Doubling each of these last three added masses to account for the other face and adding the results to the added mass of the side frames, we find the total added mass of the structure as 12,660 slugs.

Collecting together all the results above, we determine the mass M from equation (68) as

$$M = M_D + \tfrac{1}{2}M_S = 15{,}530 + \tfrac{1}{2}(21{,}645 + 12{,}660) = 32{,}680 \text{ slugs}$$

The natural frequency Ω of the structure is next determined from equation (78) as

$$\Omega = \sqrt{\frac{K}{M}} = \sqrt{\frac{1.06 \times 10^6}{3.27 \times 10^4}} = 5.69 \text{ rad/sec}$$

Finally, the wave frequency ω is determined from its period as $2\pi/9 = 0.698$ rad/sec, so that (with $\omega/\Omega = 0.122$, $\zeta/\Omega = 0.08$) the dynamic-amplification factor is calculated from equation (77) as

$$\frac{U}{U_s} = 1.02$$

This differs from unity by only 2% and hence we conclude that static analysis is appropriate for this structure. #

EXAMPLE 4.10-2. For the gravity structure shown in Fig. 4.41, determine whether static analysis is appropriate for a design wave of 8-sec period.

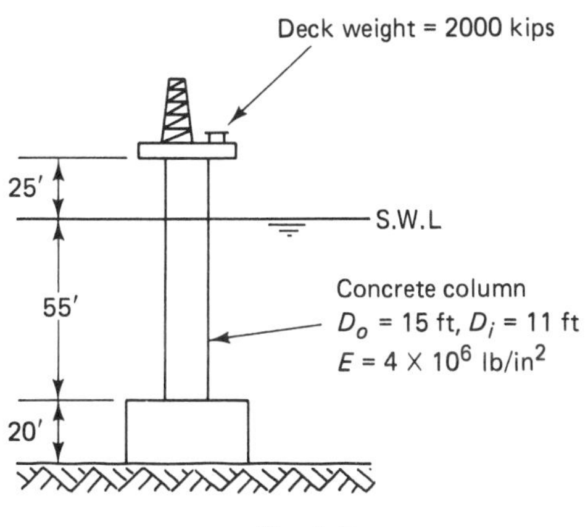

Fig. 4.41

For this simple structure, the stiffness K relating horizontal force and displacement at the top of the structure is expressible (neglecting any base movement) as

$$K = \frac{3EI}{L^3}$$

where E denotes Young's modulus of the column material, L the length of the column, and I the moment of inertia of its cross section. Hence, with $E = 4 \times 10^6$ lb/in^2 $= 5.76 \times 10^8$ lb/ft^2, $I = 1766$ ft^4, and $L = 80$ ft, we have $K = 5.96 \times 10^6$ lb/ft.

The weight of the deck is 2000 kips, corresponding to a deck mass $M_D =$ 62,100 slugs. Assuming that the concrete weighs 145 lb/ft^3, the weight of the support column is 948 kips. This corresponds to a mass of 948,000/32.2 = 29,400 slugs. The total added mass of the column resulting from its motion in the water is

$$\frac{\rho \pi D_o^2 L}{4} = \frac{(1.99)(\pi)(15^2)(80)}{4} = 28{,}100 \text{ slugs}$$

Collecting together the results above, we thus have the mass M given from equation (68) as

$$M = 62{,}100 + \tfrac{1}{2}(29{,}400 + 28{,}100) = 90{,}900 \text{ slugs}$$

The natural frequency of the structure is next determined from equation (78) as

$$\Omega = \sqrt{\frac{K}{M}} = \sqrt{\frac{5.96 \times 10^6}{9.09 \times 10^4}} = 8.10 \text{ rad/sec}$$

The frequency of the waves is $2\pi/8 = 0.785$ rad/sec, so that $\omega/\Omega = 0.097$. Assuming that $\zeta/\Omega = 0.08$, the dynamic amplification factor is determined from equation (77) as $U/U_s = 1.01$, which differs from unity by only 1%, so that static methods of analysis are appropriate. #

REFERENCES

ACI (1977). *Building Code Requirements for Reinforced Concrete*, American Concrete Institute Publication 318–77, Detroit, Mich.

ACI (1978). Guide for the Design and Construction of Fixed Offshore Concrete Structures, *Journal of the American Concrete Institute*, Vol. 75, No. 12, pp. 684–709.

AISC (1969). *Specifications for the Design, Fabrication and Erection of Structural Steel for Buildings*, American Institute of Steel Construction, New York.

API (1980). *Recommended Practice for Planning, Designing and Constructing Fixed Offshore Platforms*, American Petroleum Institute Publication RP-2A, Dallas, Tex.

Bea, R. G. (1974). Gulf of Mexico Hurricane Wave Heights, *Proceedings, Sixth Annual Offshore Technology Conference*, pp. 791–810.

Gaylord, E. H., and C. N. Gaylord (1972). *Design of Steel Structures*, McGraw-Hill Book Company, New York.

Lin, T. Y. (1963), *Design of Prestressed Concrete*, John Wiley & Sons, Inc., New York.

Maddox, N. R., and A. N. Wildenstein (1975). A Spectral Fatigue Analysis for Offshore Structures, *Proceedings, Seventh Annual Offshore Technology Conference*, pp. 185–193.

McGuire, W. (1968). *Steel Structures*, Prentice-Hall, Inc., Englewood Cliffs, N.J.

Winter, G., and A. H. Nilson (1973). *Design of Concrete Structures*, McGraw-Hill Book Company, New York.

PROBLEMS

1. A proposed offshore steel platform consists of four legs of 4-ft outside diameter and 1-in wall thickness, braced at the top with four horizontal members of 2-ft outside diameter and 0.5-in wall thickness. A side face of the structure is shown in Fig. P.1. Assuming the legs fixed against displace-

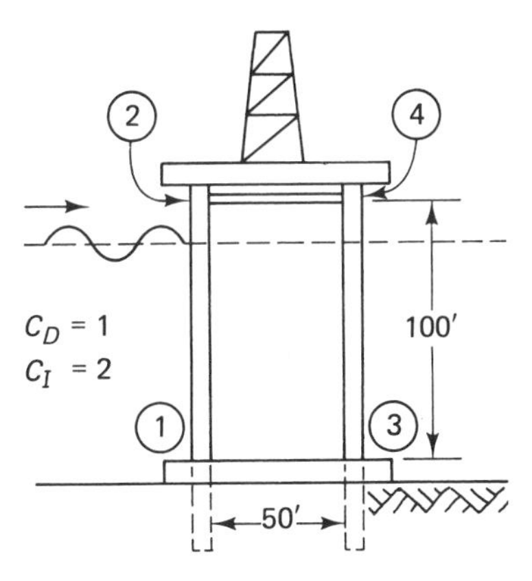

Fig. P.1

ment and rotation at joints 1 and 3, the methods of Chapter 2 provide the following (inverted) matrix equation for determining unknown displacements and rotations at joints 2 and 4 of the side face shown

$$\begin{Bmatrix} 5U_2 \\ 100V_2 \\ 100\theta_2 \\ 5U_4 \\ 100V_4 \\ 100\theta_4 \end{Bmatrix} = 10^{-7} \begin{bmatrix} 333 & & & \text{Symmetric} & & \\ 48.32 & 2256 & & & & \\ 842.0 & 19.33 & 475.2 & & & \\ 3327 & 48.32 & 840.5 & 3333 & & \\ -48.32 & 1.483 & -19.33 & -48.31 & 2256 & \\ 840.5 & 19.33 & 197.9 & 842.2 & -19.33 & 475.3 \end{bmatrix} \begin{Bmatrix} 200F_{2x} \\ 10F_{2y} \\ 10M_2 \\ 200F_{2x} \\ 10F_{4y} \\ 10M_4 \end{Bmatrix}$$

where the U and V's are expressed in feet, the θ's in radians, the F_x and F_y's in kips, and the M's in ft-kips.

(a) For Airy waves of height 20 ft and length 300 ft, in water of depth 80 ft, use the methods of Chapter 3 to determine, for the full three-dimensional structure, the maximum horizontal wave force and associated time ωt.
Ans. 64,460 lb at $\omega t = 5.8$.

(b) Using the methods of Chapter 3, determine equivalent joint loads for the wave described in part (a) at the instant ωt when the total wave force is maximum.

(c) If, in addition to the wave loading above, the side face shown carries 250 kips of deck weight acting midway between the legs and 50 kips of wind loading acting 10 ft above level 2–4, estimate the maximum (axial plus bending) stress induced in members 1–2, 3–4, and 2–4. Include contributions from member weights.
Ans. $-12{,}500$ lb/in², $-21{,}200$ lb/in², $-39{,}400$ lb/in², respectively.

(d) Estimate the dynamic amplification factor for the structure, assuming the legs flooded to the still-water level and a total deck loading on the entire three-dimensional structure of 500 kips.
Ans. 1.08 for $\zeta/\Omega = 0.08$.

2. For the steel offshore structure shown in Fig. P.2, determine the maximum (axial plus bending) stresses existing in members 2–7 and 6–7, for an Airy wave of height 25 ft and length 350 ft. All four faces of the structure are identical. The deck weight carried by the side frame shown is 300 kips. The

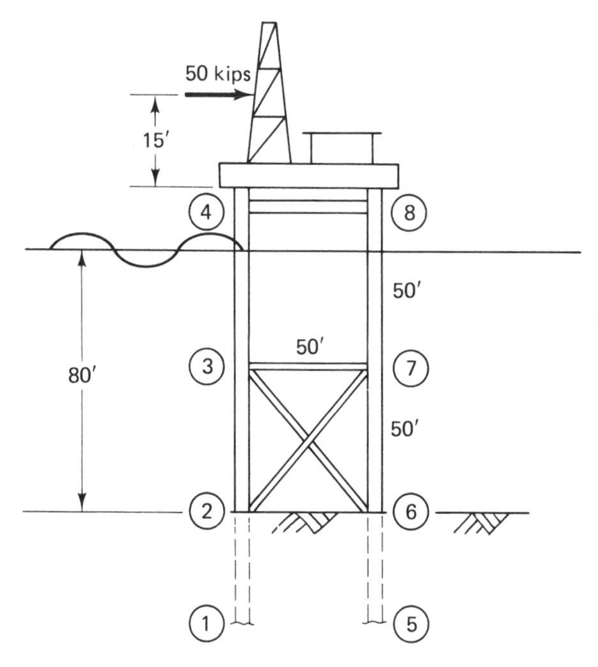

Fig. P.2

net wind force on the side face equals 50 kips and acts 15 ft above level 4–8. Vertical members above the seafloor have outside diameters of 4 ft and wall thicknesses of 1.5 in. Horizontal and diagonal members have 2-ft outside diameter and wall thicknesses of 0.5 in. Assume equivalent steel piles beneath the seafloor to have length $l = 15$ ft, area $A = 0.160$ ft², and moment of inertia $I = 2.30$ ft⁴. Horizontal and diagonal members are unflooded. (*Note:* Use the results from Problem 14 of Chapter 2 and Problems 24, 25, and 26 of Chapter 3.)

3. For the structure and wave of Problem 2, determine the dynamic amplification factor. Is a static analysis appropriate?

4. Use the displacement data of Table 4.3 and determine the maximum stress (bending plus axial) existing in member 6–7 of the structure of Example 4.2-1.

5. Repeat Problem 4 for member 4–8.

6. A diagonal bracing member of an offshore structure has an axial compressive stress of 20 kips/in². If $\sigma_e = 35$ kips/in², determine the bending-stress amplification factor, assuming that $C_m = 1.0$.
Ans. $\alpha = 2.33$.

7. Repeat Example 4.3-1 assuming the lengths of all vertical members to equal 40 ft.

8. An unflooded steel member 1–2 of an offshore structure, having outside diameter of 3 ft and wall thickness 0.75 in, is located at a depth of 800 ft, as shown in Fig. P.8. Determine the maximum hoop stress σ_θ and associated

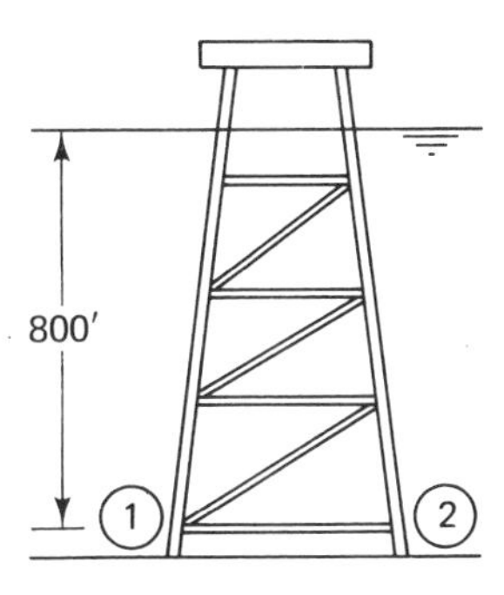

Fig. P.8

longitudinal stress σ_z in the member resulting from hydrostatic loading.
Ans. $\sigma_\theta = -8530$ lb/in², $\sigma_z = -4270$ lb/in².

9. For the member 1–2 of Problem 8, determine the greatest longitudinal compressive stress σ_z and associated hoop stress σ_θ at the ends of the member resulting from hydrostatic loading.
Ans. $\sigma_z = -17{,}400$ lb/in², $\sigma_\theta = -5240$ lb/in².

10. Re-solve Problem 9 for the greatest longitudinal tensile stress and associated hoop stress.
Ans. $\sigma_z = 8930$ lb/in², $\sigma_\theta = -2680$ lb/in².

11. Examine the acceptability of the stresses in Problem 8 if the steel has a yield stress of 36,000 lb/in².
Ans. $|\sigma_a/S_a| = 0.20$, $|\sigma_\theta/S_\theta| = 1.49$ [equation (28)].

12. Show that ring stiffeners spaced every 12 ft along member 1–2 of Problem 8 yields a stress ratio $|\sigma_\theta/S_\theta| \approx 0.8$.

13. Assume in Problem 12 that the ring stiffeners have a thickness of 3.75 in. Determine the necessary width.
Ans. $d = 5.0$ in.

14. Estimate the punching shear in a main leg of an offshore structure, of wall thickness 1 in, if a diagonal member of 1-ft radius frames into it at $\theta = 55°$ and exerts a normal force of 200 kips and a moment of 400 ft-kips.
Ans. $\tau_p = 10{,}700$ lb/in².

15. Estimate the fatigue life of a joint having stress history indicated below. Assume a stress concentration factor of 2.0.

H (ft)	T (sec)	P (%)	σ_a (kips/in²)
40–60	12	0.01	5.0
20–40	10	0.03	3.0
10–20	9	0.10	1.5
5–10	7	1.00	0.6
3–5	4	10.00	0.5
0–3	3	88.86	0.2

Ans. $T \approx 20$ years.

16. Re-solve Problem 15 assuming that the stresses for periods 4 sec or less are increased 50% because of dynamic effects.
Ans. $T \approx 14$ years.

17. A small concrete gravity structure consists of a single column, of inside radius 4.4 ft and outside radius 6.0 ft, supporting a deck and equipment in 60 ft of water. Internal forces and moments acting at the base of the column are as indicated in Fig. P.17.

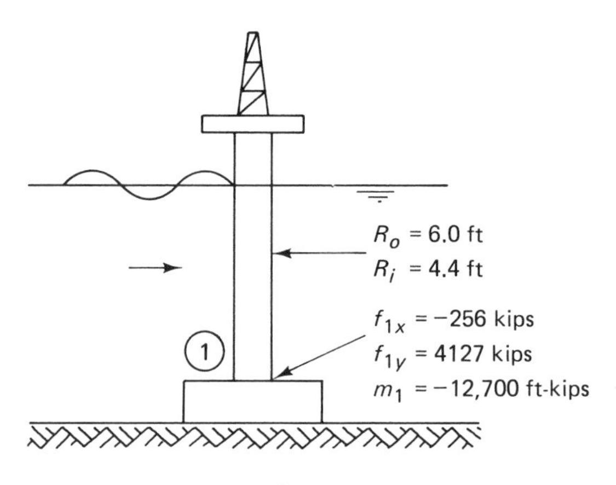

Fig. P.17

(a) Determine the net area of symmetrically located prestress tendons, with design prestress of 150 kips/in², needed to eliminate bending tensile stresses in the column.
Ans. $A_s = 9.22$ in².

(b) Determine the cracking moment.
Ans. $M_c = 21{,}900$ ft-kips.

(c) Determine the ultimate moment that the column can withstand without crushing failure of the concrete, assuming tendons grouped as in Example 4.8-2.
Ans. $M_u = 33{,}100$ ft-kips.

18. Reconsider Example 4.9-1 assuming a net external pressure of 200 lb/in².

19. Reconsider Example 4.9-1 assuming a net external pressure of 200 lb/in² and ellipsoidal end caps having ratio of height to radius of $\frac{1}{2}$.

20. An offshore gravity (Fig. P.20) structure consists of four concrete columns supporting a deck and equipment weighting 20,000 kips. The inside and outside diameters of the columns are 48 and 50 ft, respectively. Assuming

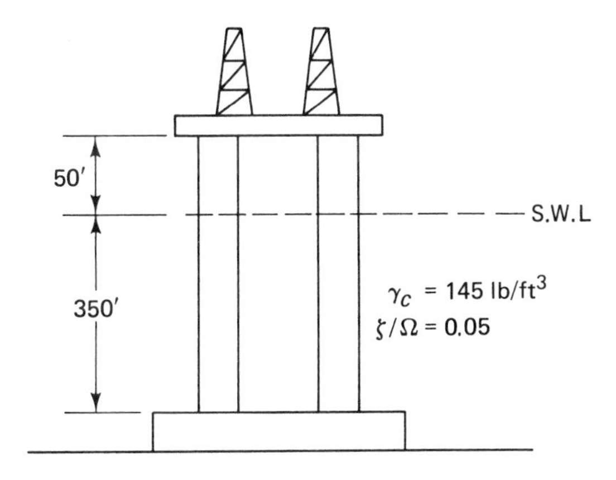

Fig. P.20

the columns unflooded, determine the dynamic amplification factor for waves of 10-sec period. Calculations show that a total force of 100 kips acting at the deck level on the entire three-dimensional structure will cause a deck deflection of 0.060 in.
Ans. $U_0/U_s = 1.08$.

21. Estimate the dynamic amplification factor for the structure and design wave of Example 4.2-1. Vertical members are assumed flooded to the still-water level; all other members are assumed unflooded. For the equivalent pile stiffnesses used with the design wave, calculations for a side face show that a 1-kip horizontal force at the deck level will cause a horizontal deflection at this level of 0.024 in.
Ans. $U_0/U_s = 1.02$ for $\zeta/\Omega = 0.08$.

Foundations for offshore structures. Top: Segment of a pile being driven through the leg of a small template substructure. (Courtesy of the U.S. Navy.) Bottom: Concrete gravity structure under construction, with cylindrical foundation elements complete and main legs nearing completion. (Courtesy of the Mobil Oil Corporation and the American Petroleum Institute.)

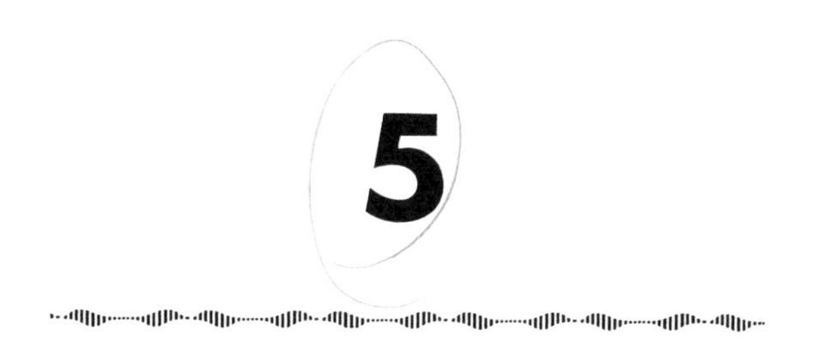

5 FOUNDATION ANALYSIS

IN CHAPTER 4 we concentrated attention on the structural analysis of offshore platforms and showed how the stresses within the individual members could be computed for given environmental conditions. The present chapter extends this analysis to include consideration of the foundation of the structure. For template structures, this involves study of the ultimate capacity and response of support piles driven into the seafloor and, for gravity structures, it involves study of the ultimate capacity and response of base elements used to distribute the loading of the structure over the seafloor.

5.1 SOIL CHARACTERISTICS

Soils, often referred to as *sediments* in the marine environment, consist generally of particles, or grains, of disintegrated rock and possibly organic material of varying sizes and varying aggregate characteristics. They are classified broadly in terms of their particle size and their plastic or nonplastic aggregate behavior in the presence of water, that is, their ability or inability to be molded without cracking or breaking. Two major categories of soils are *sand* and *clay*. Sand is characterized typically as a nonplastic soil having particle sizes in the range between about 0.2 in and 0.003 in. Clay, on the other hand, is characterized as a plastic soil having particle sizes less than about 0.003 in. A third soil category of interest in the marine environment is *silt*, which is a relatively nonplastic soil with particle sizes less then about 0.003 in. Most sediment deposits consist of a mixiture of these three soil types. However, for engineering purposes, a sediment can generally be classed broadly as either clay or sand, depending on its plastic or nonplastic behavior.

Soil deposits found at and below the seafloor are usually assumed to have all the voids, or pores, between the soil particles completely filled with water. The *total stress* at any point in this saturated soil mass may thus be regarded as the sum of the stress carried by the soil particles and the pressure carried by the pore water.

When a soil sample is subjected to a uniform gradually increasing stress, it first exhibits an elastic response followed by shear failure (large-scale displacement) at a critical stress level. This failure stress is generally assumed to be that at which the magnitude of the shearing stress in the sample reaches a critical value S described empirically by the *Coulomb equation*,

$$S = C + \sigma \tan \phi \tag{1}$$

where C and ϕ denote constants for a given soil and σ denotes the *effective stress* normal to the shear plane, that is, the stress carried by the soil particles in the direction normal to the shear plane.

By considering the definition of the effective stress, as given above, we may also express equation (1) in terms of the applied stress normal to the shear failure plane. Two limiting cases present themselves: that where the applied stress is carried entirely by the pore water and that where it is carried entirely by the soil particles. In the first case, the effective stress is zero and, in the second, it is equal to the applied stress. The significance of these cases depends on the soil type and loading times involved.

Take first sand soils. These soils possess high permeability; hence, any tendency for the pore water to carry the applied stress is quickly relieved by flow of water out of the soil mass. The effective stress may thus generally be assumed equal to the applied stress. Moreover, the shear strength for sands is known experimentally to be proportional directly to the effective stress so that the appropriate form of equation (1) is simply

$$S = \sigma_A \tan \phi \tag{2}$$

where σ_A denotes the applied stress normal to the shear failure plane and ϕ denotes the *angle of friction* of the sand, as determined for a given soil sample by laboratory test. Values of ϕ vary somewhat with the density of the sand deposit but typically fall within the range 30 to 35°.

Next consider clay soils. In contrast with sand, these soils possess low permeability, and hence a part of the applied stress can be carried for an appreciable time by the pore water before drainage finally allows the applied stress to be transferred completely to the soil particles as effective stress. Because of the low strength of the soil relative to the nearly incompressible pore water, almost all of the applied stress will be carried initially by the pore water, so that both of the limiting cases referred to above can be identified: the undrained case where the effective stress is zero and the drained case where the effective stress is equal to the applied stress. In the latter case, experiment shows that the shearing strength may be assumed directly proportional to the applied stress σ_A normal to the shear failure plane. The appropriate form of equation (1) for the undrained case is accordingly

$$S = C \tag{3}$$

and that for the drained case is

$$S = \sigma_A \tan \phi_D \tag{4}$$

where C and ϕ_D are referred to, respectively, as the *cohesion* and *effective angle of friction* of the clay.

Values of C and ϕ_D may be determined from standard laboratory tests on soil samples taken at various depths in a soil deposit (see, for example, Terzaghi and Peck, 1967). The cohesion of a given soil sample, as measured under undrained conditions, depends on the amount of water content present in the soil and hence on the amount of prior compression and drainage, or *consolidation*, experienced by the soil. Values of the cohesion can range from near zero to 4000 lb/ft^2 or more. Appropriate ranges for clays whose consistency is designated as very soft, soft, medium, etc., are listed in Table 5.1. Values of the effective

Table 5.1 RANGE OF COHESION FOR CLAYS OF VARIOUS DESIGNATIONS OF CONSISTENCY

Consistency	Cohesion (lb/ft^2)
Very soft	Less than 250
Soft	250–500
Medium	500–1000
Stiff	1000–2000
Very stiff	2000–4000
Hard	Greater than 4000

angle of friction ϕ_D vary with the degree of plasticity exhibited by the clay, but fall typically within the range 20 to 40°.

If, in a natural soil deposit, the soil has become completely consolidated under its own weight loading, the deposit is said to be *normally consolidated.* On the other hand, if the soil deposit is relatively new, it may be *underconsolidated,* in which case all the water drainage needed to transfer the weight loading completely to the soil particles has not yet taken place. Finally, if a normally consolidated soil deposit has suffered erosion at its surface, or if it has previously been subjected to compression loading, the degree of consolidation existing will be greater than that expected from its present weight loading and the deposit is then said to be *overconsolidated.*

5.2 PILES FOR TEMPLATE STRUCTURES

As noted earlier, template-type offshore structures rely mainly on steel pipe piles driven through the columns of the structure into the seafloor to provide support for deck loadings and resistance to movement during storm conditions. Under compression loading, the piles derive their support capacity from vertical shear forces exerted on the lateral surface of the pile by the surrounding soil and from vertical normal forces exerted on their base end by the soil there. In most cases, it is the vertical shear force that provide the majority of a pile's load capacity, and, as this force increases with increasing lateral surface area, deeply driven piles are normally needed to carry the large loads exerted by the overhead structure.

The size of the piles and the depth to which they are driven vary, of course, from one structure to another and depend on the number employed, the loading anticipated and the subsoil conditions. Typically, though, piles of outside diameter in the range 2 to 5 ft with wall thicknesses in the range 0.5 to 1 in are not uncommon, with these being driven to depths of 200 ft or more. In certain cases where very soft soil conditions exist, additional *skirt piles* may also be employed. These are piles driven around the base of the structure and attached there in order to provide further support. Figure 5.1 illustrates both the conventional pile system and that where skirt piles are used.

Under design conditions such as illustrated in Fig. 5.2, the piles are subjected to large compressive loads, often of the order of 1000 kips, or more. Because of high overturning moments exerted on the structure by wind and wave action, the piles may also be subjected to large tensile loads of this same order. The wind and wave forces can also induce large horizontal forces and moments on the piles at the groundline, often of the order, respectively, of 100 kips and 1000 ft-kips or more. These can, in turn, cause significant movement of the piles at the groundline and hence of the attached overhead structure.

In view of these loading conditions, we may conveniently divide the study of pile foundations for template structures into three major categories:

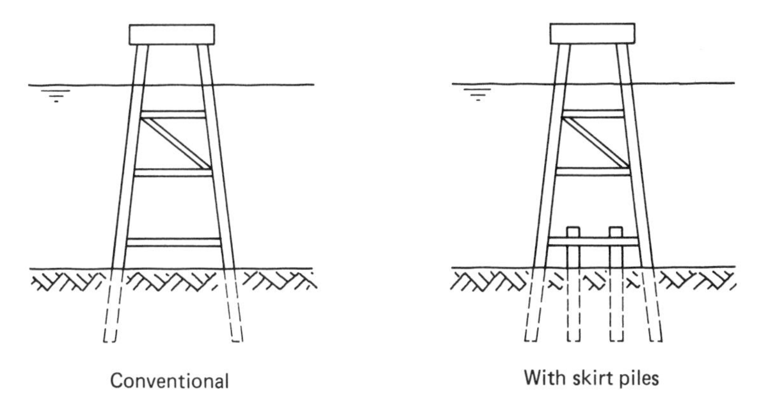

Fig. 5.1. Conventional and skirt piles for offshore template structures.

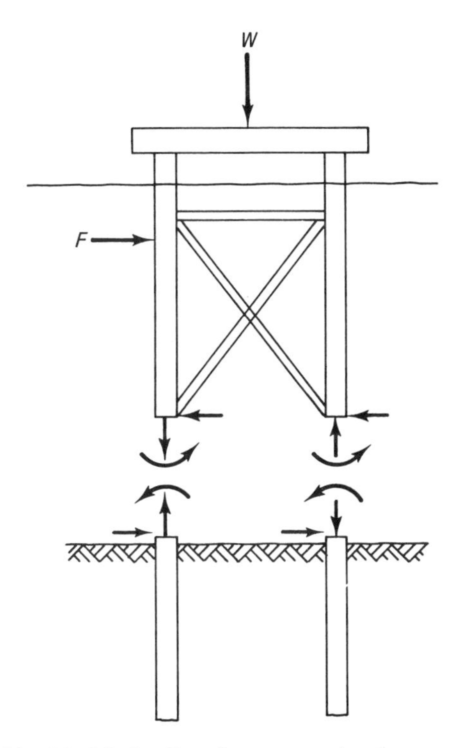

Fig. 5.2. Pile loadings from an overhead structure.

1. The prediction of the axial capacity of piles driven to a certain depth
2. The prediction of the elastic axial response of piles loaded to levels less than ultimate capacity
3. The prediction of the lateral response of piles to applied horizontal loading and moment at the groundline

The first category is important for determining the depth that support piles must be driven into the seafloor in order to provide adequate support for the applied axial loads. Such knowledge is, of course, needed for installation of a properly designed structure. It is also necessary for evaluating a proposed structural design since the possibility exists that the depth needed is greater than that which could be achieved with available pile-driving hammers. If this were the case, the design would then have to be modified to include more piles so that the design depth of each could be reduced. The second and third categories are important for determining the response to the piles at the groundline to design-level loadings on the structure. This response knowledge is needed to establish the stiffness properties of equivalent free-standing piles for use in the analysis of the overhead structure, as discussed in Chapter 4. The last category is also of major importance in the design of the piles themselves, since lateral loadings generally induce large bending stresses in the piles at some distance below the seafloor.

The spacings of the piles of an offshore template structure are usually sufficiently great (greater, say, than 5 to 10 pile diameters) to allow neglect of any interaction between them through the adjacent soil. Their loading response is thus adequately represented by considering each pile separately. Also, settlement of the piles resulting from compression of the soil can usually be neglected, partly because the loadings expected on them most, if not all, of the time are many times smaller than the extreme environmental loadings they are designed to withstand, and partly because deeply driven piles transfer their loadings to the soil mainly through vertical shear forces which are not effective in compressing the soil.

5.3 PREDICTION OF AXIAL PILE CAPACITY

The resistance of a solid cylindrical pile to large-scale vertical movement under axial compressive loading is customarily assumed to result from the combined effects of a shear yield force acting on the lateral surface of the pile and a normal yield force acting over the entire base end of the pile. For an open-ended pipe pile this same assumption is also generally valid since, on driving, soil is forced up into the interior of the pile, forming a solid soil plug whose resistance to movement under static loading is usually much greater than that of the soil at the base of the pile. Thus, for open-ended pipe piles such as commonly used in offshore work, the ultimate axial capacity Q, like that of a solid pile, is expressible as

$$Q = Q_s + Q_p \tag{5}$$

where, as indicated in Fig. 5.3, Q_s denotes the total shear yield force acting on the outside lateral surface area of the pile and Q_p denotes the total end normal yield force.

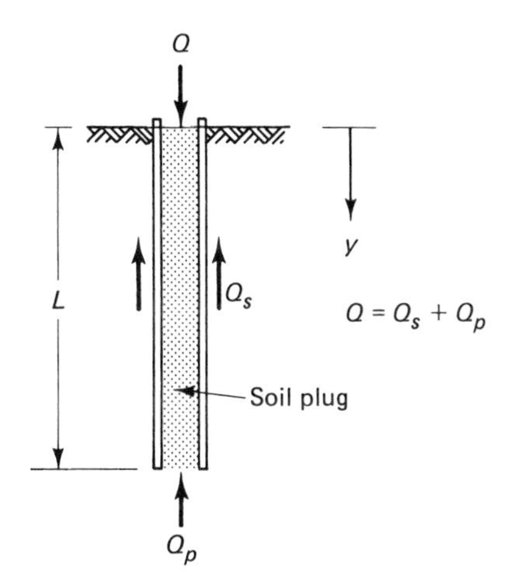

Fig. 5.3. Forces on a pile.

To calculate the total shear force Q_s in the equation above, we introduce a shear yield force f per unit of lateral surface area of the pile and integrate this over the entire embedded pile length. Assuming f to depend, in general, on the depth y below the groundline, we thus have

$$Q_s = \pi D_o \int_0^L f\,dy \tag{6}$$

where D_o denotes the outside diameter of the pile and L denotes the depth to which it is driven in the soil. Similarly, with q denoting the normal end yield force per unit of pile-end area, we also have

$$Q_p = q\frac{\pi D^2}{4} \tag{7}$$

where q can, in general, be expected to depend on the pile depth L. Finally, if F denotes critical axial loading on the pile at the groundline, and w_p denotes the combined submerged weight per unit of length of the pile and soil plug, the axial capacity Q can be expressed as

$$Q = F + w_p L \tag{8}$$

Substituting the expressions above into equation (5) and rearranging, we have the ultimate compressive loading capacity F_c of the pile expressible as

$$F_c = \pi D_o \int_0^L f\,dy + q\frac{\pi D_o^2}{4} - w_p L \tag{9}$$

The tensile capacity F_T of the pile is similarly calculated by the formula

$$F_T = \pi D_o \int_0^L f\,dy + w_p L \tag{10}$$

The condition that the soil plug in the pile does not move, as assumed above, is easily seen to be expressible by the relation

$$q\frac{\pi D_i^2}{4} < \pi D_i \int_0^L f\,dy + w_s L \tag{11}$$

where D_i denotes the inside diameter of the pile and w_s denotes the submerged weight per unit length of the soil plug.

In using the expression above, it is, of course, necessary to specify the unit forces f and q for a given soil deposit. For the case of clay soils, these are related to the measurable cohesion (or undrained shear strength) C of the soil by the following relations:

$$f = \alpha C, \qquad q = N_c C \tag{12}$$

where α and N_c are dimensionless coefficients. For the case of sand soils, f and q are, within limits, related to the weight of the overhead soil and the friction angle δ between the pile and sand particles by

$$f = K\gamma_s y \tan\delta, \qquad q = N_q \gamma_s L \tag{13}$$

where γ_s denotes the submerged specific weight of the sand, and K and N_q denote dimensionless coefficients.

Table 5.2 lists typical design values of the parameters above. Notice that,

Table 5.2 TYPICAL DESIGN PARAMETERS FOR COMPUTING PILE CAPACITY

Clay	Sand
$\alpha = 1.0, 0 \leq C \leq 0.5$ kip/ft^2	$\delta = 30°$, $\gamma_s = 40$–70 lb/ft^3
$\alpha = 1.25 - 0.5C, 0.5 < C < 1.5$ kips/ft^2	$K = 0.7, f(\max) = 2$ kips/ft^2
$\alpha = 0.5, C > 1.5$ kip-ft^2	$N_q = 40, q(\max) = 200$ kips/ft^2
All clays, $N_c = 9$	

Source: Based on criteria given by McClelland et al. (1967) and the American Petroleum Institute (API, 1980).

for sand, maximum values of f and q are specified so that equation (13) applies only up to these limits, after which f and q are assumed constant.

EXAMPLE 5.3-1. For a clay soil having undrained shear strengths negligible at the mudline and increasing linearly to about 3.6 kips/ft² at 400-ft penetration, as shown in Fig. 5.4, determine the depth to which a 4-ft-diameter steel pile of 1-in wall thickness

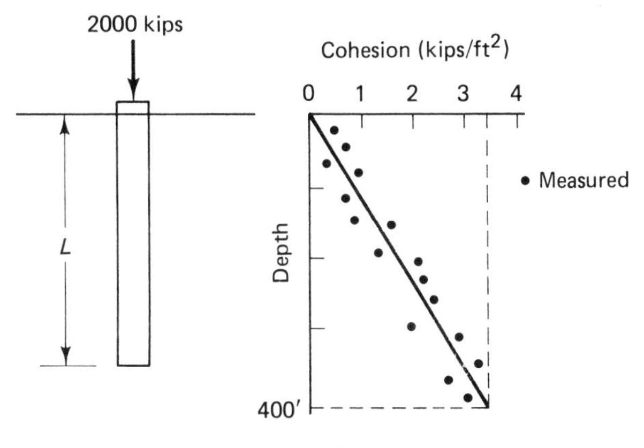

Fig. 5.4

must be driven to provide an ultimate compressive load capacity of 2000 kips. The specific weight (in air) of steel is 0.484 kip/ft³ and that of the soil is 0.099 kip/ft³.

We have the shear strength C expressible in terms of depth y from the mudline as

$$C = by, \qquad b = 0.009 \text{ kip/ft}^3$$

From Table 5.2, we thus have the shear force $f = \alpha C$ expressible as

$$f = f_1 = by, \qquad 0 < y \leq 55.6 \text{ ft}$$
$$f = f_2 = 1.25by - 0.5(by)^2, \qquad 55.6 < y \leq 167 \text{ ft}$$
$$f = f_3 = 0.5by, \qquad y > 167 \text{ ft}$$

and the normal force q at $y = L$ as

$$q = 9C = 9bL$$

The total submerged weight, per unit of length, of the pile and soil plug is calculated from the equation

$$w_p = \gamma_{st}A_{st} + \gamma_s A_s$$

where γ_{st} and γ_s denote, respectively, the submerged specific weights of the steel and soil, and A_{st} and A_s denote the cross-sectional areas of the steel and soil plug. The submerged specific weights are calculated from the corresponding specific weights in air by subtracting the specific weight of water ($\gamma = 0.064$ kip/ft^3) from these. Thus, from the data given, we have

$$\gamma_{st} = 0.484 - 0.064 = 0.420 \text{ kip/ft}$$
$$\gamma_s = 0.099 - 0.064 = 0.035 \text{ kip/ft}^3$$

and

$$w_p = 0.420\frac{\pi}{4}(4^2 - 3.83^2) + 0.035\frac{\pi}{4}(3.82^2) = 0.842 \text{ kip/ft}$$

With the results above, we may solve for the required depth L of the pile using equation (9). Assuming, tentatively, that $L > 167$ ft, we have, in view of the representations of f above,

$$\int_0^L f\,dy = \int_0^{55.6} f_1\,dy + \int_{55.6}^{167} f_2\,dy + \int_{167}^{L} f_3\,dy$$

which gives

$$\int_0^L f\,dy = 30 + 2.25 \times 10^{-3}L^2$$

Putting this in equation (9) and using

$$q = 9C = 0.081L$$

we find, with $F_c = 2000$, $w = 0.842$, the expression

$$0.0283L^2 + 0.175L - 1623 = 0$$

from which we determine

$$L = 236 \text{ ft}$$

If we wish to include a factor of safety of, say, 1.5 into the problem, we simply reduce the capacity Q given by equation (5) by 1.5. From equation (8), this can be seen to be equivalent to increasing P and w_p in equation (9) by 1.5. Thus, re-solving with $F_c = 3000$ and $w_p = 1.26$, we find

$$L = 300 \text{ ft}$$

Notice that this solution, as well as the preceding one, satisfies the tentative initial assumption that $L > 167$ ft. If this has not been the case, we would have to re-solve the problem with the assumption that $55.6 < L < 167$ ft and with the integral in equation (9) involving only the first two representations $f = f_1$ and $f = f_2$ given above, with the integral of f_1 having limits 0 and 55.6 and the integral of f_2 having limits 55.6 and L. Of course, if this solution yielded a value of

$L < 55.6$ ft, we would have to re-solve the problem again using only the first representation $f = f_1$ with the integral having limits of 0 and L.

The condition for rigid-plug behavior as given by relation (12) is easily satisfied since (for $L = 300$ ft)

$$q\frac{\pi D_i^2}{4} = 89 \text{ kips}$$

$$\pi D_i \int_0^{300} f\,dy = 3340 \text{ kips}$$

$$w_s L = 10.5 \text{ kips}$$

and $89 < 3340 + 10.5$. #

EXAMPLE 5.3-2. For the offshore structure and loading shown in Fig. 5.5, determine the depth that the piles should be driven to provide adequate support. Assume the same soil properties as in Example 5.3-1. Note that a worst-case condition is considered where the wind and wave loading act at 45° to the faces of the structure. The pile loadings are also idealized as indicated.

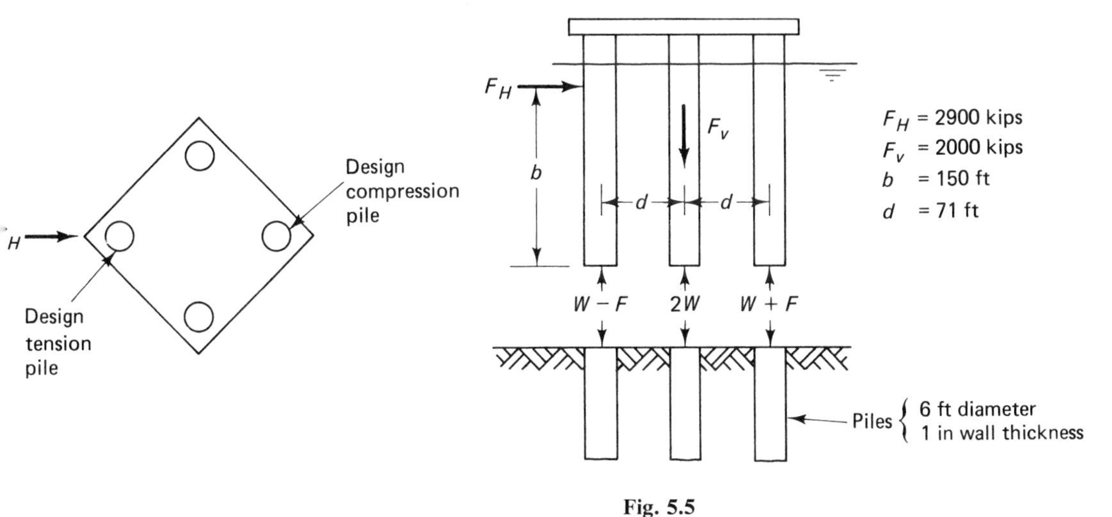

Fig. 5.5

From the balance of forces in the vertical, we have

$$4W = F_v$$

and from the balance of moments about the left-hand pile, we have

$$d(2W) = 2d(W + F) = bF_H + dF_v$$

The solution of these two equations accordingly gives

$$W = \frac{F_v}{4}, \qquad F = \frac{bF_H}{2d}$$

Substituting numerical values, we find $W = 500$ kips, $F = 3060$ kips. The maximum compressive force (on the right-hand pile) is thus

$$F_c = W + F = 3560 \text{ kips}$$

and the maximum tensile force (on the left-hand pile) is

$$F_T = W - F = -2560 \text{ kips}$$

Consideration of the maximum compression loading in the manner given in the preceding example shows, for a factor of safety of 1.5, that the piles should be driven to a depth of 337 ft. From equation (10), the tensile capacity of the piles is calculated as 6180 kips, corresponding to a factor of safety of 6180/2560 = 2.4. #

5.4 ELASTIC RESPONSE OF PILES TO AXIAL LOADING

The discussion above was concerned with the ultimate capacity of a pile to withstand axial loading. When the applied loading is such that the axial capacity is not overcome, as under design conditions, the vertical pile movement is then restrained chiefly by elastic shearing forces acting on the lateral surface area of the pile. Some resistance is, of course, also provided in compression by normal forces acting on the base of the pile. However, because the lateral surface area of a deeply driven pile is much greater than its base area, the contribution of this latter force to the pile resistance is of little importance and can be ignored.

The following discussion gives a simplified treatment of the elastic response of piles to axial tension or compression loadings. A more detailed treatment has been given by Coyle and Reese (1966).

Figure 5.6(a) illustrates the general problem where an embedded pile of length L is subjected to an axial loading F_0 at the groundline that does not exceed the ultimate capacity of the pile.

As indicated, an elastic shear force f_a per unit of pile surface area is assumed to provide resistance to movement of the pile. We assume this force to depend on the outside pile diameter D_o, the axial deflection v, and an elastic soil modulus k_a having units of force per unit area. In dimensionless form, we thus have, for

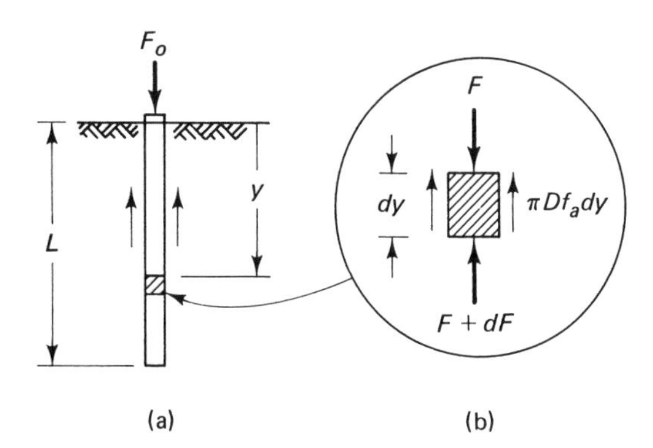

Fig. 5.6. Shear force from adjacent soil on a pile element.

linear elastic response, $f_a/k_a = v/D_o$, or, on multiplying through by k_a,

$$f_a = \frac{k_a v}{D_o} \tag{14}$$

where f_a is assumed positive upward when v is positive downward.

Referring to the pile element of Fig. 5.6(b), the equilibrium equation for an element of the pile is easily seen to be expressible as

$$\frac{dF}{dy} = -\pi D_o f_a \tag{15}$$

where F denotes the axial force (positive in compression) in the pile at distance y below the groundline. If A denotes the cross-sectional area of the pipe carrying the applied loading, and $\epsilon = -dv/dy$ denotes the strain (assumed positive in compression) in the pile, then from Hooke's law of elasticity, we have

$$F = EA\epsilon = -EA\frac{dv}{dy} \tag{16}$$

Combining equations (14)–(16), we thus obtain the following differential equation for determining the pile deflection:

$$\frac{d^2v}{dy^2} - \frac{\pi k_a}{EA}v = 0 \tag{17}$$

The solution of this equation is expressible as

$$v = C_1 \cosh \alpha y + C_2 \sinh \alpha y \tag{18}$$

where

$$\alpha = \sqrt{\frac{\pi k_a}{EA}} \tag{19}$$

and C_1 and C_2 denote arbitrary constants. These may be determined by the conditions

$$F = F_0, \qquad y = 0$$
$$F = 0, \qquad y = L$$

from which we find, with the help of equation (16),

$$C_1 = \frac{F_0}{EA\alpha} \coth \alpha L, \qquad C_2 = -\frac{F_0}{EA\alpha}$$

Substituting these into equation (18), we thus have the equation for the pile deflection expressible as

$$v = \frac{F_0}{EA\alpha} (\coth \alpha L \cosh \alpha y - \sinh \alpha y) \tag{20}$$

Of chief interest in this result is the force–deflection relation at the groundline. This is determined from equation (20), with $y = 0$, as

$$F_0 = K_0 v \tag{21}$$

where the effective axial stiffness K_0 of the pile is given by

$$K_0 = EA\alpha \tanh \alpha L \tag{22}$$

This last equation is of value in defining the effective axial stiffness of a pile at the groundline for structural analysis purposes. To use the equations, we must, of course, have available an appropriate value of the elastic soil modulus k_a introduced above. Fortunately, this can be determined directly from application of equations (19) and (22) to load–deflection field tests on embedded piles. In particular, for a test pile of known sectional dimensions driven to a known depth, the value of K_0 can be taken directly from the load–deflection measurements and the value of the elastic modulus k_a then determined by solution of equations (19) and (22). Table 5.3 gives typical values for clay and sand, as

Table 5.3 TYPICAL VALUES OF THE ELASTIC MODULUS k_a FOR CLAY AND SAND

Soil	k_a (kips/ft^2)
Clay	160
Sand	190

Source: Data for clay from Cox et al. (1979); for sand, from Reese and Cox (1976).

estimated from load tests. Figure 5.7 shows the calculated responses of the test piles used in these studies, based on the data of Table 5.3, and contrasts these with the range of measured values.

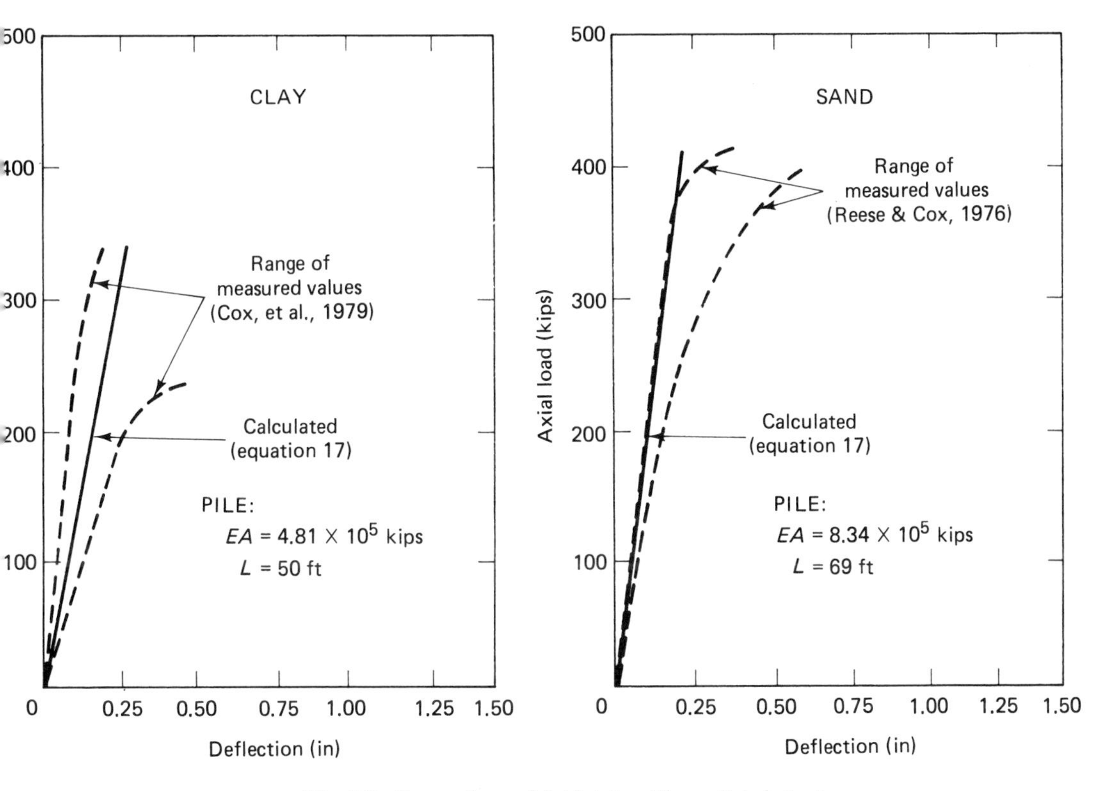

Fig. 5.7. Comparison of field data with predicted elastic response.

EXAMPLE 5.4-1. Determine the effective axial stiffness of a 4-ft-diameter pipe pile having wall thickness of 1 in when the pile is driven 200 ft in clay soil.

We have (with $E = 3 \times 10^7$ lb/in², $A = 148$ in²)

$$EA = 4.4 \times 10^9 \text{ lb}$$

so that, with $k_a = 1.6 \times 10^5$ lb/ft² from Table 5.3, we find

$$\alpha = \sqrt{\frac{\pi k_a}{EA}} = 0.011$$

From equation (22), we then have

$$K_0 = 4.7 \times 10^7 \text{ lb/ft} \quad \#$$

5.5 RESPONSE OF PILES TO CYCLIC LATERAL LOADING

When, as illustrated in Fig. 5.8, an offshore support pile is subjected to cyclic lateral force and moment at the groundline from wave action on the overhead structure, the surrounding soil along the pile length exerts a restraining force that is greatest near the groundline and that decreases with depth as the tendency of the pile to deflect decreases. To determine the pile response under these conditions, it is necessary to represent the soil reaction in an analytical form that includes not only the upper layers of the soil, where yielding and large lateral displacements may occur, but also the lower layers where only elastic response of the soil is involved.

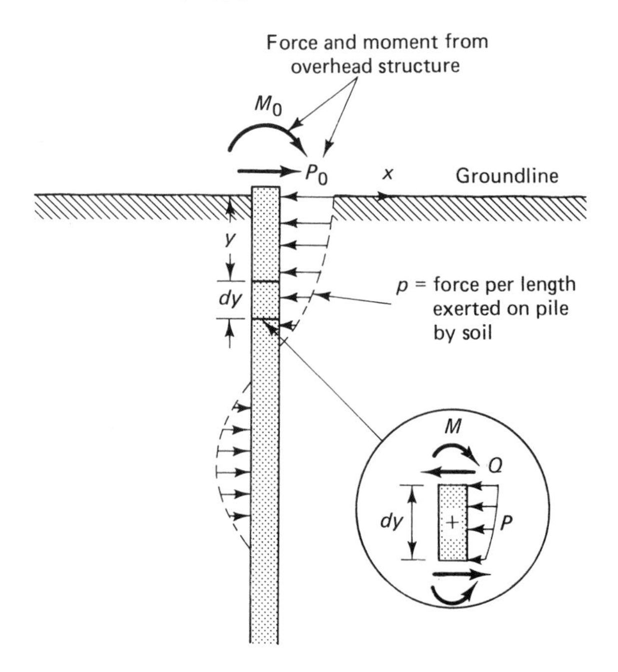

Fig. 5.8. Forces on a laterally loaded pile.

The following discussion gives a simplified representation of this soil reaction, together with the attendant solution for the lateral pile response. More detailed treatments of the problem have been given by Matlock (1970) and Reese et al. (1974, 1975).

In fundamental terms, we assume the lateral soil resistance p per unit of pile length to depend, in general, on the pile diameter D, the lateral deflection u, a characteristic strength parameter σ of the soil having dimensions of stress, and an effective elastic modulus k of the soil, also having dimensions of stress.

In dimensionless form, we accordingly have the relation

$$\frac{p}{kD} = f\left(\frac{u}{D}, \frac{\sigma}{k}\right) \tag{23}$$

where f denotes an arbitrary function. This relation is further restricted to have the simple form shown in Fig. 5.9, which, in algebraic terms, is expressible

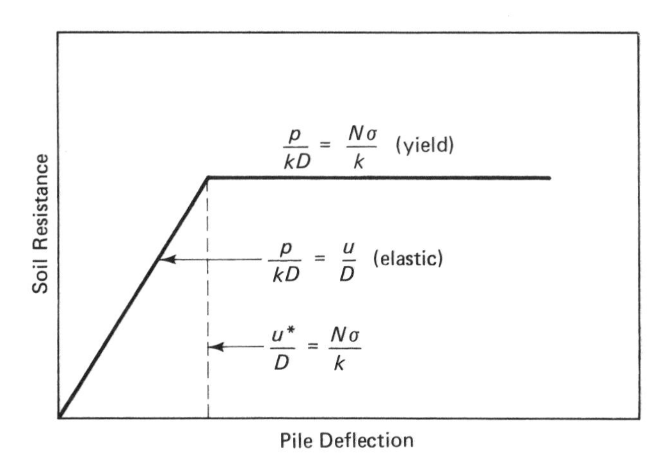

Fig. 5.9. Idealized soil response to a laterally loaded pile.

(for positive u) as

$$\begin{aligned} p &= ND\sigma, & \frac{u}{D} &> \beta \\ p &= ku, & \frac{u}{D} &\leq \beta \end{aligned} \tag{24}$$

where N denotes a coefficient and β is determined by the intersection of the two relations, that is,

$$\beta = \frac{N\sigma}{k} = \frac{u^*}{D} \tag{25}$$

with u^* denoting the pile deflection at the transition between the linear (elastic) response and the uniform (yield) response.

In the relations above, we allow the strength of the soil as represented by σ to vary with depth, as it generally does for soil deposits; however, for simplicity of analysis, we assume the effective elastic modulus k of the soil to be constant for a reasonably homogeneous soil deposit.

Restriciting attention now to the case of clay soils, we assume the strength to be characterized adequately by the cohension C of the clay. For the upper layers of most soil deposits (where significant lateral movement of the pile occurs), this may further be assumed to vary linearly with depth y below the groundline, so that we have

$$\sigma = C = a + by \tag{26}$$

where a and b denotes constants for a given soil. Substituting this relation into equations (24) and (25), we then have the resistive force per unit length of pile expressible as

$$\begin{aligned} p &= ND(a + by), & \frac{u}{D} &> \beta \\ p &= ku, & \frac{u}{D} &\leq \beta \end{aligned} \tag{27}$$

with β given by

$$\beta = \frac{N(a + by)}{k} = \frac{u^*}{D} \tag{28}$$

Next consider the case of sands. For this case, we assume that the lateral stress exerted on the soil mass at any depth y by the pile is the maximum principal stress existing there, in which case the strength of the soil is then expressible as (see Terzaghi and Peck, 1967)

$$\sigma = K_p \gamma_s y \tag{29}$$

where γ_s denotes the submerged specific weight of the sand and K_p is defined in terms of the angle of friction ϕ of the sand by

$$K_p = \frac{1 + \sin \phi}{1 - \sin \phi} \tag{30}$$

Substituting equation (29) into equations (24) and (25), we thus have the final form of the assumed soil resistance for sands expressible as

$$\begin{aligned} p &= NDK_p \gamma_s y, & \frac{u}{D} &> \beta \\ p &= ku, & \frac{u}{D} &\leq \beta \end{aligned} \tag{31}$$

with

$$\beta = \frac{NK_p \gamma_s y}{k} = \frac{u^*}{D} \tag{32}$$

To analyze the pile response to cyclic lateral loading using the relations above, we consider the pile–soil system and assumed applied maximum loadings and directions shown earlier in Fig. 5.8. In the general case, we assume the upper layers of the soil to have yielded under the pile loading to a depth L_1. The governing equation for the pile deflection is the usual beam equation

$$EI\frac{d^4u}{dy^4} = -p \tag{33}$$

where EI denotes the flexural rigidity of the pile, u is measured positive to the right, and p is positive for positive deflection u.

For the yield zone ($y < L_1$) we have for either the clay or sand case, the equation (u being positive under the assumed loadings)

$$EI\frac{d^4u}{dy^4} = -P_1 - P_2y \tag{34}$$

where, in the case of clay,

$$P_1 = NDa, \qquad P_2 = NDb \tag{35}$$

and, in the case of sand,

$$P_1 = 0, \qquad P_2 = NDK_p\gamma_s \tag{36}$$

Integrating equation (34) and using boundary conditions at $y = 0$ given by

$$EI\frac{d^2u}{dy^2} = M_0, \qquad EI\frac{d^3u}{dy^3} = P_0 \tag{37}$$

where M_0 and P_0 denote the maximum applied moment and lateral force at the ground line, we have (for constant EI)

$$EIu = -\frac{P_1y^4}{24} - \frac{P_2y^5}{120} + \frac{P_0y^3}{6} + \frac{M_0y^2}{2} + C_1y + C_2 \tag{38}$$

where C_1 and C_2 denote arbitrary constants.

Next consider the elastic zone ($y \geq L_1$). For either the clay or sand case, we have the equation

$$EI\frac{d^4u}{dy^4} + ku = 0 \tag{39}$$

whose solution for deeply driven piles may be written as

$$EIu = e^{-\alpha y'}(C_3 \cos \alpha y' + C_4 \sin \alpha y') \tag{40}$$

where $y' = y - L_1$ denotes distance from the transition depth $y = L_1$, α is defined by

$$\alpha = \left(\frac{k}{EI}\right)^{1/4} \tag{41}$$

and where the assumption of deeply driven piles implies that

$$L - L_1 \geq \frac{3}{\alpha} \tag{42}$$

with L denoting the total pile length.

There remain five constants to be determined in the solutions above, namely C_1, C_2, C_3, C_4, and L_1. These may be evaluated using equation (28) for the case of clays or equation (32) for sands, together with the conditions that the deflection, slope, internal moment, and shear force at $y = L_1$, in the yield-zone solution be equal to those at $y' = 0$ in the elastic-zone solution.

Noting that the internal moment M and shear force Q are given by the relations

$$EI\frac{d^2u}{dy^2} = M, \qquad EI\frac{d^3u}{dy^3} = -Q \tag{43}$$

we easily find from the yield-zone solution, the moment M^* and the shear force Q^* at the transition level $y = L_1$ to be given by

$$M^* = -\frac{P_1L_1^2}{2} - \frac{P_2L_1^3}{6} + P_0L_1 + M_0 \tag{44}$$

$$Q^* = P_1L_1 + \frac{P_2L_1^2}{2} - P_0 \tag{45}$$

From the elastic-zone solution we then have, on equating the internal moment and shear at $y' = 0$ to the expressions above,

$$C_3 = \frac{\alpha M^* - Q^*}{2\alpha^3} \tag{46}$$

$$C_4 = -\frac{M^*}{2\alpha^2} \tag{47}$$

Next, using equation (28) or (32), we have, with the help of equations (40) and (41),

$$2(P_1 + P_2L_1) = M^*\alpha^2 - Q^*\alpha \tag{48}$$

which, with equations (44) and (45), may be solved by trial and error to find the value of L_1 defining the depth of the yield zone. Clearly, if equation (48) provides a value of L_1 less than or equal to zero, no yield zone exists and the entire pile response is described by the elastic-zone solution, with L_1 set equal to zero.

Assuming a nonzero, positive value of L_1, the remaining two constants C_1 and C_2 can be determined from the condition that the deflection and slope at $y = L_1$ as given by the yield-zone solution are the same as those given by the elastic zone solution at $y' = 0$. We find, in terms of M^* and V^* given by equations (44) and (45)

$$C_1 = \frac{Q^* - 2\alpha M^*}{2\alpha^2} + \frac{P_1 L_1^3}{6} + \frac{P_2 L_1^4}{24} - \frac{P_0 L_1^2}{2} - M_0 L_1 \tag{49}$$

and

$$C_2 = \frac{(1 + 2\alpha L_1)M^*}{2\alpha^2} - \frac{(1 + \alpha L_1)Q^*}{2\alpha^3} - \frac{P_1 L_1^4}{8} - \frac{P_2 L_1^5}{30} + \frac{P_0 L_1^3}{3} + \frac{M_0 L_1^2}{2} \tag{50}$$

This completes the analytical solution of the problem for either the clay on sand case, since all constants in the yield and elastic-zone solutions have now been evaluated.

Estimates of the coefficients N and modulus k appearing in the equations above may be determined by trial-and-error comparisons with field data from instrumented piles where groundline deflections and moment variations along the piles are measured for applied groundline forces and moments. Typical values are shown in Table 5.4 for soft clay in the region of significant pile deflection, as determined from field measurements of Matlock (1970), and for

Table 5.4 TYPICAL SOIL PARAMETERS FOR CYCLIC LATERAL LOADING

Soft Clay	Sand
$N = 3.5$	$N = 1.8$
$k = 3.2$ kips/in^2	$k = 1.2$ kips/in^2

Source: Data from Dawson (1980), as amended.

sand, as determined from field measurements of Reese et al. (1974). Figure 5.10 on page 250 shows the agreement between the measured responses of the test piles and the calculated responses using the values of Table 5.4 and the theory above.

EXAMPLE 5.5-1. Consider the typical pipe pile of an offshore structure shown in Fig. 5.11. The pile has a diameter of 4 ft and a wall thickness of 1 in and is driven to a depth of 250 ft with an 8° batter. Cyclic wave forces on the overhead structure induce estimated cyclic lateral force P_0 at the groundline of 150 kips maximum intensity and estimated cyclic moment M_0 at the groundline of 2000 ft-kips maximum intensity, the moment being counterclockwise when the lateral force is to the right, as shown. The soil consists of clay having shear strength that is negligible at the groundline and that increases linearly with depth at a rate of 8.7 lb/ft^2 per foot for the first 100 ft of depth. We wish to determine the deflection and rotation of the pile at the groundline and the maximum stress in the pile.

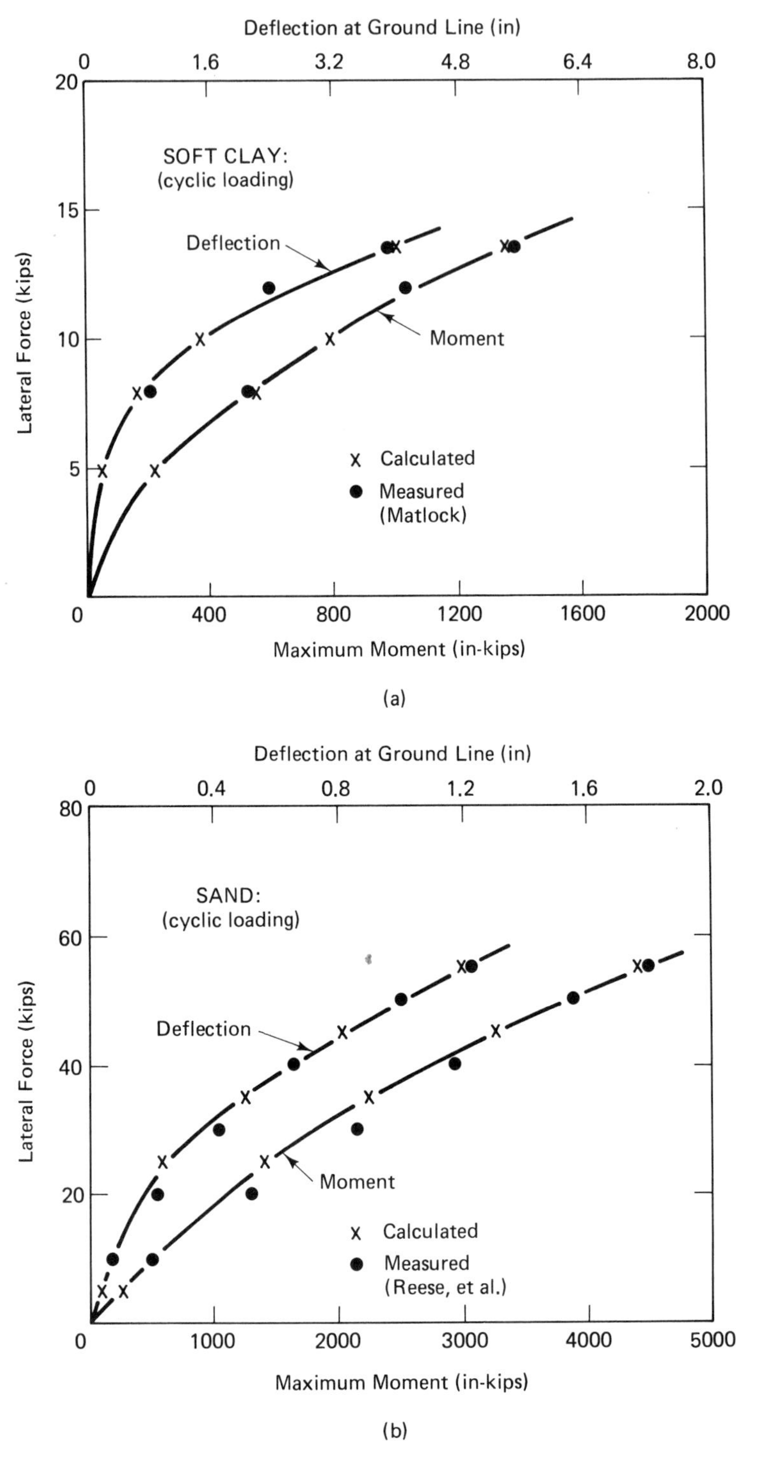

Fig. 5.10. Comparison of field measurements with the predicted response of a pile to lateral loading.

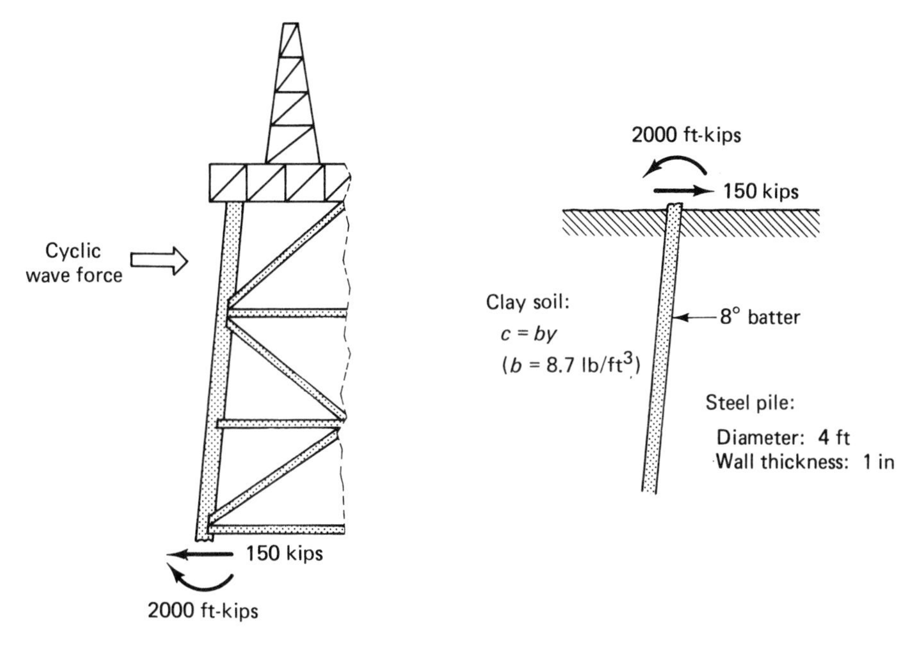

Fig. 5.11

The batter of the pile is sufficiently small (cos 8° = 0.9903) to allow analysis of the pile as a vertical pile. The soil strength C is characterized by equation (26), which here takes the form

$$C = by, \qquad b = 8.7 \text{ lb/ft}^3$$

Assuming the soil parameters listed in Table 5.4 for clay to be applicable, we thus have from equation (35)

$$P_1 = 0, \qquad P_2 = 122 \text{ lb/ft}^2$$

and, with $EI = 8.5 \times 10^9$ lb-ft², we have, from equation (41),

$$\alpha = 0.0858 \text{ ft}^{-1}$$

We first determine the depth L_1 of the yield zone associated with the loading. By trial-and-error solution of equation (48), we find

$$L_1 = 59.0 \text{ ft}$$

The associated moment and shear force at this depth are determined from equations (44) and (45), with $P_0 = 150$ kips, $M_0 = -2000$ ft-kips, as

$$M^* = 2670 \text{ ft-kips}$$
$$Q^* = 62.3 \text{ kips}$$

The constants C_3 and C_4 appearing in the elastic solution are next determined from equations (46) and (47) as

$$C_3 = 1.32 \times 10^8 \text{ lb-ft}^3$$
$$C_4 = -1.81 \times 10^8 \text{ lb-ft}^3$$

From equations (49) and (50), we similarly determine the constants C_1 and C_2 in the yield-zone solution as

$$C_1 = -1.08 \times 10^8 \text{ lb-ft}^2$$
$$C_2 = 5.60 \times 10^9 \text{ lb-ft}^3$$

With these last constants known, we may now determine the deflection u and rotation

$$\theta = -\frac{du}{dy} \quad (\theta \text{ positive clockwise})$$

at the groundline using equation (38) with $y = 0$. We find

$$u = 0.659 \text{ ft}, \qquad \theta = 0.0128 \text{ rad}$$

Finally, the internal moment in the pile may be investigated using the first of equation (43) together with equation (38) for $y < L_1$ and equation (40) for $y > L_1$. By numerical examination, we find the maximum moment to occur within the yield zone at a depth of approximately 40 ft and to be given by

$$M(\max) = 2960 \text{ ft-kips}$$

The maximum bending stress in the pile is thus determined as

$$\sigma = \frac{MR_0}{I} = \frac{(2960)(2)}{1.97} = 3000 \text{ kips/ft}^2 \text{ (20.9 kips/in}^2\text{)}$$

To ensure that the solution above is consistent with the assumption of a deeply driven pile, we use the condition of equation (42) requiring that the depth $L - L_1$ below the transition level L_1 be equal to or greater than the ratio $3/\alpha$. This condition is easily seen to be met in the present problem, since $L - L_1 = 250 - 59.0 = 191$ ft and $3/\alpha = 35$ ft. #

5.6 STIFFNESS PROPERTIES OF EQUIVALENT FREE-STANDING PILES

It was noted earlier in Chapter 4 that the inclusion of pile response in the stress analysis of an overhead structure is conveniently carried out by replacing the actual embedded piles by equivalent free-standing ones, fixed at their base and having stiffness properties at the groundline approximating those of the actual piles. The determination of these equivalent stiffness properties can be made using methods given in the preceding two sections of this chapter.

The problem is illustrated in Fig. 5.12(a), where the actual support piles of the structure have been replaced by equivalent ones, the pile on the left being denoted by end coordinates 1 and 2.

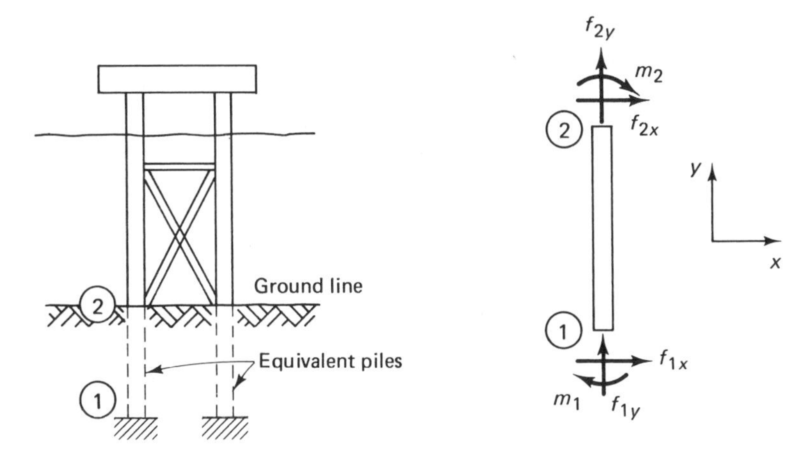

Fig. 5.12. Equivalent piles.

From the results of Chapter 2, we have the stiffness equation of pile 1–2 expressible, in terms of the horizontal and vertical displacements U_2, V_2 and rotation θ_2 at the groundline, as

$$\begin{Bmatrix} f_{1x} \\ f_{1y} \\ m_1 \\ f_{2x} \\ f_{2y} \\ m_2 \end{Bmatrix} = \begin{bmatrix} -k_1 & 0 & k_4 \\ 0 & k_2 & 0 \\ -k_4 & 0 & \frac{1}{2}k_3 \\ k_1 & 0 & -k_4 \\ 0 & k_2 & 0 \\ -k_4 & 0 & k_3 \end{bmatrix} \begin{Bmatrix} U_2 \\ V_2 \\ \theta_2 \end{Bmatrix} \tag{51}$$

where f_{1x}, f_{1y}, m_1 and f_{2x}, f_{2y}, m_2 denote internal forces and moments acting at the ends 1 and 2, as indicated in Fig. 5.12(b), and k_1, k_2, k_3, and k_4 denote stiffness components. The task is to determine these latter components for any given structure, loading, and pile system.

Using equation (51), we may easily write the reduced stiffness equation expressing pile response at the groundline as

$$\begin{Bmatrix} f_{2x} \\ f_{2y} \\ m_2 \end{Bmatrix} = \begin{bmatrix} k_1 & 0 & -k_4 \\ 0 & k_2 & 0 \\ -k_4 & 0 & k_3 \end{bmatrix} \begin{Bmatrix} U_2 \\ V_2 \\ \theta_2 \end{Bmatrix} \tag{52}$$

When the stiffness components are specified, this relation, together with similar ones for the remaining piles, can be incorporated into the stiffness equation of the overhead structure and the structure analyzed following procedures described in Chapter 2 (see Example 2.4-3.).

It is easily seen that the stiffness component k_2 relating axial force and displacement is given directly by equation (22). The determination of the remaining components is, however, not so simple since, unlike the axial response, the lateral response of the pile under design loads is generally nonlinear in nature because of the soil yielding as described in the preceding section. This nonlinearity has the effect of making the components k_1, k_3, and k_4 depend, in general, on the magnitude of the lateral loading and moment at the groundline.

To develop equations for determining these components, we note first from the results of Chapter 2 that they are expressible in terms of the length l and flexural rigidity EI of the equivalent pile by the relations

$$k_1 = \frac{12EI}{l^3}, \qquad k_3 = \frac{4EI}{l}, \qquad k_4 = \frac{6EI}{l^2} \tag{53}$$

so that

$$k_3 = \frac{l^2}{3}k_1, \qquad k_4 = \frac{l}{2}k_1 \tag{54}$$

From equation (52), we thus have

$$f_{2x} = k_1 U_2 - k_1 \frac{l}{2}\theta_2 \tag{55}$$

$$m_2 = -k_1 \frac{l}{2} U_2 + k_1 \frac{l^2}{3}\theta_2 \tag{56}$$

Eliminating k_1 between these equations, we find

$$l^2 - \frac{3}{2}\left(\frac{U_2}{\theta_2} - \frac{m_2}{f_{2x}}\right)l - \frac{3m_2}{f_{2x}}\frac{U_2}{\theta_2} = 0$$

so that a solution for l is expressible as

$$l = \frac{3}{4}\left(\frac{U_2}{\theta_2} - \frac{m_2}{f_{2x}}\right) + \sqrt{\frac{9}{16}\left(\frac{U_2}{\theta_2} - \frac{m_2}{f_{2x}}\right)^2 + 3\frac{m_2}{f_{2x}}\frac{U_2}{\theta_2}} \tag{57}$$

The corresponding solution for k_1 is also determined from equation (55) as

$$k_1 = \frac{f_{2x}}{U_2 - (l/2)\theta_2} \tag{58}$$

Knowing k_1 and l, the remaining components k_3 and k_4 can then be determined from equation (54).

The equations above have been developed using the implicit assumption that the force f_{2x} and moment m_2 at the groundline are known. When this is the case, the deflection U_2 and rotation θ_2 at the groundline can be determined from the pile analysis of Section 5.5 and these then used with the results above to determine directly the stiffness components of the equivalent pile. In the more usual case, however, the internal force and moment at the groundline are initially unknown and must be determined simultaneously with the stiffness components from analysis of both the response of the overhead attached structure and the actual embedded pile.

We illustrate this procedure below, first for the case where f_{2x} and m_2 are specified, and second, for the case where f_{2x} and m_2 are initially unknown.

EXAMPLE 5.6-1. For the pile and loading considered in Example 5.5-1 of the preceding section, determine the lateral stiffness components of an equivalent free-standing pile fixed at its base.

From Example 5.5-1, we have

$$f_{2x} = 150 \text{ kips}, \qquad m_2 = -2000 \text{ ft-kips}$$

and

$$U_2 = 0.659 \text{ ft}, \qquad \theta_2 = 0.0128 \text{ rad}$$

Substituting into equation (57), we find

$$l = 66.1 \text{ ft}$$

and, from equations (54) and (58), we find

$$k_1 = 6.36 \times 10^2 \text{ kips/ft}, \qquad k_3 = 9.26 \times 10^5 \text{ kip-ft},$$
$$k_4 = 2.10 \times 10^4 \text{ kips} \quad \#$$

EXAMPLE 5.6-2. Consider the offshore steel structure and loading of Example 4.2-1 and determine stiffness components for equivalent piles. A side face of the structure is shown in Fig. 5.13 together with joint loading. The actual piles have an outside

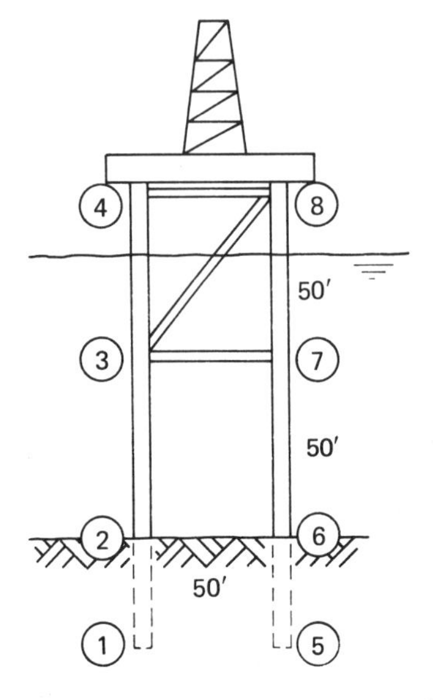

Joint	F_x	F_y	M
2	2.5	−18.4	21.6
3	15.7	−31.9	25.9
4	30.5	−134.6	−34.4
6	3.4	−18.4	29.0
7	12.4	−31.5	31.6
8	28.1	−166.5	−71.4

Forces in units of kips
Moments in units of ft-kips

Fig. 5.13

diameter of 4 ft and a wall thickness of 1 in and are driven to a depth of 200 ft. The soil is assumed to consist of soft clay with cohesion proportional to depth below the groundline, the proportionality factor being 8.7 lb/ft³.

The axial stiffness component of the equivalent pile is determined using methods outlined in Section 5.4. We find (see Example 5.3-1)

$$k_2 = 4.7 \times 10^4 \text{ kips/ft}$$

To determine the lateral stiffness components of the equivalent pile, we use an iterative procedure wherein values of the internal force and moment at the top of each pile are assumed and the stiffness components calculated from the actual pile deflection as in the preceding example. These stiffness components are next incorporated into the overall stiffness matrix of the structure and the deflections calculated at joints 2 and 6 using the joint forces shown in Fig. 5.13. The associated internal force and moment at the top of each pile are then calculated from these deflections, using equation (52). These new values are then used to compute new values of the stiffness components and the process repeated until the calculated force and moment at the top of each pile equal the assumed values used for the stiffness calculations.

The iterative procedure can be started by assuming the total horizontal force acting on the frame of Fig. 5.13 to be divided equally between the two piles. The total horizontal force is seen to equal 92.6 kips so that the assumed internal horizontal force acting at the top of each pile is 46.3 kips.

As a first trial, we accordingly assume that $f_{2x} = f_{6x} = 46.3$ kips and arbitrarily assume moments $m_2 = m_6 = 2000$ ft-kips. The results of the iteration are listed in Table 5.5.

Table 5.5

Trial	Type	f_{2x} (kips)	m_2 (ft-kips)	f_{6x} (kips)	m_6 (ft-kips)
1	Assumed	46.3	2000	46.3	2000
	Calculated	46.6	1128	46.0	1106
2	Assumed	46.6	1128	46.0	1106
	Calculated	46.4	1090	46.2	1090
.	.	.	.	.	.
.	.	.	.	.	.
.	.	.	.	.	.
8	Assumed	46.5	1090	46.1	1080
	Calculated	46.5	1090	46.1	1080

Agreement between the assumed and calculated values of the forces and moments at the top of the piles is seen to occur with the eighth trial calculations. The associated stiffness components are found as

Member 1–2	Member 5–6
$k_1 = 4.80 \times 10^4$ kips/ft	$k_1 = 4.93 \times 10^4$ kips/ft
$k_3 = 2.87 \times 10^6$ ft-kips	$k_3 = 2.91 \times 10^6$ ft-kips
$k_4 = 3.21 \times 10^5$ kips	$k_4 = 3.28 \times 10^5$ kips

#

5.7 FOOTINGS FOR OFFSHORE STRUCTURES

Footings are foundational elements used to distribute the loading from a structure over a sufficient area to prevent failure of the underlying soil. In offshore work, they are used primarily with gravity-type structures which rest directly on the seafloor. They are, however, sometimes used also with template-type structures located on very soft soils for the purpose of providing temporary bearing support of the substructure until piles are driven through its legs to secure it. Examples of these footings are illustrated in Fig. 5.14. As indicated,

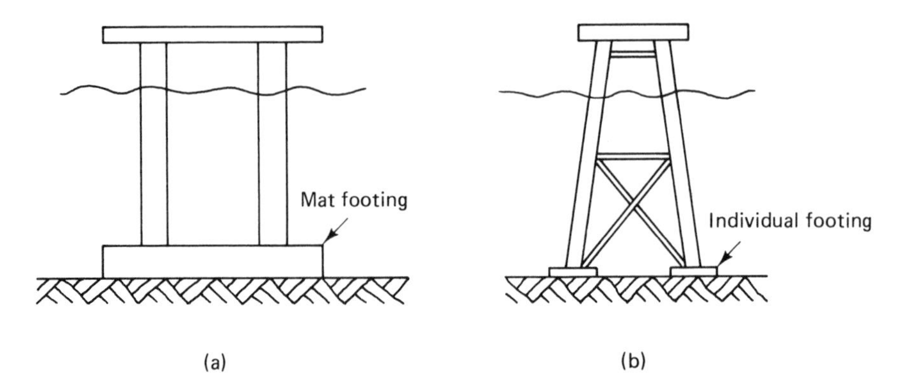

Fig. 5.14. Footings for offshore structures.

when the footing covers the entire area under the structure, it is referred to as a *mat footing*, and when it supports only a single element of the structure, it is referred to as an *individual footing*.

5.8 BEARING CAPACITY OF FOOTINGS

When a footing is centrally and vertically loaded, the soil tends to be pushed out from beneath it in the manner indicated in Fig. 5.15. Failure results when

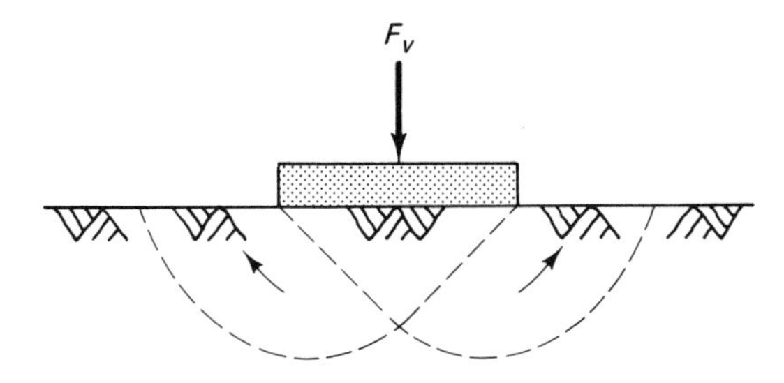

Fig. 5.15. Failure motion of soil beneath an axially loaded footing.

large-scale movement of the soil takes place. This occurs when the shearing stress at points in the soil mass reaches the critical value discussed earlier in Section 5.1. For the general form of the failure relation of that section [equation (1)], the ultimate bearing capacity q_u per unit area of the footing is known from soil mechanics to be expressible approximately as

$$q_u = N_c C + \tfrac{1}{2} N_\gamma \gamma_b B \tag{59}$$

where C denotes the cohesion of the soil, γ_b denotes its submerged specific weight, B denotes a characteristic dimension of the footing, and N_c and N_γ denote coefficients dependent on the angle of friction ϕ of the soil. For a square footing, B is the width of a side; for a rectangular footing, B is the smaller side dimension; and for a circular footing, B is its radius. Typical values of the coefficients N_c and N_γ are listed in Table 5.6.

Table 5.6 VALUES OF BEARING-CAPACITY COEFFICIENTS

ϕ	N_c	N_γ	ϕ	N_c	N_γ
0	5.1	0	25°	20.7	10.9
5°	6.4	0.5	30°	30.1	22.4
10°	8.3	1.2	35°	46.1	48.0
15°	11.0	2.7	40°	75.3	109.4
20°	14.8	5.4	45°	133.9	271.7

As discussed earlier in Section 5.1, for sand soils we have $C = 0$, so that equation (59) reduces to

$$q_u = \tfrac{1}{2}N_\gamma \gamma_b B \tag{60}$$

For clay soils, we take for the initial undrained case, $\phi = 0$, $N_c = 5.1$, $N_\gamma = 0$, and for the long-term drained case, we take $C = 0$, so that equation (59) reduces to

$$\begin{aligned} q_u &= 5.1C \quad \text{(undrained)} \\ q_u &= \tfrac{1}{2}N_\gamma \gamma_b B \quad \text{(drained)} \end{aligned} \tag{61}$$

The undrained case is normally associated with the end-of-construction phase of an offshore structure where inadequate time has existed to permit relief of elevated pore-water pressure by flow of water out of the soil mass. The drained case, on the other hand, is associated with the long-term capacity of the footing after drainage has taken place.

Values of the ultimate bearing capacity determined from equations (60) or (61) are normally reduced by a factor of safety in the range 2.5 to 3 for design purposes.

EXAMPLE 5.8-1. Consider a circular footing of 20-ft radius and determine its ultimate bearing capacity for a normally consolidated clay having undrained shear strength (cohesion) C that varies linearly with depth y below the seafloor according to the relation

$$C = \alpha y, \qquad \alpha = 12 \text{ lb/ft}^3$$

The effective angle of friction ϕ_D of the clay is 20° and its submerged specific weight is 40 lb/ft³.

For the initial undrained case, we estimate the cohesion by taking the value of C at a depth of 10 ft [$B/2$ in equation (59)]. Thus, $C = 120$ lb/ft², and from the first of equations (61), we have

$$q_u = (5.1)(120) = 612 \text{ lb/ft}^2$$

For the long-term drained case, we have from Table 5.6 that $N_\gamma = 5.4$, so that, using the second of equations (61), we find

$$q_u = \tfrac{1}{2}(5.4)(40)(20) = 2160 \text{ lb/ft}^2$$

The ultimate bearing capacity of the soil is thus seen to increase with consolidation and drainage of the soil. The smaller (undrained) value, together with a suitable factor of safety, must, of course, be used in design. #

EXAMPLE 5.8-2. Reconsider Example 5.8-1 for the case where the clay soil is overconsolidated, having an undrained shear strength of 500 lb/ft² at a depth of 10 ft.

For the initial, undrained case, we have

$$q_u = (5.1)(500) = 2550 \text{ lb/ft}^2$$

and for the long-term drained case, we have, as before,

$$q_u = 2160 \text{ lb/ft}^2$$

Thus, for this soil, the long-term capacity of the footing is less than its initial value, and hence must be the controlling capacity in the design. #

5.9 RESISTANCE OF FOOTINGS TO SLIDING

Footings may be subjected to sliding under horizontal loading if the penetration or depth of the foundation is relatively small and if the horizontal loading is sufficiently large. To estimate the lateral load capacity f_u per unit of area of a footing not deeply embedded, we may appeal directly to the Coulomb failure relation of equation (1). Assuming the seafloor to be the plane of shear failure when sliding occurs, we have immediately that

$$f_u = C + \frac{F_v}{A} \tan \phi \qquad (62)$$

where A denotes the area of the footing and F_v denotes the vertical loading. When the soil under the footing is sand, we have $C = 0$ and when it is clay we may consider both the undrained case ($\phi = 0$) and the drained case ($C = 0$)

and choose the smaller as the appropriate value. Again factors of safety of 2.5 to 3 are normally considered for design purposes.

5.10 DESIGN OF FOOTINGS SUBJECTED TO GENERAL LOADING CONDITIONS

When a footing is subjected to a central vertical loading, the pressure exerted on the seafloor by the footing may be assumed to be uniformly distributed over the footing area and the results of Section 5.8 apply. In particular, if this pressure is less than the working value of the bearing capacity per unit of footing area, the footing may be assumed adequate in size; otherwise, not so.

Footings of offshore structures are, however, generally subjected not only to a vertical force from weight loading, but also a horizontal force and moment resulting from wind and wave action on the overhead structure. The horizontal force causes a shearing stress on the soil beneath the footing while the vertical force and moment cause a nonuniform bearing pressure. For a satisfactory footing design, the shearing stress from the horizontal loading must be less than the working value of the sliding resistance of the footing as determined from equation (62). To examine the adequacy of the bearing capacity of the footing under these conditions, we may, as a conservative approximation, imagine the maximum value of the bearing pressure to be uniformly distributed over the base of the footing. For a satisfactory footing design, this value must then be less than the working value of the bearing capacity as determined using equations (60) or (61).

The maximum value of the bearing pressure exerted on the soil beneath the footing may be calculated assuming the pressure to vary linearly over the footing as indicated in Fig. 5.16(a). The balance of vertical force and moment about the left-hand edge of the footing requires that

$$\int_A P\,dA = F_v, \qquad \int_A xP\,dA = \frac{B}{2}F_v + bF_H \tag{63}$$

where A denotes the area of the footing. Taking the case of a rectangular footing of width B and length L, and assuming the pressure exists over the entire area of the footing [Fig. 5.16(b)], we easily find

$$P_1 = \frac{F_v}{BL} - \frac{F_H bB}{2I}, \qquad P_2 = \frac{F_v}{BL} + \frac{F_H bB}{2I} \tag{64}$$

where $I = LB^3/12$ and b denotes the moment arm of the horizontal force.

If the minimum value of the pressure P_1 determined from the equation above is nonnegative, the maximum value of the pressure P_2 may be calculated

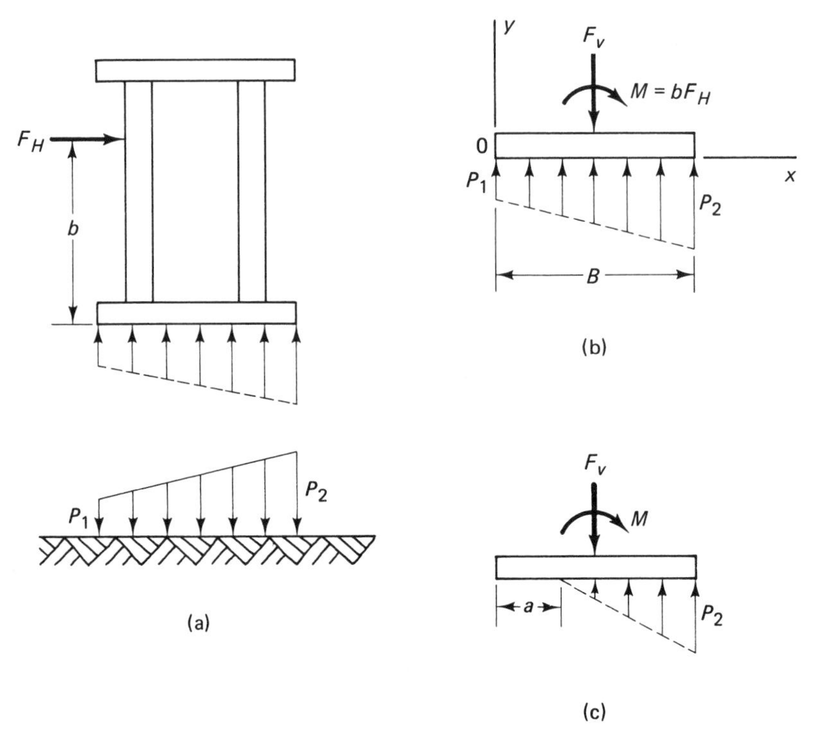

Fig. 5.16. Idealized load distribution under a footing subjected to overturning moment and vertical force.

directly from the second of these equations. For a sufficiently large horizontal force, it can, however, happen that the calculated minimum pressure P_1 will be negative, in which case the solution above has to be discarded since the footing is incapable of exerting a tensile force on the soil. In this case, an alternative solution must be considered based on the linear pressure distribution of Fig. 5.16(c). Using equations (63), we find in this case

$$P_2 = \frac{4}{3L} \frac{F_v^2}{F_v B - 2bF_H} \tag{65}$$

$$a = \frac{3bF_H}{F_v} - \frac{B}{2} \tag{66}$$

It is interesting to note that the condition for impending overturn of the structure occurs when the soil reaction is concentrated at the right-hand edge of the footing. Putting $a = B$ in equation (66), we thus find $F_H = F_v B/2b$ as the horizontal force needed to cause impending overturn. From equation (65), we also see that the pressure P_2 is infinite for this case. Thus, if P_2 is limited by bearing-capacity considerations of the footing, there is no need to consider separately the possibility of overturning the structure from wind and wave loadings.

Equations (64)–(66) have been developed for a rectangular or square footing. Similar, but more complicated relations, can be worked out for a circular footing. For an approximate analysis, however, the circular footing may simply be replaced by an equivalent square footing of equal area and the equations employed above. For more complex footing geometries, the maximum pressure may similarly be determined by approximating the actual geometry by a number of rectangular sections and using equation (63) with an assumed linear distribution of the pressure (or a suitable approximation of it) in a manner analogous to that employed above. In certain problems, it is possible to retain the geometry of the footing and approximate only the linear pressure distribution (Example 5.10-2). An alternative procedure for calculating the capacity of footings of various shapes and loadings is to incorporate directly into equation (59) empirically determined shape and loading coefficients (API, 1980).

EXAMPLE 5.10-1. Consider the offshore gravity structure shown in Fig. 5.17 and determine if the mat footing is adequate for the indicated loadings and soil conditions. The total submerged weight of the structure, including the base, is 5000 kips.

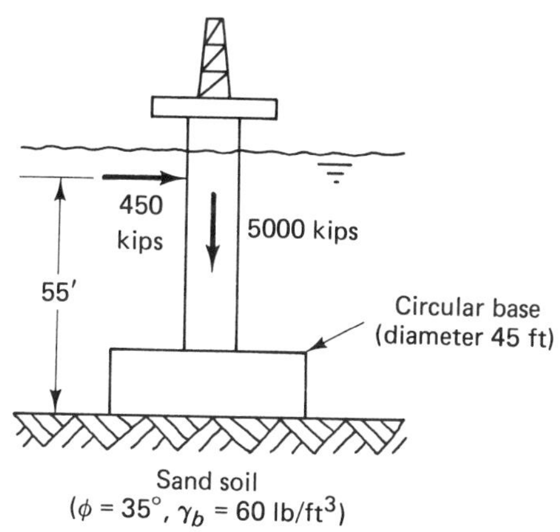

Fig. 5.17

We use the first of equations (64) to find the minimum pressure exerted on the soil. The area of the circular base is 1590 ft², corresponding to an equivalent square base of sides equal approximately to 40 ft. With $I = (40)^4/12 = 2.13 \times 10^5$, we have

$$P_1 = \frac{5000}{(40)^2} - \frac{(450)(55)(40)}{2(2.13 \times 10^5)} = 0.81 \text{ kip/ft}^2$$

Since this is positive, we accordingly may determine the maximum pressure P_2 from the second of equations (64) as

$$P_2 = 5.45 \text{ kips/ft}^2$$

The bearing capacity for the sand soil is given by equation (60). With $\phi = 35°$, we find from Table 5.6 that $N_\gamma = 48$. Hence,

$$q_u = \tfrac{1}{2}(48)(60)(22.5) = 32.4 \text{ kips/ft}^2$$

Using a factor of safety of 2.5, we have a working value of the footing capacity of $32.4/2.5 = 13.0$ kips/ft². Since this is considerably greater than the maximum pressure exerted on the soil, we thus conclude that the footing dimensions are more than adequate insofar as bearing is concerned.

To examine the adequacy of the footing to prevent sliding, we use equation (62). We have

$$f_u = \frac{5000}{1590} \tan 30° = 1.82 \text{ kips/ft}^2$$

The applied lateral force per unit of area is $450/1590 = 0.28$ kip/ft², so that the footing is adequate to prevent sliding.

EXAMPLE 5.10-2. Reconsider Example 5.10-1 for the case where the footing has the shape shown in Fig. 5.18.

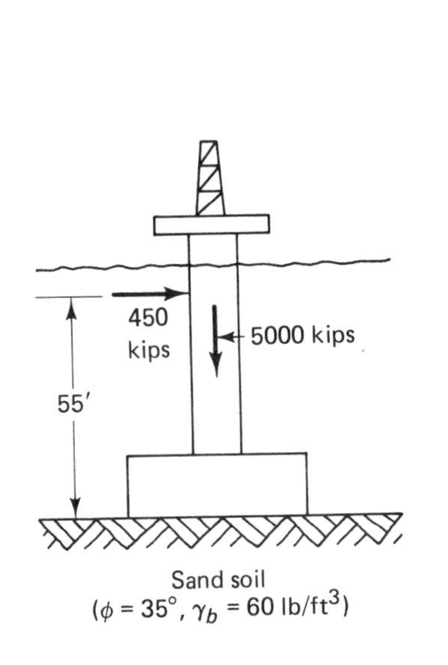

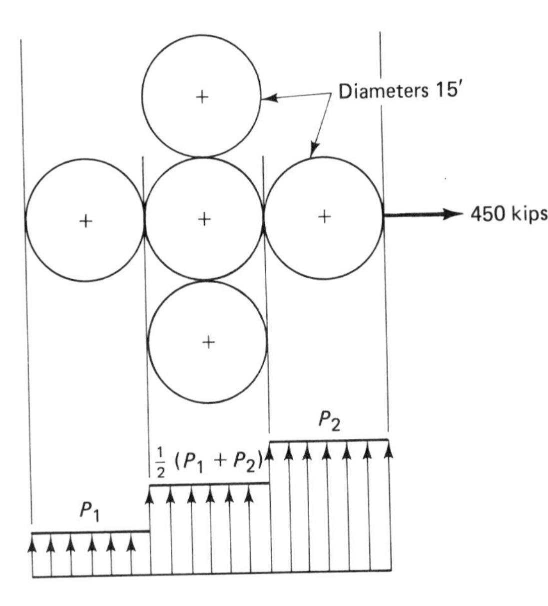

Fig. 5.18

As an approximation, the linear variation of pressure over the footing is represented by the distribution shown in Fig. 5.18.

If A denotes the base area of each of the cylindrical elements, the balance of forces in the vertical direction gives

$$P_2 A + \frac{P_1 + P_2}{2} 3A + P_1 A = F_v$$

where F_v denotes the applied vertical loading. Similarly, with R denoting the radius of each cylindrical element, the balance of moments about the left-hand edge of the footing gives

$$6RP_2 A + 4R\left(\frac{P_1 + P_2}{2}\right) 3A + RP_1 A = M$$

where M denotes the applied moment about the left-hand edge. If F_H denotes the applied horizontal force and b its moment arm from the base of the structure, we have

$$M = bF_H + 3RF_v$$

Combining the three equations above, we easily find the relations

$$P_1 = \frac{2F_v}{10A} - \frac{F_H b}{4RA}$$

$$P_2 = \frac{2F_v}{10A} + \frac{F_H b}{4RA}$$

Substituting $F_v = 5000$ kips, $F_H = 450$ kips, $b = 55$ ft, $R = 7.5$ ft, $A = 177$ ft^2, we find

$$P_1 = 0.99 \text{ kip/ft}^2, \qquad P_2 = 10.3 \text{ kips/ft}^2$$

Since P_1 is positive, the solution for P_2 is valid. Considering the outer element where this pressure acts, we find from equation (60) that the ultimate bearing capacity is

$$q_u = \tfrac{1}{2}(48)(60)(7.5) = 10.8 \text{ kips/ft}^2$$

This is only about 5% greater than the maximum bearing pressure and we must thus conclude that the footing design is unsatisfactory since no margin is available for a suitable factor of safety. One possible remedy is to attach a circular slab of 45 ft diameter to the base of the cylindrical foundation elements, in which case the analysis of Example 5.10-1 applies. #

5.11 ELASTIC RESPONSE OF FOOTINGS

When a properly sized footing is subjected to design-level loads, the soil beneath the footing responds in an elastic manner. To include this response into the analysis of the overhead structure, we may consider an equivalent footing fixed at its base and having stiffness properties at its upper surface approximating those of the actual footing (Fig. 5.19).

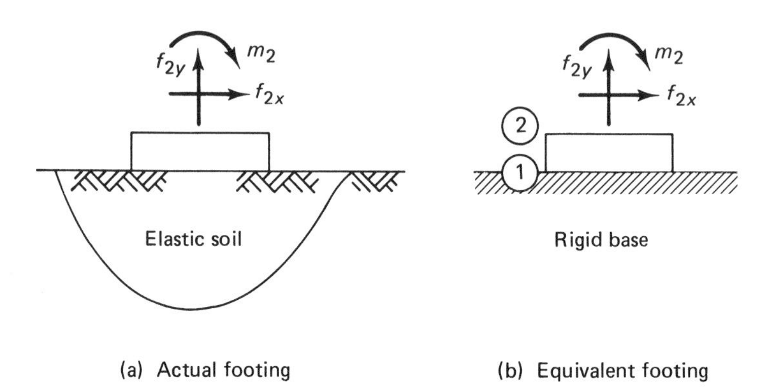

Fig. 5.19. Equivalent footing for use in estimating stiffness properties: (a) actual footing; (b) equivalent footing.

The stiffness equation for the equivalent footing is customarily assumed expressible as

$$\begin{Bmatrix} f_{2x} \\ f_{2y} \\ m_2 \end{Bmatrix} = \begin{bmatrix} k_1 & 0 & 0 \\ 0 & k_2 & 0 \\ 0 & 0 & k_3 \end{bmatrix} \begin{Bmatrix} U_2 \\ V_2 \\ \theta_2 \end{Bmatrix} \tag{67}$$

where f_{2x}, f_{2y}, and m_2 denote the horizontal and vertical force and the moment at the top of the footing and U_2, V_2, and θ_2 denote the corresponding displacements. For a rigid circular footing resting on a homogeneous elastic soil, we have (see Funston and Hall, 1967)

$$k_1 = \frac{32(1-\nu)\mu R}{7-8\nu}, \qquad k_2 = \frac{4\mu R}{1-\nu}, \qquad k_3 = \frac{8\mu R^3}{3(1-\nu)} \tag{68}$$

where R denotes the radius of the footing and μ and ν denote the elastic shear modulus and Poisson's ratio of the soil. For clay soils, Poisson's ratio may be assumed equal to 0.5, and for sand soils, it may be assumed to be about 0.25. Values of the shear modulus for a soil deposit must be estimated from laboratory tests on samples taken at the site. Since the shear modulus can generally be expected to vary with depth, the values used should be an appropriate average value over depths equal to two or three times the diameter of the footing.

Although strictly applicable only to a circular footing, the stiffness relations of equation (68) may also be employed in an approximate manner for square footings, with the area of the equivalent circular footing chosen equal to the area of the square footing.

EXAMPLE 5.11-1. For the offshore structure shown in Fig. 5.20, determine the deck deflection for a concentrated horizontal deck loading of 100 kips.

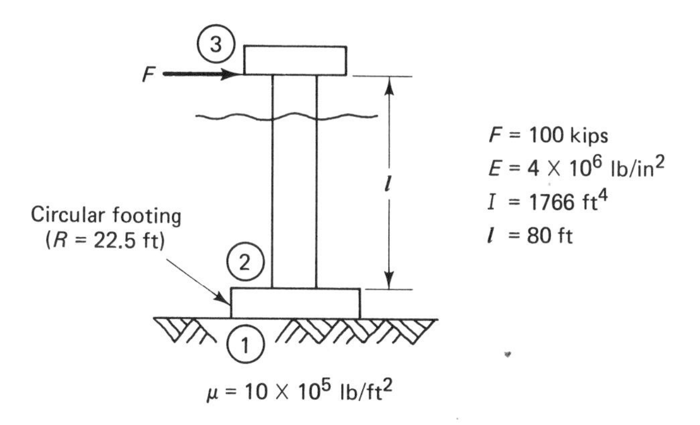

Fig. 5.20

Assuming vertical displacements zero, the reduced stiffness equation for determining the deflections at 2 and 3 is expressible using equation (67) and the direct stiffness method of Chapter 2 as (see Example 2.4-3)

$$\begin{Bmatrix} F_{2x} \\ M_2 \\ F_{3x} \\ M_3 \end{Bmatrix} = \begin{bmatrix} (a+k_1) & b & -a & b \\ b & (2c+k_3) & -b & c \\ -a & -b & a & -b \\ b & c & -b & 2c \end{bmatrix} \begin{Bmatrix} U_2 \\ \theta_2 \\ U_3 \\ \theta_3 \end{Bmatrix}$$

where F_{2x}, M_2, F_{3x}, and M_3 denote horizontal forces and moments at joints 2 and 3; U_2, θ_2, U_3, and θ_3 denote corresponding displacements and rotations; a, b, c and d are defined as

$$a = \frac{12EI}{l^3}, \qquad b = \frac{6EI}{l^2}, \qquad c = \frac{2EI}{l}$$

and k_1 and k_3 are as defined by equation (68). The boundary conditions are

$$F_{2x} = M_2 = M_3 = 0, \qquad F_{3x} = 100 \text{ kips}$$

Substituting numerical values and solving the equation above, we find

$$U_2 = 8.39 \times 10^{-4} \text{ ft}, \qquad \theta_2 = 1.33 \times 10^{-4} \text{ rad}$$

$$U_3 = 2.82 \times 10^{-2} \text{ ft}, \qquad \theta_3 = 4.47 \times 10^{-4} \text{ rad}$$

The deck deflection is accordingly equal to 2.82×10^{-2} ft or 0.338 in. This may be contrasted with the value 0.201 found by assuming both a rigid footing and rigid soil. #

EXAMPLE 5.11-2. For the structure and footing shown in Fig. 5.21(a), determine the stiffness matrix governing movement of the base of the column.

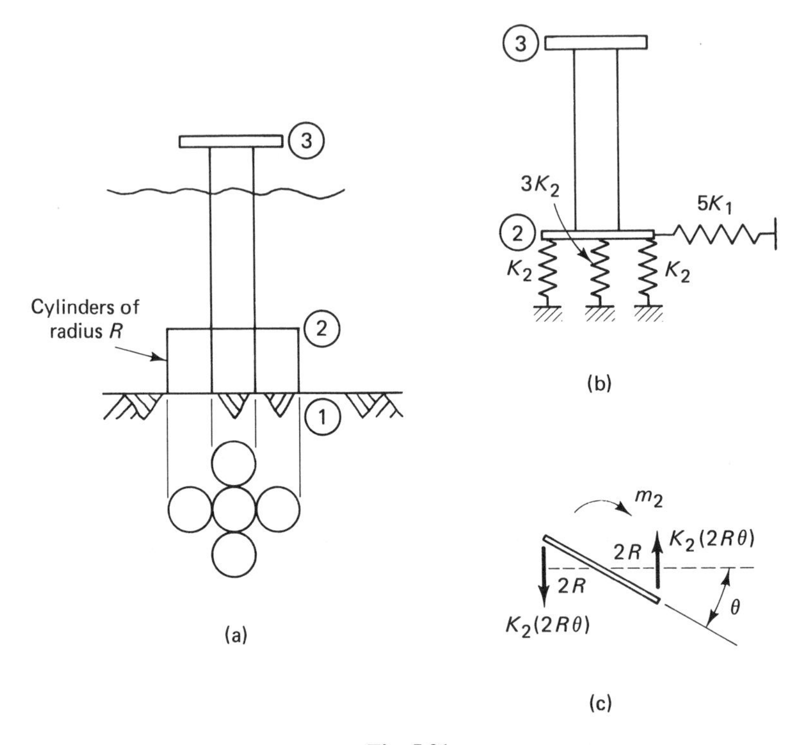

Fig. 5.21

We assume the response of the base elements to be equivalent to that of elastic springs of stiffnesses K_1 and K_2, as indicated in Fig. 5.21(b), where the K's are defined by equation (68). The connection between the horizontal force f_{2x} and displacement U_2 at joint 2 is thus

$$f_{2x} = 5K_1U_1$$

The connection between vertical force f_{2y} and displacement V_2 is similarly

$$f_{2y} = 5K_2V_2$$

Finally, for a moment m_2 applied at the center of the footing, we have from equilibrium considerations applied to the free body of Fig. 5.21(c),

$$m_2 = 4RK_2\theta_2$$

Hence, the matrix stiffness equation governing the movement of the base of the column is

$$\begin{Bmatrix} f_{2x} \\ f_{2y} \\ m_2 \end{Bmatrix} = \begin{bmatrix} 5K_1 & 0 & 0 \\ 0 & 5K_2 & 0 \\ 0 & 0 & 4RK_2 \end{bmatrix} \begin{Bmatrix} U_2 \\ V_2 \\ \theta_2 \end{Bmatrix}$$

As in the example above, this may be used with the stiffness equation for the column to determine the overall structural response. #

5.12 SETTLEMENT OF FOUNDATIONS

Foundation footings for offshore structures are subject to settlement as a result of compression of the soil by the dead-weight loading of the structure. Uniform settlement of a foundation generally causes no adverse effects on the footing or overhead structure but unequal settlement, called *differential settlement*, can lead to unacceptable stresses in the footing or structure. The amount of differential settlement that can be tolerated by a structure and footing vary, of course, with the particular design. As a general rule, however, differential settlements greater than about 1 in are usually considered excessive.

If a circular or square footing is located on a relatively deep layer of sand (depth equal to at least three or four times the diameter or width of the footing), the maximum settlement will generally be small and can be estimated using the elastic solution given in the preceding section. Thus, if ΔH denotes the settlement, we have from equation (67), with $\nu = 0.25$,

$$\Delta H = \frac{W}{k_2} = \frac{3W}{16\mu R} \tag{69}$$

where R denotes the radius of a circular footing or one-half the width of a square footing, W denotes the vertical loading, and μ the elastic shear modulus of the sand. The value given by this equation represents the maximum uniform settlement that can be expected. If the foundation consists of several footings, each carrying different loads, the differential settlement can be estimated as the difference in the values given by equation (69) when applied to each footing. For a mat footing, the maximum differential settlement can similarly be estimated by assuming the value given by equation (69) to represent the average settlement, with the maximum at the center of the footing being twice that at the edges. In this case, the settlement at the center will be a third greater than the average and that at the edges will be a third less than it. The differential settlement is then two-thirds of the average value.

For a footing on a relatively deep layer of clay, the settlement calculation is more complicated since it must include the initial type of elastic deformation and also the long-term deformation associated with compression and drainage of the clay. From the discussion of Section 5.1, it will be recalled that sand drains very quickly under compression loading, so that no such effect need be considered for settlement of footings on sand soils. Because of their low permeability, clay soils drain slowly and such effects therefore have a significant role in the settlement of footings on clay soils.

The settlement of a footing on a clay soil is usually estimated using compression stress–strain data taken from a sample of the sediment. The compression test (known as a consolidation test) is carried out in the laboratory, with full drainage of the sample allowed at each increment of the applied stress.

Figure 5.22 shows results from such compression tests for samples taken

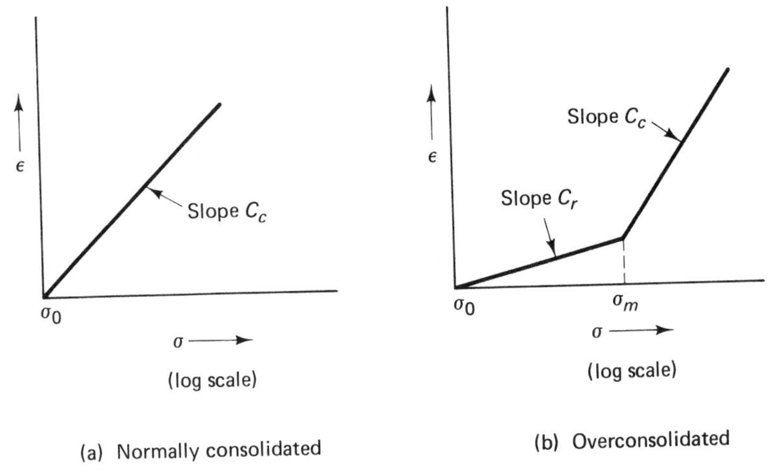

Fig. 5.22. Stress-strain compression curves for (a) normally consolidated and (b) overconsolidated clays.

from normally consolidated and overconsolidated clay deposits (see Section 5.1). As is customary, the strain (change in length per unit of length) is plotted against the logarithm (base 10) of the stress, resulting in linear or bilinear curves. In these graphs, σ_0 denotes the stress to which the sample had been subjected in the soil deposit by the weight of the overhead soil (submerged specific weight of the soil times the distance below the seafloor where the sample was taken). For the normally consolidated sediment, this stress is the greatest stress ever experienced by the sample while in the soil deposit. For the overconsolidated case, however, the sample had, at some time in its history, been subjected in the soil deposit to a stress σ_m greater than σ_0, and this is revealed in the compression test by the change in slope of the stress–strain data at stress σ_m.

Since the soil samples are assumed representative of the soil behavior at any depth, the strain ϵ at any depth associated with a stress σ (equal to σ_0 at that depth plus an applied stress $\Delta\sigma$) is determined for a normally consolidated clay simply as

$$\epsilon = C_c(\log \sigma - \log \sigma_0) \tag{70}$$

where C_c is the slope of the linear curve shown in Fig. 5.22(a). Similarly, for an overconsolidated clay, we have for $\sigma \leq \sigma_m$,

$$\epsilon = C_r(\log \sigma - \log \sigma_0) \tag{71a}$$

and for $\sigma > \sigma_m$,

$$\epsilon = C_r(\log \sigma_m - \log \sigma_0) + C_c(\log \sigma - \log \sigma_m) \tag{71b}$$

where the slopes C_r and C_c are as defined in Fig. 5.22(b). Note that the stress σ_m can vary with depth and this variation must therefore be determined from compression tests on samples taken at various distances below the seafloor.

To calculate the applied stress $\Delta\sigma$ resulting from a vertical loading W applied to a circular or square footing of diameter or width D, it is customary to use the results from a theoretical solution (the Boussinesq solution) that treats the soil as a homogeneous elastic solid. Figure 5.23 on page 272 shows the stress so calculated at the center and edges of circular and square footings as a function of depth below the seafloor. In this figure the stress $\Delta\sigma$ is expressed per unit of the uniform stress $q = W/A$ acting at the seafloor, where A denotes the area of the footing. The depth y is also expressed per unit of footing diameter or width.

Using the theory above, the maximum settlement at the center of a footing, as well as that at its edges, can be calculated. For several individual footings, the differential settlement can be estimated as the difference in the average values of each. For a mat footing, the differential settlement is estimated simply as the distance between the settlements at the center and edges.

EXAMPLE 5.12-1. Estimate the settlement of the circular foundation footing shown in Fig. 5.24. The underlying soil is sand with a shear modulus as indicated. The weight of the structure and footing, allowing for buoyancy, is 5000 kips.

Since the soil under the foundation is sand, we may use equation (69), which gives for the average settlement

$$\Delta H = \frac{(3)(5000)}{(16)(10 \times 10^2 \times 20)} = 0.047 \text{ ft} = 0.56 \text{ in}$$

The differential settlement is estimated as being $(0.67)(0.56) = 0.37$ in. #

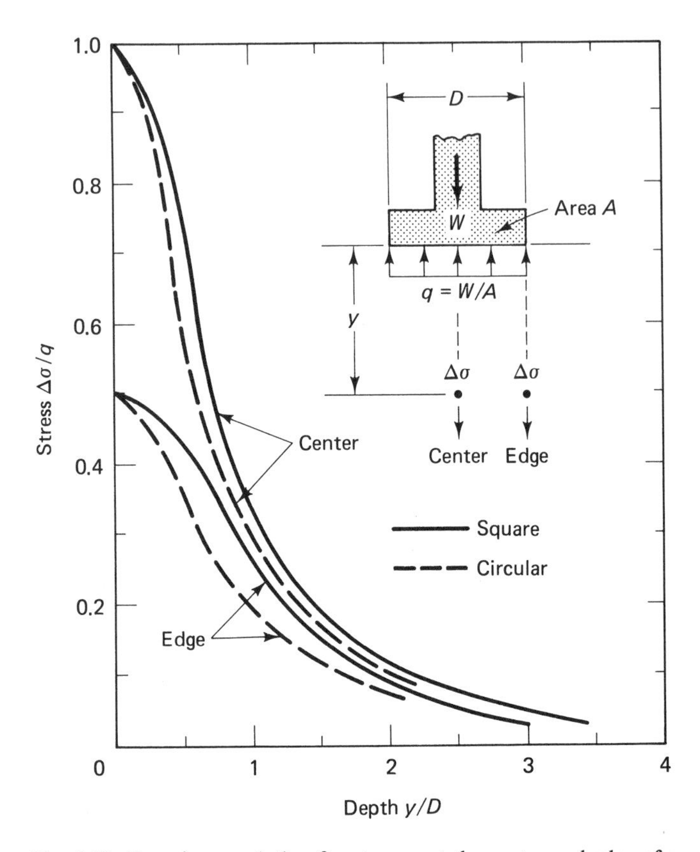

Fig. 5.23. Boussinesq solution for stresses at the center and edge of a circular or square footing.

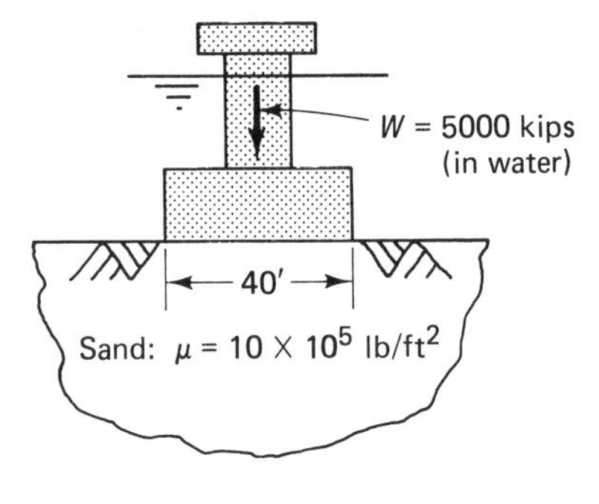

Fig. 5.24

EXAMPLE 5.12-2. Estimate the maximum differential settlement of the foundation of Fig. 5.25. The underlying sediment is an overconsolidated clay having compression properties $C_r = 0.02$, $C_c = 0.18$, and σ_m values as indicated. The submerged

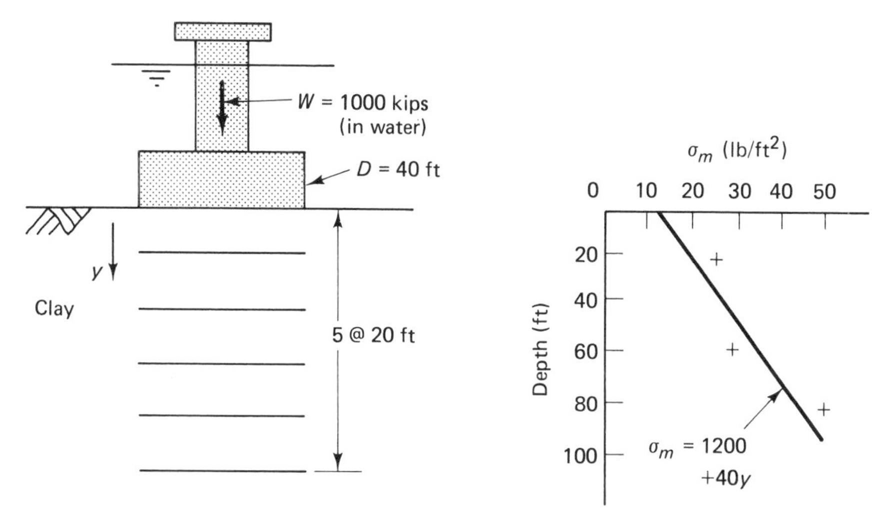

Fig. 5.25

specific weight of the clay is 40 lb/ft². The net weight of the structure and footing, allowing for buoyancy, is 1000 kips.

The settlement is calculated using equation (71). The soil mass is arbitrarily divided into five layers of 20-ft thickness for calculation purposes. The stress $\sigma_0 = \gamma_s y$ at the middepth of each layer is calculated and the additional applied stress $\Delta\sigma$ at each middepth is determined for the center and edge of the footing from the graph of Fig. 5.23, with $q = W/A = 1 \times 10^6/1257 = 796$ lb/ft². The corresponding strains are determined from equation (71). The results are tabulated in Table 5.7.

Table 5.7

y (ft)	σ_0 (lb/ft²)	$\Delta\sigma$ (lb/ft²)	σ (lb/ft²)	σ_m (lb/ft²)	ϵ
		Center of Footing			
10	400	716	1116	1600	8.91×10^{-3}
30	1200	334	1534	2400	2.13×10^{-3}
50	2000	178	2178	3200	7.41×10^{-4}
70	2800	103	2903	4000	3.14×10^{-4}
90	3600	64	3664	4800	1.53×10^{-4}
		Edge of Footing			
10	400	358	758	1600	6.52×10^{-3}
30	1200	207	1407	2400	1.38×10^{-3}
50	2000	119	2119	3200	5.02×10^{-4}
70	2800	80	2880	4000	2.45×10^{-4}
90	3600	48	3648	4800	1.15×10^{-4}

The settlement ΔH is calculated from the strains as

$$\Delta H = \int_0^\infty \epsilon \, dy = H \sum \epsilon_i$$

where H denotes the thickness of each layer and ϵ_i denotes the strain in the ith layer. The sum of the strains at the center and edge of the footing are determined from the results of Table 5.7 as 0.0122 and 0.00876, respectively. The settlement at the center of the footing is thus (0.0122)(20) = 0.245 ft and that at its edge is (0.00876)(20) = 0.175 ft. The maximum differential settlement is accordingly 0.245 − 0.175 = 0.07 ft or about 0.8 in. #

REFERENCES

API (1980). *Recommended Practice for Planning, Designing and Constructing Fixed Offshore Platforms*, American Petroleum Institute Publication RP-2A, Dallas, Tex.

Cox, W. R., L. M. Kraff, and E. A. Verner (1979). Axial Load Tests on 14-Inch Pipe Piles in Clay, *Proceedings, Eleventh Annual Offshore Technology Conference*, pp. 1147–1151.

Coyle, H. M., and L. C. Reese (1966). Load Transfer for Axially Loaded Piles in Clay, *Journal of the Soil Mechanics and Foundation Division*, ASCE, Vol. 92, pp. 1–26.

Dawson, T. H. (1980). Simplified Analysis of Offshore Piles under Cyclic Lateral Loads, *Ocean Engineering*, Vol. 7, pp. 553–562.

Funston, N. E., and W. J. Hall (1967). Footing Vibration with Non-Linear Subgrade Support, *Journal of the Soil Mechanics and Foundation Division*, ASCE, Vol. 93, pp. 191–211.

Matlock, H. (1970). Correlations for Design of Laterally Loaded Piles in Soft Clay, *Proceedings, Second Annual Offshore Technology Conference*, pp. 577–587.

McClelland, B., J. A. Focht, Jr., and W. J. Emrich (1967). Problems in Design and Installation of Heavily Loaded Pipe Piles, *Proceedings, Conference Civil Engineering in the Oceans*, ASCE, pp. 601–634.

Reese, L. C., and W. R. Cox (1976). Pullout Tests of Piles in Sand, *Proceedings*, Eighth Annual Offshore Technology Conference, pp. 527–338.

Reese, L. C., W. R. Cox, and F. D. Koop (1974). Analysis of Laterally Loaded Piles in Sand, *Proceedings, Sixth Annual Offshore Technology Conference*, pp. 473–483.

Reese, L. C., W. R. Cox, and F. D. Koop (1975). Field Testing and Analysis of Laterally Loaded Piles in Stiff Clay, *Proceedings, Seventh Annual Offshore Technology Conference*, pp. 671–675.

Terzaghi, K., and R. B. Peck (1967). *Soil Mechanics in Engineering Practice*, John Wiley & Sons, Inc., New York.

PROBLEMS

1. A clay soil deposit has undrained shear strength (cohesion) C varying with depth y as indicated in Fig. P.1. Determine the depth L that a steel pile of 6-ft outside diameter and 2-in wall thickness needs to be driven to carry an

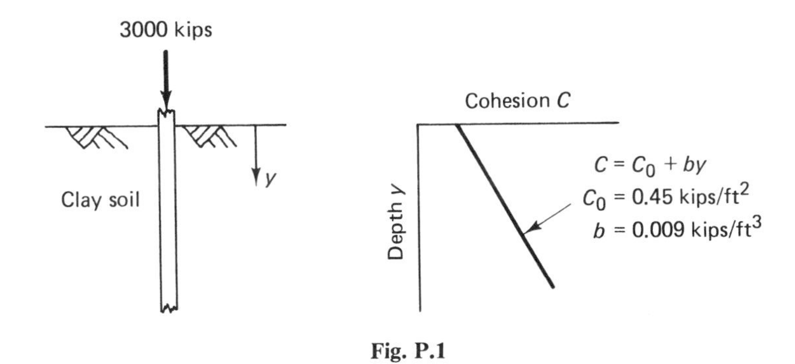

Fig. P.1

axial compressive load of 3000 kips. Assume a factor of safety of 1.5. Assume that the submerged weight of soil is 0.035 kip/ft³ and that of steel is 0.420 kip/ft³.
Ans. $L = 274$ ft.

2. A sand soil deposit has specific submerged weight of 50 lb/ft³. Determine the depth L that a 4-ft-diameter steel pile of wall thickness 0.75 in must be driven to support an applied compressive load of 2000 kips. Use the data of Table 5.2. Assume that the submerged weight of steel is 420 lb/ft³ and use a factor of safety of 1.5.
Ans. $L = 86$ ft.

3. The side frame shown in Fig. P.3 consists of two main legs and two skirt piles. The spacing between the piles is 50 ft. For the loading shown, estimate the vertical force in each pile, assuming a linear variation across the base.
Ans. $P_1 = -1150$ kips, $P_2 = 450$ kips, $P_3 = 2050$ kips, $P_4 = 3650$ kips.

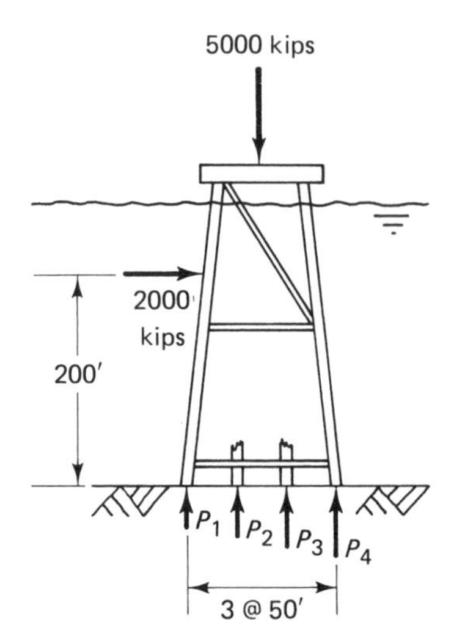

Fig. P.3

4. Solve Problem 3, assuming that the skirt piles are not present.
Ans. $P_1 = -167$ kips, $P_4 = 5167$ kips.

5. Determine the effective axial stiffness of a 6-ft-diameter steel pipe pile of wall thickness 1 in when driven 250 ft in soft clay.
Ans. $K_0 = 5.65 \times 10^7$ lb/ft.

6. Determine the yield depth L_1, the deflection U, and the rotation θ at the groundline for the steel pile and loading of Fig. P.6.
Ans. $L_1 = 12.8$ ft, $U = 17.12 \times 10^{-3}$ ft, $\theta = 9.19 \times 10^{-4}$ rad.

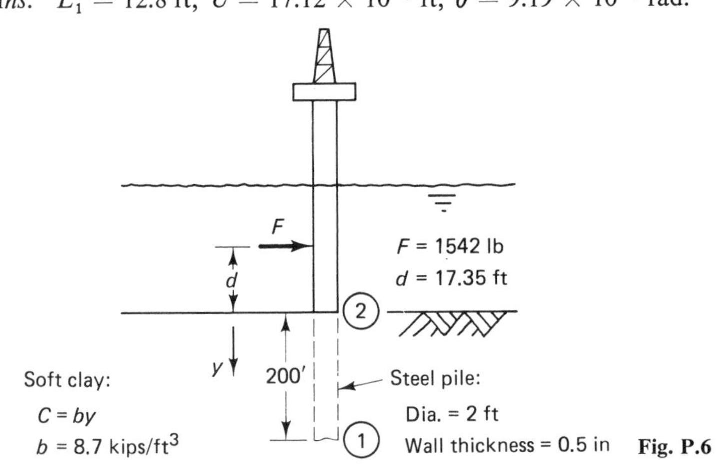

Fig. P.6

7. For the pile and loading of Problem 6, determine the maximum moment in the pile, the depth that it occurs, and the associated bending stress.
Ans. $M = 34{,}100$ ft-lb, $y = 7.0$ ft, $\sigma_b = 1920$ lb/in^2.

8. For the pile and loading of Problem 6, determine stiffness properties of an equivalent free-standing pile fixed at its base.

Ans. $k_1 = 2.47 \times 10^6$ lb/ft, $k_2 = 2.36 \times 10^7$ lb/ft, $k_3 = 1.64 \times 10^8$ lb-ft, $k_4 = 1.74 \times 10^7$ lb.

9. The simple steel frame shown in Fig. P.9 consists of two vertical piles of 4-ft outside diameter and 1-in wall thickness driven 200 ft into the seafloor and rising 50 ft above it to a horizontal bracing of 2-ft outside diameter and 0.5-in wall thickness. For the horizontal loadings shown, determine the stiffness properties of equivalent free-standing piles 1–2 and 4–5.

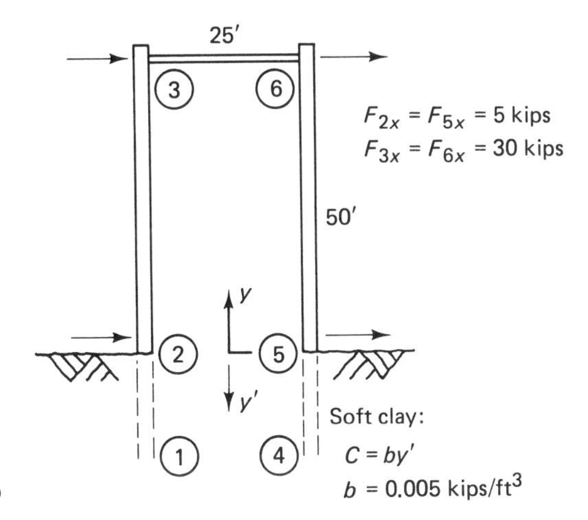

Fig. P.9

10. Assume a clay soil having $C = 1$ kip/ft², $\phi_D = 20°$, $\gamma_b = 40$ lb/ft³ and determine initial and long-term bearing capacities of the foundation footing of Fig. P.10. Assume a factor of safety of 2.5.

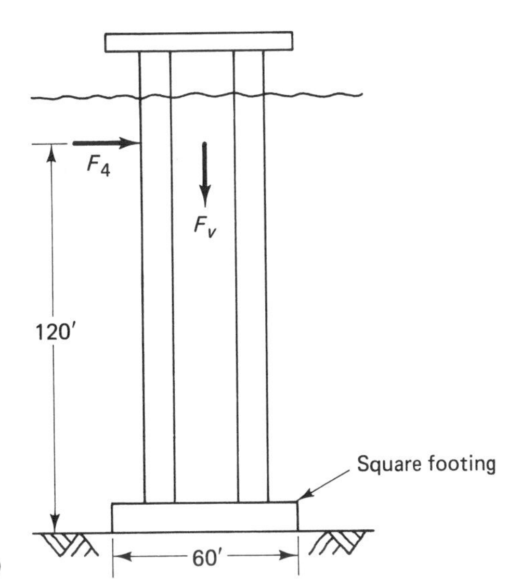

Fig. P.10

11. Determine the capacity of the footing of Problem 10 if the soil is sand with $\phi = 30°$, $\gamma_b = 60$ lb/ft³. Assume a factor of safety of 2.5.
Ans. 16.1 kips/ft².

12. For the capacity of Problem 11, determine maximum and minimum values of F_v for the structure of Problem 10, assuming $F_H = 500$ kips.
Ans. 52,000 kips, 2100 kips.

13. Assume the soil data of Example 5.12-2 and estimate the maximum differential settlement of a circular foundation of 60 ft diameter, carrying a net vertical force of 2250 kips.
Ans. 1.25 in.

Exxon's HONDO platform off California in the Santa Barbara Channel in 850 feet of water. Because of the depth of water and the resulting tall, slender structure, dynamic effects from wave action had to be considered in its design. Dynamic loadings from earthquakes also had to be considered because of the location of the structure in an active seismic area. (Courtesy of McDermott Incorporated.)

6

DYNAMICS OF STRUCTURES

WHEN AN OFFSHORE STRUCTURE is subjected to periodic wave forces, it experiences corresponding periodic displacements of the same frequency as that of the wave loading. The associated back-and-forth accelerations of the structural mass induce, in turn, dynamic inertia forces on the structure. The effect of these forces is generally to cause an increase in the displacements over those caused by the wave forces. A simplified analysis including these forces was given in Chapter 4. If the natural response frequency of the structure is considerably greater than that of the wave loading, this analysis showed that the dynamic amplification of the deflections will be small. Such is often the case for relatively short, stiff offshore structures placed in water depths of, say, 300 ft or less, and is the basis for using the static method of structural analysis described in Chapter 4.

In the case of taller structures, or in the case of structures having considerable flexibility because of their particular form, it can, however, happen that the natural frequency of the structure will be close to that of the wave loading, thus resulting in appreciable dynamic amplification. In such circumstances, a structure designed on the basis of static considerations must be checked for possible overstress from dynamic effects. The present chapter describes basic procedures involved in such an analysis. These same procedures are also appropriate for analysis when earthquake-induced loadings must be considered for structures located in areas of seismic activity. This topic is also discussed briefly in the present chapter.

6.1 LUMPED DESCRIPTION OF WAVE FORCES

In describing the static method of analysis of offshore structures in Chapter 4, we considered wave-induced joint loadings only for the instant when the total horizontal wave force on the structure was at its maximum value. This is normally sufficient when dynamic inertial loadings are negligible because the overall structural response will then vary directly with the total wave force acting on it. Its maximum response will therefore occur at about the same instant as that when the total horizontal wave force is maximum. When dynamic effects are important, this is, however, no longer the case, since the structural mass cannot respond instantly to the applied force. For this reason, a dynamic analysis of the structure must consider joint loadings not only for the instant of maximum wave force but also for other times in the wave cycle.

Because of the complexity of the analysis in this case, a simplified treatment of the wave forces, more approximate than that used in a static analysis, is customarily employed. This treatment involves lumping the areas and volumes of the individual structural members into areas and volumes at the joints of the structure and calculating the horizontal wave forces acting on these assumed concentrated bodies. Also, to keep the analysis manageable, attention is generally restricted to linear (Airy) wave theory for determining the water motion to be used in the wave-force calculations.

Consider, for instance, the structure shown below in Fig. 6.1(a). To calculate the joint forces on this structure arising from the wave loading, we suppose the presented areas and volumes of the individual cylindrical members to be lumped into concentrated joint areas A_1, A_2, etc., and concentrated joint volumes B_1, B_2, etc., as shown in Fig. 6.1(b). The presented areas and volumes of the individual members are conveniently calculated by considering the actual outside diameters of the members and their projected lengths normal to the direction of wave propagation, with only those members, or member segments, that are beneath the still-water level being considered. For example, at joint 2, the concentrated area A_2 includes one-half the presented areas of members 1–2, 2–3, 2–5, and 2–7 of the side face shown, together with similar contributions from members on the front face that frame into the joint. Analogous

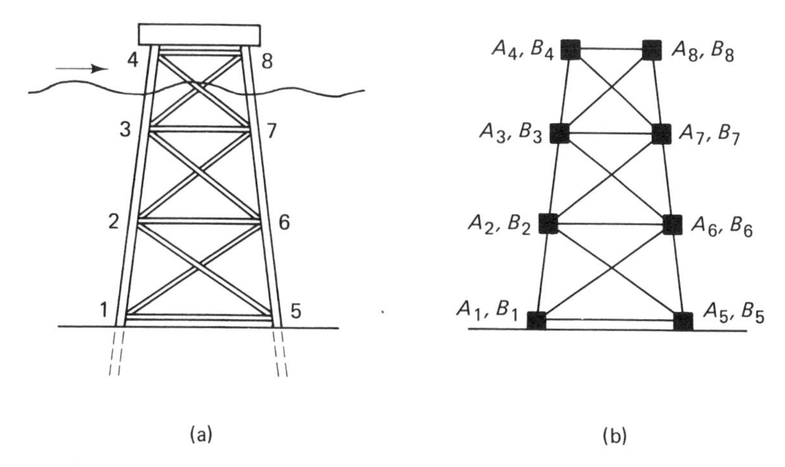

Fig. 6.1. Lumped area and volume model of an offshore structure for wave force calculations.

lumping for the volumes of the members (based on actual outside diameters and projected lengths) framing into the joint yields the concentrated volume B_2. Notice that the horizontal member 2–6 contributes nothing to the concentrated areas and volumes since its projected length is essentially zero. This is consistent with the more exact wave-force considerations described earlier.

Having lumped the presented areas and volumes of the members in the foregoing manner, the wave force F_p acting on the pth joint can then be calculated from a Morison-type equation (including the effects of relative motion) of the form

$$F_p = \tfrac{1}{2}\rho C_D A_p |u'_p| u'_p + \rho C_I B_p a_{px} - \rho(C_I - 1)B_p \ddot{U}_p \tag{1}$$

where u'_p denotes the horizontal water velocity relative to the joint, expressible in terms of the horizontal water velocity u_p and the horizontal joint velocity $\dot{U}_p$ as

$$u'_p = u_p - \dot{U}_p \tag{2}$$

and where a_{px} denotes the horizontal acceleration of the water at the joint, and $\ddot{U}_p$ denotes the acceleration of the joint. Here, also, ρ denotes the water density and C_D and C_I denote drag and inertia coefficients as discussed earlier.

Assuming further that the joint velocity is small in comparison with the water velocity there, we have, approximately,

$$|u'_p| u'_p = |u_p| u_p - 2|u_p| \dot{U}_p$$

If, as an additional approximation, we replace $|u_p|$ by an appropriate average value $\hat{u}_p$, independent of time, we may write the force equation above in terms of the actual water velocity u_p as

$$F_p = \tfrac{1}{2}\rho C_D A_p \hat{u}_p u_p + \rho C_I B_p a_{px} - \rho C_D A_p \hat{u}_p \dot{U}_p - \rho(C_I - 1)B_p \ddot{U}_p \tag{3}$$

Using Airy wave theory and assuming a wave of height H, frequency ω, and wavenumber k in water of depth h, we have the horizontal velocity and acceleration at joint p expressible as

$$u_p = E_p \cos(kx_p - \omega t) \tag{4}$$

$$a_{px} = \omega E_p \sin(kx_p - \omega t) \tag{5}$$

where E_p is given by

$$E_p = \frac{\omega H}{2} \frac{\cosh ky_p}{\sinh kh} \tag{6}$$

and x_p and y_p denote x and y coordinates of joint p, with the origin of the coordinates taken at the seafloor. Note that for joints with $y > h$, we may use $y = h$ in the formula above to account approximately for the wave forces acting on the submerged parts of the members lumped at these joints. Substituting these equations into the force formula above, we may write the resulting expression as

$$F_p = F_{0p} \sin(kx_p - \omega t + \phi_p) - \rho C_D A_p \hat{u}_p \dot{U}_p - \rho(C_I - 1)B_p \ddot{U}_p \tag{7}$$

where F_{0p} and ϕ_p are defined by the equations

$$F_{0p} = E_p[(\tfrac{1}{2}\rho C_D A_p \hat{u}_p)^2 + (\rho C_I B_p \omega)^2]^{1/2} \tag{8}$$

$$\tan \phi_p = \frac{C_D A_p \hat{u}_p}{2C_I B_p \omega}, \qquad 0 \leq \phi_p \leq \frac{\pi}{2} \tag{9}$$

It remains to determine an appropriate value for the term $\hat{u}_p$. For regular Airy waves, we accordingly choose $\hat{u}_p$ such that the difference between $|u_p|u_p$ and $\hat{u}_p u_p$ is minimized in a least-square integral sense, that is, such that the integral

$$I = \int_0^{2\pi} (|u_p|u_p - \hat{u}_p u_p)^2 \, d\omega t \tag{10}$$

is a minimum. Using equation (4), we find

$$\hat{u}_p = \frac{8}{3\pi} E_p = 0.849 E_p \tag{11}$$

EXAMPLE 6.1-1. Consider the offshore structure shown in Fig. 6.2 and determine the expression for the force at joint 2, assuming a wave of height 20 ft and length 300 ft. The water depth is 75 ft, as indicated. The vertical members have an outside diameter

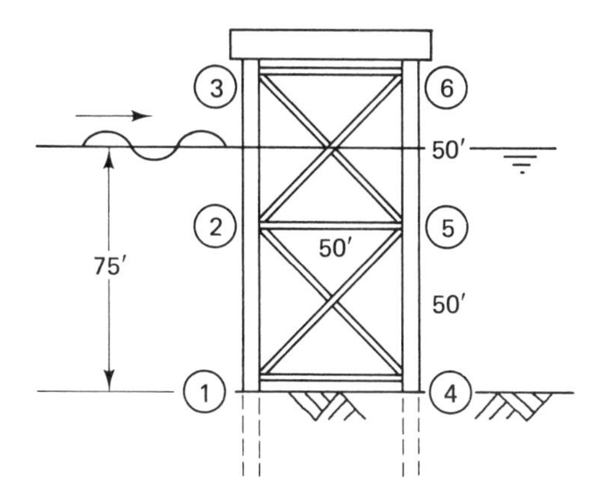

Fig. 6.2

of 4 ft and the horizontal and diagonal members have an outside diameter of 2 ft. All four sides of the structure are identical.

The lumped area A_2 is calculated as follows: For members 1–2 and 2–3, we have the following areas presented to the waves (neglecting that part of member 2–3 above the still-water level):

$$A_{1\text{-}2} = 4(50) = 200 \text{ ft}^2, \qquad A_{2\text{-}3} = (4)(25) = 100 \text{ ft}^2$$

For the diagonal member 2–4, we have similarly, treating it as an equivalent vertical member of height 50 ft.

$$A_{2\text{-}4} = (50)(2) = 100 \text{ ft}^2$$

For the horizontal member 2–5, we have $A_{2\text{-}5} = 0$, since its projected length onto the wave direction is essentially zero. For the diagonal member 2–6, treating it as an equivalent vertical member of height 25 ft, we have

$$A_{2\text{-}6} = (25)(2) = 50 \text{ ft}^2$$

For the lower diagonal of the front face, say 2–7, framing into joint 2, we have

$$A_{2\text{-}7} = (70.7)(2) = 141 \text{ ft}^2$$

For the horizontal member on the front face, say 2–8, framing into joint 2, we have

$$A_{2\text{-}8} = (50)(2) = 100 \text{ ft}^2$$

Finally, for the upper diagonal member on the front face, say 2–9, framing into joint 2, we have

$$A_{2\text{-}9} = (35.4)(2) = 70.8 \text{ ft}^2$$

Taking one-half of each of the areas calculated above and adding them together, we thus obtain the lumped area A_2 as

$$A_2 = 331 \text{ ft}^2$$

The lumped volume B_2 is similarly determined by calculating effective volumes of each of the members framing into joint 2. We have

$$B_{1\text{-}2} = \frac{(50)(\pi)(4^2)}{4} = 628 \text{ ft}^3$$

$$B_{2\text{-}3} = \frac{(25)(\pi)(4^2)}{4} = 314 \text{ ft}^2$$

$$B_{2\text{-}4} = \frac{(50)(\pi)(2^2)}{4} = 157 \text{ ft}^3$$

$$B_{2\text{-}5} = 0$$

$$B_{2\text{-}6} = \frac{(25)(\pi)(2^2)}{4} = 78.5 \text{ ft}^3$$

$$B_{2\text{-}7} = \frac{(70.7)(\pi)(2^2)}{4} = 222 \text{ ft}^3$$

$$B_{2\text{-}8} = \frac{(50)(\pi)(2^2)}{4} = 157 \text{ ft}^3$$

$$B_{2\text{-}9} = \frac{(35.4)(\pi)(2^2)}{4} = 111 \text{ ft}^3$$

Taking one-half of each of the above and adding, we thus find

$$B_2 = 834 \text{ ft}^3$$

For an Airy wave of height 20 ft and length 300 ft in 75 ft of water, we may determine the wavenumber k and frequency ω as

$$k = 0.0209 \text{ ft}^{-1}, \qquad \omega = 0.786 \text{ sec}^{-1}$$

The force F_2 is then determined from equation (7). Choosing $y_2 = 50$ ft, we have from equation (6) the water-velocity amplitude E_2 given as

$$E_2 = 5.48 \text{ ft/sec}$$

From equation (11) we also have

$$\hat{u}_2 = 4.65 \text{ ft/sec}$$

The amplitude F_{02} of the wave force acting at joint 2 and the phase angle ϕ_2 are determined from equations (8) and (9) as (choosing $\rho = 1.99$ slugs/ft^3,

$C_D = 1$ and $C_I = 2$)

$$F_{02} = 3030 \text{ lb}$$
$$\phi_2 = 0.531 \text{ rad} = 30.4°$$

The force F_2 (in pounds) is thus expressible, on choosing the origin of x such that $x_2 = 0$, as

$$F_2 = 3030 \sin(-\omega t + 0.531) - 3060\dot{U}_2 - 1660\ddot{U}_2$$

where $\dot{U}$ and $\ddot{U}$ are expressed, respectively, in units of ft/sec and ft/sec², and ω equals 0.786 sec^{-1}, as determined above. #

6.2 GOVERNING DYNAMIC EQUATIONS

To illustrate the development of dynamic equations governing the motion of an offshore structure, we consider the side frame of a simple structure as shown in Fig. 6.3. The wave force acting at each joint of the frame is assumed to be

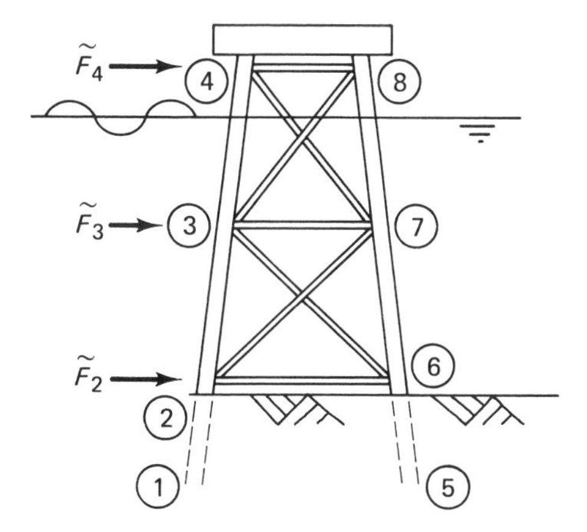

Fig. 6.3. Side frame of offshore structure having wave forces acting at its movable levels.

horizontal and described by equation (7), taking into account contributions from front face members. Under these loads, the frame can then be expected to experience primarily horizontal motions. We accordingly restrict attention to the movable levels 2–6, 3–7, and 4–8 and note that, because of the relatively stiff horizontal members present, each level will move horizontally as a nearly rigid section, with horizontal joint displacements at a level being equal, or nearly so.

Now, for horizontal motion of a rigid body, it is the total horizontal force acting on the body that is important rather than its distribution. Thus, under the restriction of rigid movement of each level, we need consider only the sum of the wave forces acting on all joints at that level. We denote the total wave force acting at level p–q by $\tilde{F}_p$, where, for example, $\tilde{F}_2$ denotes the sum of the forces acting on joints 2 and 6. From equation (7), we thus have (with $U_p = U_q$)

$$\tilde{F}_p = F_{0p} \sin (kx_p - \omega t + \phi_p) + F_{0q} \sin (kx_p + kL_p - \omega t + \phi_q) \\ - \rho C_D \hat{u}_p \tilde{A}_p \dot{U}_p - \rho (C_I - 1) \tilde{B}_p \ddot{U}_p \qquad (12)$$

where $L_p = x_q - x_p$ denotes the horizontal distance between the joints p and q and

$$\tilde{A}_p = A_p + A_q, \qquad \tilde{B}_p = B_p + B_q \qquad (13)$$

With the help of appropriate trigonometric identities, equation (12) can be written more simply as

$$\tilde{F}_p = F^*_{0p} \sin (kx_p - \omega t + \phi^*_p) - \rho C_D \hat{u}_p \tilde{A}_p \dot{U}_p - \rho (C_I - 1) \tilde{B}_p \ddot{U}_p \qquad (14)$$

where ϕ^*_p and F^*_{0p} are defined by

$$\tan \phi^*_p = \frac{F_{0p} \sin \phi_p + F_{0q} \sin (\phi_q + kL_p)}{F_{0p} \cos \phi_p + F_{0q} \cos (\phi_q + kL_p)} \qquad (15)$$

and

$$F^*_{0p} = \frac{F_{0p} \cos \phi_p + F_{0q} \cos (\phi_q + kL_p)}{\cos \phi^*_p} \qquad (16)$$

In addition to the wave force acting at any level, there will also exist an inertia force associated with the acceleration of the mass at that level. To calculate this force, we suppose the mass of the individual members framing into each joint of the side face (including contributions from the front faces) to be lumped into concentrated masses at the joints, with the lumped masses at joints 4 and 8 also including contributions from the deck mass. The total mass $\tilde{M}_p$ at level p–q then consists of the sum of the lumped masses at joints p and q and the inertia force at that level is given by $-\tilde{M}_p \ddot{U}_p$, where overhead dots denote time derivatives.

Finally, to include approximately the effect of damping forces arising from internal friction in the structure, we may assume ficticious resistive forces at level p–q of the form $-\tilde{C}_p \dot{U}_p$, where $\tilde{C}_p$ denotes a constant damping coefficient.

Collecting together the results above, we have the total force $\tilde{F}^T_p$ acting at level p–q described by the equation

$$\tilde{F}^T_p = \tilde{F}_p - \tilde{M}_p \ddot{U}_p - \tilde{C}_p \dot{U}_p \qquad (17)$$

Assuming level 1–5 to be fixed against displacement, the total forces acting at the other three levels may thus be written in matrix form as

$$\{\tilde{F}^T\} = \{\tilde{F}\} - [\tilde{M}]\{\ddot{U}\} - [\tilde{C}]\{\dot{U}\} \tag{18}$$

where

$$\{\tilde{F}^T\} = \begin{Bmatrix} \tilde{F}_2^T \\ \tilde{F}_3^T \\ \tilde{F}_4^T \end{Bmatrix}, \qquad \{\tilde{F}\} = \begin{Bmatrix} \tilde{F}_2 \\ \tilde{F}_3 \\ \tilde{F}_4 \end{Bmatrix}, \qquad \{U\} = \begin{Bmatrix} U_2 \\ U_3 \\ U_4 \end{Bmatrix}$$

$$[\tilde{M}] = \begin{bmatrix} \tilde{M}_2 & 0 & 0 \\ 0 & \tilde{M}_3 & 0 \\ 0 & 0 & \tilde{M}_4 \end{bmatrix}, \qquad [\tilde{C}] = \begin{bmatrix} \tilde{C}_2 & 0 & 0 \\ 0 & \tilde{C}_3 & 0 \\ 0 & 0 & \tilde{C}_4 \end{bmatrix}$$

and $\{\dot{U}\}$, $\{\ddot{U}\}$ are determined from $\{U\}$ by taking first and second derivatives, respectively, of the individual components.

Having the forces above, we may next relate these to the horizontal displacements U_2, U_3, U_4 at levels 2–6, 3–7, and 4–8 using the methods of Chapter 2. As in the static analysis of Chapter 4, the actual embedded piles are replaced with equivalent free-standing ones, fixed at their base ends and having stiffness properties at the groundline approximating those of the actual piles.

In particular, if we determine the reduced matrix equation connecting forces and displacements at joints 2, 3, 4 and 6, 7, 8 of the side frame shown in Fig. 6.3, we may determine the average displacements U_2, U_3, U_4 associated with net horizontal forces $F_2 + F_6$, $F_3 + F_7$, and $F_4 + F_8$ applied to the side frame and write the result as

$$\{U\} = [K]^{-1}\{\tilde{F}^T\} \tag{19}$$

Here, the first column of matrix $[K]^{-1}$ is determined by setting $F_2 = F_6 = \frac{1}{2}$, so that $F_2^T = 1$, and setting all other forces and moments at the joints equal to zero. The solution for the (average) displacements at the three levels 2–6, 3–7, and 4–8 will then yield the values for the first column of $[K]^{-1}$. Similar treatments will yield values for the remaining two columns. Taking the inverse of the matrix $[K]^{-1}$ so determined, we thus obtain the stiffness matrix relating horizontal forces with displacements for the side face, that is,

$$\{\tilde{F}^T\} = [K]\{U\} \tag{20}$$

Combining equations (18) and (20) and using equation (14) with $C_D = 1$, $C_I = 2$, we thus find the equations of motion of the structure expressible as

$$[M^*]\{\ddot{U}\} + [C^*]\{\dot{U}\} + [K]\{U\} = \{F^*\} \tag{21}$$

where $[M^*]$, $[C^*]$, and $\{F^*\}$ are defined as

$$[M^*] = \begin{bmatrix} \tilde{M}_2 + \rho\tilde{B}_2 & 0 & 0 \\ 0 & \tilde{M}_3 + \rho\hat{B}_3 & 0 \\ 0 & 0 & \tilde{M}_4 + \rho\tilde{B}_4 \end{bmatrix} \tag{22}$$

$$[C^*] = \begin{bmatrix} \tilde{C}_2 + \rho\tilde{A}_2\hat{u}_2 & 0 & 0 \\ 0 & \tilde{C}_3 + \rho\tilde{A}_3\hat{u}_3 & 0 \\ 0 & 0 & \tilde{C}_4 + \rho\tilde{A}_4\hat{u}_4 \end{bmatrix} \tag{23}$$

$$\{F^*\} = \begin{Bmatrix} F_{02}^* \sin(kx_2 - \omega t + \phi_2^*) \\ F_{03}^* \sin(kx_3 - \omega t + \phi_3^*) \\ F_{04}^* \sin(kx_4 - \omega t + \phi_4^*) \end{Bmatrix} \tag{24}$$

Since dots over the variables U_2, U_3, U_4 in the equation above imply time derivatives, the matrix equation (21) can be seen to represent three scalar differential equations whose solution yield the horizontal displacements at the levels 2–6, 3–7, and 4–8 of the structure shown in Fig. 6.3. Observing that the stiffness matrix $[K]$ will generally involve nonzero off-diagonal terms, we see that the equations are *coupled* to one another in the sense that the differential equation for, say, U_2 will have terms containing U_3 and U_4 as well as U_2.

We also note that the individual damping constants $\tilde{C}_2$, $\tilde{C}_3$, and $\tilde{C}_4$ appearing in the matrix $[C^*]$ are generally not known and that, to avoid having to specify each constant, it is customary to assume *Rayleigh damping*, where the constants are such that $[C^*]$ is proportional to $[M^*]$ or $[K]$ or a linear combination of these. In the present formulation where $[C^*]$ is diagonal, we accordingly assume $\tilde{C}_2$, $\tilde{C}_3$, and $\tilde{C}_4$ to be such that the diagonal terms of $[C^*]$ are proportional to the corresponding diagonal terms of $[M^*]$. Mathematically, we assume that

$$[C^*] = 2\zeta[M^*] \tag{25}$$

where ζ denotes a single damping constant to be assigned.

EXAMPLE 6.2-1. Consider the steel structure shown in Fig. 6.4 and determine the wave force matrix $\{F^*\}$ for levels 2–6, 3–7, and 4–8, assuming a wave of height 40 ft and length 600 ft. All four faces of the structure are identical. The members have outside diameter (OD) and inside diameters (ID) as indicated:

Member	OD (ft)	ID (ft)
Verticals	4	3.75
Horizontals	2	1.92
Diagonals	2	1.92

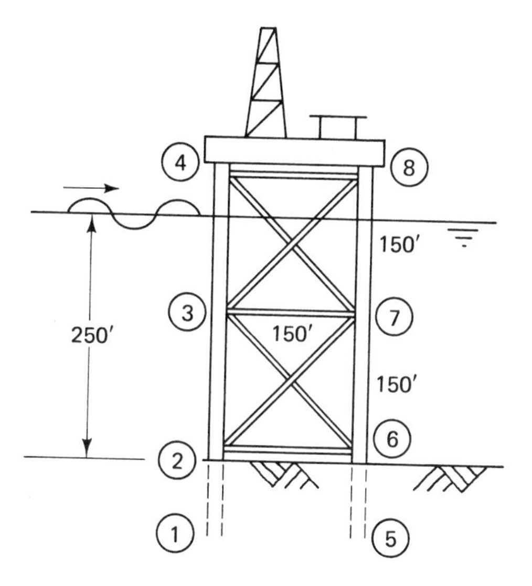

Fig. 6.4

The amplitude F_{0p} and phase angle ϕ_p of the wave force acting at each joint are calculated according to procedures outlined in Section 6.1. The results are tabulated in Table 6.1.

Table 6.1

Joints	A_p	B_p	y_p	E_p	u_p	F_{0p}	ϕ_p
2,6	812	1750	0	1.70	1.44	7,124	0.281
3,7	1253	2759	150	4.26	3.62	33,180	0.618
4,8	441	1009	250	11.68	9.82	57,620	1.081
units: ft, sec, lb, rad							

The components of $\{F^*\}$ are given by the first part of equation (14),

$$F_p^* = F_{0p}^* \sin(kx_p - \omega t + \phi_p^*)$$

with ϕ_p^* and F_{0p}^* given by equations (15) and (16). Choosing the origin of x at the leg 2–3–4, we have $x_2 = x_3 = x_4 = 0$ and, from the results of Table 6.1, we find

$$\{F^*\} = 10^4 \begin{Bmatrix} 1.01 \sin(-\omega t + 1.07) \\ 4.69 \sin(-\omega t + 1.40) \\ -8.15 \sin(-\omega t - 1.28) \end{Bmatrix} \quad \#$$

EXAMPLE 6.2-2. Reconsider the steel structure of the preceding example and determine the mass matrix $[M^*]$ associated with the dynamic response of levels 2–6, 3–7, and 4–8. Assume a deck weight of 1500 kips.

The nonzero diagonal components M_1^*, M_2^*, M_3^* of the mass matrix are expressible in terms of the lumped masses $\tilde{M}_2$, $\tilde{M}_3$, $\tilde{M}_4$ and volumes $\tilde{B}_2, \tilde{B}_3, \tilde{B}_4$ at the three levels as (equation 22)

$$M_2^* = \tilde{M}_2 + \rho\tilde{B}_2, \qquad M_3^* = \tilde{M}_3 + \rho\tilde{B}_3, \qquad M_4^* = \tilde{M}_4 + \rho\tilde{B}_4$$

where ρ denotes the water density.

To calculate the lumped masses, we first determine the masses of the vertical, horizontal, and diagonal members. The members are steel with a mass of 15.0 slugs/ft³. In addition, the lower members are assumed fully flooded and the upper members are assumed flooded to the still-water level. Table 6.2 summarizes

Table 6.2

Member	Position	Mass (slugs)
Vertical	Lower	6230
	Upper	5630
Horizontal	Lower	1440
	Upper	578
Diagonal	Lower	2040
	Upper	1630

the masses of the individual members. To determine the mass $\tilde{M}_2$, we take one-half the sum of the masses of all members framing into joints 2 and 6. For the side-face members framing into joint 2, we have

$$\tfrac{1}{2}(6230) + \tfrac{1}{2}(1440) + \tfrac{1}{2}(2040) = 4855 \text{ slugs}$$

For the front-face members, we also have

$$\tfrac{1}{2}(1440) + \tfrac{1}{2}(2040) = 1740 \text{ slugs}$$

Thus, the total lumped mass at joint 2 (neglecting any contribution from the imbedded pile) is 6595 slugs. This is the same for joint 6, so that

$$\tilde{M}_2 = 2(6595) = 13{,}190 \text{ slugs}$$

Similar calculations for joints 3 and 7 give

$$\tilde{M}_3 = 2(11{,}040) = 22{,}080 \text{ slugs}$$

Finally, considering joints 4 and 8, we find the total lumped mass from members framing into these joints to equal 10,046 slugs. To this must be added one-half the total deck mass, that is,

$$\frac{1}{2}\left(\frac{1.5 \times 10^6}{32.2}\right) = 23{,}292 \text{ slugs}$$

so that

$$\tilde{M}_4 = 10{,}046 + 23{,}292 = 33{,}338 \text{ slugs}$$

Next we calculate the lumped volumes at each level of the structure. These are equal to the sum of the lumped volumes at each joint of a level. From the results of Example 6.2-1, we thus have

$$\tilde{B}_2 = B_2 + B_6 = 3500 \text{ ft}^3$$
$$\tilde{B}_3 = B_3 + B_7 = 5518 \text{ ft}^3$$
$$\tilde{B}_4 = B_4 + B_8 = 2018 \text{ ft}^3$$

Combining the results above, we have

$$M_2^* = 13{,}190 + (1.99)(3500) = 20{,}155 \text{ slugs}$$
$$M_3^* = 22{,}080 + (1.99)(5518) = 33{,}061 \text{ slugs}$$
$$M_4^* = 33{,}338 + (1.99)(2018) = 37{,}354 \text{ slugs}$$

so that the mass matrix is expressible as

$$[M^*] = 10^4 \begin{bmatrix} 2.016 & 0 & 0 \\ 0 & 3.306 & 0 \\ 0 & 0 & 3.735 \end{bmatrix} \qquad \#$$

EXAMPLE 6.2-3. A static analysis of the structure of Example 6.2-1 gives, for unit applied forces $F_2^T = F_2 + F_6$, etc., the average horizontal displacements U_2, U_3, U_4 indicated below at levels 2–6, 3–7, and 4–8.

F_2^T (lb)	F_3^T (lb)	F_4^T (lb)	U_2 (10^{-6} ft)	U_3 (10^{-6} ft)	U_4 (10^{-6} ft)
1	0	0	3.004	3.050	3.051
0	1	0	3.050	3.336	3.403
0	0	1	3.051	3.403	3.755

Using these data, determine the stiffness matrix $[K]$ governing the dynamic lateral displacements of the three movable levels of the structure.

From the results above, we have the deflection-force relation

$$\begin{Bmatrix} U_2 \\ U_3 \\ U_4 \end{Bmatrix} = 10^{-6} \begin{bmatrix} 3.004 & 3.050 & 3.051 \\ 3.050 & 3.338 & 3.403 \\ 3.051 & 3.403 & 3.755 \end{bmatrix} \begin{Bmatrix} F_2^T \\ F_3^T \\ F_4^T \end{Bmatrix}$$

Inverting this equation, we find the stiffness matrix connecting forces F_2^T, F_3^T, F_4^T with displacements U_2, U_3, U_4 expressible as

$$[K] = 10^6 \begin{bmatrix} 4.87 & -5.47 & 1.00 \\ -5.47 & 10.07 & -4.69 \\ 1.00 & -4.69 & 3.70 \end{bmatrix} \quad \#$$

6.3 MODAL METHOD OF SOLUTION

Equation (21) gives the matrix equation governing the motion of an offshore structure having three movable levels. In the general case when n movable levels are involved, procedures analogous to those described above will lead to an identical matrix equation, with the matrices $[M^*]$, $\{U\}$, etc., simply increased in size to account for the additional movable levels. Thus, in general, the matrix equation governing the motion of an offshore structure may be written, with the help of the damping assumption of equation (25), as

$$[M^*]\{\ddot{U}\} + 2\zeta[M^*]\{\dot{U}\} + [K]\{U\} = \{F^*\} \tag{26}$$

To solve the set of governing differential equations represented by this equation, we may use *modal analysis* to obtain an associated set of uncoupled equations, each of which may be solved separately and the results then combined to obtain the solution to the original set of equations. To illustrate, we restrict attention again to the case where equation (26) represents the governing equations for a structure with only three movable levels.

We first consider the special case of equation (26) for undamped free vibration, namely

$$[M^*]\{\ddot{U}\} + [K]\{U\} = 0 \tag{27}$$

This matrix equation represents three scalar equations whose solutions may be assumed periodic and of the form

$$\{U\} = \{U_0\} \sin \omega t \tag{28}$$

where $\{U_0\}$ is a column matrix having components U_{02}, U_{03}, U_{04}. Substituting this assumed solution into equation (27), we find

$$-\omega^2[M^*]\{U_0\} + [K]\{U_0\} = 0 \tag{29}$$

This equation represents a set of three homogeneous algebraic equations involving U_{02}, U_{03}, and U_{04}. Written out, we have

$$\begin{aligned}
(k_{22} - M_2^*\omega^2)U_{02} + k_{23}U_{03} + k_{24}U_{04} &= 0 \\
k_{32}U_{02} + (k_{33} - M_3^*\omega^2)U_{03} + k_{34}U_{04} &= 0 \\
k_{42}U_{02} + k_{43}U_{03} + (k_{44} - M_4^*\omega^2)U_{04} &= 0
\end{aligned} \tag{30}$$

Now, for a nonzero solution of these equations, the determinant of the coefficients of U_{02}, U_{03}, and U_{04} must vanish, as required by Cramer's rule for simultaneous solution of linear algebraic equations. This leads to a cubic equation in ω^2 having roots λ_1, λ_2, and λ_3 such that

$$\lambda_1 = \omega_1^2, \qquad \lambda_2 = \omega_2^2, \qquad \lambda_3 = \omega_3^2 \tag{31}$$

with $\lambda_1 \leq \lambda_2 \leq \lambda_3$. The corresponding positive solutions $\omega_1, \omega_2, \omega_3$ are the frequencies of the so-called first, second, and third modes of vibration. For each of these solutions, we may use any two of the three equations above to solve for the ratios U_{02}/U_{04}, U_{03}/U_{04}. Setting $U_{04} = 1$, the solution U_{02}, U_{03} for each root λ may thus be obtained. These define the *mode shapes* associated with λ_1, λ_2, and λ_3.

Using equation (27) and the fact that $[M^*]$ and $[K]$ are symmetric, it can be found that the following relations exist between the mode shapes associated with $\lambda = \lambda_i$ and $\lambda = \lambda_j$:

$$\begin{aligned}
\{U_0(\lambda_i)\}^T[M^*]\{U_0(\lambda_j)\} &= 0 \\
\{U_0(\lambda_i)\}^T[K]\{U_0(\lambda_j)\} &= 0
\end{aligned} \tag{32}$$

provided that $\lambda_i \neq \lambda_j$.

If we now form the *modal matrix* $[P]$ as

$$[P] = \begin{bmatrix} U_{02}(\lambda_1) & U_{02}(\lambda_2) & U_{02}(\lambda_3) \\ U_{03}(\lambda_1) & U_{03}(\lambda_2) & U_{03}(\lambda_3) \\ U_{04}(\lambda_1) & U_{04}(\lambda_2) & U_{04}(\lambda_3) \end{bmatrix} \tag{33}$$

it follows, by expansion, from equation (32) that

$$\begin{aligned}
[P]^T[M^*][P] &= [M'] \\
[P]^T[K][P] &= [K']
\end{aligned} \tag{34}$$

where $[M']$ and $[K']$ are diagonal matrices, with $[K']$ having the additional property that

$$[K'] = [M']\{\lambda\} \tag{35}$$

with $\{\lambda\}$ denoting a diagonal matrix having diagonal components λ_1, λ_2, λ_3, as defined earlier.

Using the modal matrix, we now introduce new variables $\{Y\}$ such that

$$\{Y\} = [P]^{-1}\{U\}, \qquad \{U\} = [P]\{Y\} \tag{36}$$

where $\{Y\}$ denotes a column matrix having components Y_1, Y_2, Y_3. Substituting into equation (26), we thus find

$$[M^*][P]\{\ddot{Y}\} + 2\zeta[M^*][P]\{\dot{Y}\} + [K][P]\{Y\} = \{F^*\}$$

Finally, premultiplying both sides of this equation by $[P]^T$ and using equation (34), we have

$$[M']\{\ddot{Y}\} + 2\zeta[M']\{\dot{Y}\} + [K']\{Y\} = [P]^T\{F^*\} \tag{37}$$

Since $[M']$ and $[K']$ are diagonal, it is easily seen that the three scalar equations reprsented by this last equation are uncoupled and of the form

$$\begin{aligned} M'_1\ddot{Y}_1 + 2\zeta M'_1\dot{Y}_1 + K'_{11}Y_1 &= F'_1 \\ M'_2\ddot{Y}_2 + 2\zeta M'_2\dot{Y}_2 + K'_{22}Y_2 &= F'_2 \\ M'_3\ddot{Y}_3 + 2\zeta M'_3\dot{Y}_3 + K'_{33}Y_3 &= F'_3 \end{aligned} \tag{38}$$

where M'_1, K'_{11}, F'_1, etc., denote respectively the components of $[M']$, $[K']$, and $[P]^T\{F^*\}$. Making use of equation (35), we may rewrite these last equations as

$$\begin{aligned} \ddot{Y}_1 + 2\zeta\dot{Y}_1 + \lambda_1 Y_1 &= \frac{F'_1}{M'_1} \\ \ddot{Y}_2 + 2\zeta\dot{Y}_2 + \lambda_2 Y_2 &= \frac{F'_2}{M'_2} \\ \ddot{Y}_3 + 2\zeta\dot{Y}_3 + \lambda_3 Y_3 &= \frac{F'_3}{M'_3} \end{aligned} \tag{39}$$

The components F'_1, F'_2, F'_3 denote known functions of time, associated with the wave force through the transformation $[P]^T\{F^*\}$. By expanding and using trigonometric functions of sums of angles, these force components can be written in the form

$$\begin{aligned} F'_1 &= F'_{01}\sin(-\omega t + \phi'_1) \\ F'_2 &= F'_{02}\sin(-\omega t + \phi'_2) \\ F'_3 &= F'_{03}\sin(-\omega t + \phi'_3) \end{aligned} \tag{40}$$

and the steady-state solutions for Y_1, Y_2, Y_3 expressible as

$$\begin{aligned} Y_1 &= Y_{01}\sin(-\omega t + \phi_1' + \phi_1'') \\ Y_2 &= Y_{02}\sin(-\omega t + \phi_2' + \phi_2'') \\ Y_3 &= Y_{03}\sin(-\omega t + \phi_3' + \phi_3'') \end{aligned} \tag{41}$$

where ϕ_1'', ϕ_2'', ϕ_3'' and Y_{01}, Y_{02}, Y_{03} are given by

$$\tan\phi_n'' = \frac{2\zeta\omega}{\lambda_n - \omega^2} \tag{42}$$

$$Y_{0n} = \frac{F_{0n}'/M_n'}{(\lambda_n - \omega^2)\cos\phi_n'' + 2\zeta\omega\sin\phi_n''} \tag{43}$$

with $n = 1, 2, 3$, respectively.

Having the solutions for Y_1, Y_2, Y_3, the displacements U_2, U_3, U_4 can then be determined using the second of equations (36) which, when expanded, becomes

$$\begin{aligned} U_2 &= U_{02}(\lambda_1)Y_1 + U_{02}(\lambda_2)Y_2 + U_{02}(\lambda_3)Y_3 \\ U_3 &= U_{03}(\lambda_1)Y_1 + U_{03}(\lambda_2)Y_2 + U_{03}(\lambda_3)Y_3 \\ U_4 &= U_{04}(\lambda_1)Y_1 + U_{04}(\lambda_2)Y_2 + U_{04}(\lambda_3)Y_3 \end{aligned} \tag{44}$$

These last equations show that the solutions for the dynamic deflections at each level of the structure can be regarded as a weighted sum of the solutions Y_1, Y_2, Y_3, the weightings being the mode-shape factors $U_{02}(\lambda_1)$, $U_{02}(\lambda_2)$, etc. An examination of the relative contributions to the deflection values for any particular case will, in fact, usually show that most is provided by the first mode and that a good approximation to the deflection values can often be obtained by neglecting altogether contributions from the second and third modes and writing equation (44) simply as

$$\begin{aligned} U_2 &= U_{02}(\lambda_1)Y_1 \\ U_3 &= U_{03}(\lambda_1)Y_1 \\ U_4 &= U_{04}(\lambda_1)Y_1 \end{aligned} \tag{45}$$

Of course, a somewhat better approximation would be obtained by considering contributions from the first and second modes and neglecting only the third mode.

It should be noted that only three vibration modes arise in the present illustration because there are only three movable levels in the structure we are considering. More generally, when n movable levels exist, solution procedures analogous to those just described will lead to n vibration modes. In most cases, the remarks above will, however, continue to apply and a very good description of the deflections can be obtained by considering only the first two or three modes in a series expansion analogous to equation (44).

It may also be noted that the lowest (first mode) frequency ω_1 determined above will represent the *fundamental frequency* of the structure and will correspond approximately to the natural frequency Ω found by the simplified analysis of Section 4.10. Consistent with that earlier treatment, the damping factor ζ may be chosen as a certain fraction of the fundamental frequency ω_1, typically about 0.05 to 0.10, i.e., about 5 to 10% of critical damping of the fundamental mode of vibration.

EXAMPLE 6.3-1. The mass and stiffness matrices associated with a particular offshore structure are shown in Fig. 6.5. Using these, determine the horizontal displacements of the three movable levels of the structure for the wave forces indicated. Assume that wave frequency $\omega = 0.785$ rad/sec.

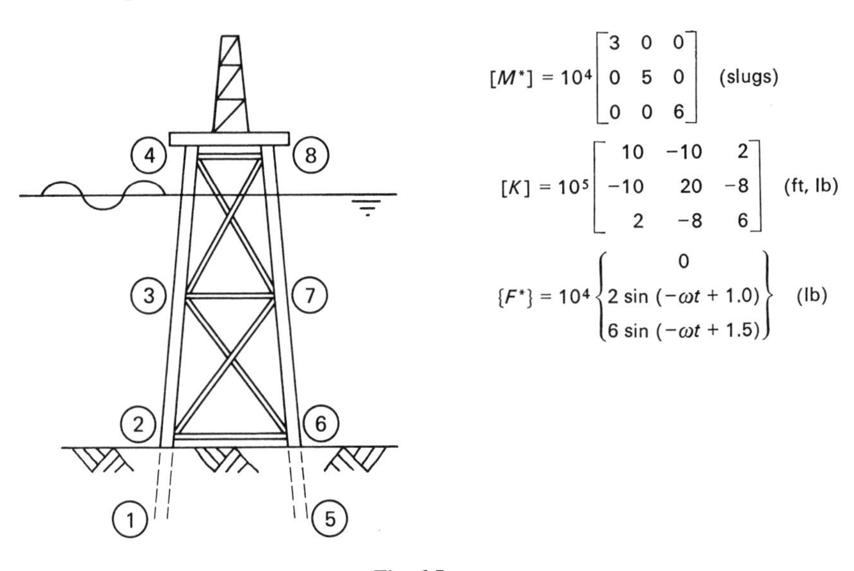

Fig. 6.5

We first determine the free-vibration frequencies of the structure using equation (28). Writing $\lambda = \omega^2$ and setting the determinant of the coefficients of U_{02}, U_{03}, and U_{04} equal to zero, we find the following equation ($M_2^* = 3 \times 10^4$, $K_{22} = 10 \times 10^5$, etc.):

$$\lambda^3 - 83.3\lambda^2 + 1164\lambda - 2222 = 0$$

The solutions, as determined by numerical investigation, are

$$\lambda_1 = 2.27, \qquad \lambda_2 = 14.8, \qquad \lambda_3 = 66.2$$

from which the vibration frequencies $\omega_1 = \sqrt{\lambda_1}$, etc., are determined as

$$\omega_1 = 1.51 \text{ rad/sec}, \qquad \omega_2 = 3.85 \text{ rad/sec}, \qquad \omega_3 = 8.14 \text{ rad/sec}$$

We next determine the mode shapes for each value of λ using any two of the three relations of equation (30). The results are tabulated in Table 6.3. The

Table 6.3

λ	U_{02}	U_{03}	U_{04}
2.27	0.559	0.720	1.00
14.8	−1.832	−0.819	1.00
66.3	3.530	−3.293	1.00

modal matrix is accordingly determined from equation (33) as

$$[P] = \begin{bmatrix} 0.559 & -1.832 & 3.530 \\ 0.720 & -0.819 & -3.293 \\ 1.000 & 1.000 & 1.000 \end{bmatrix}$$

Next, we determined the elements of the transformed mass matrix $[M']$ and force matrix $\{F'\}$. These are given by

$$[M'] = [P]^T[M^*][P]$$
$$\{F'\} = [P]^T\{F^*\}$$

Carrying out the multiplications, we find

$$M'_1 = 9.53 \times 10^4 \text{ slugs}, \qquad M'_2 = 19.4 \times 10^4 \text{ slugs}, \qquad M'_3 = 97.6 \times 10^4 \text{ slugs}$$

and (in units of 10^4 lb)

$$F'_1 = 1.44 \sin(-\omega t + 1) + 6 \sin(-\omega t + 1.5)$$
$$F'_2 = -1.64 \sin(-\omega t + 1) + 6 \sin(-\omega t + 1.5)$$
$$F'_3 = -6.59 \sin(-\omega t + 1) + 6 \sin(-\omega t + 1.5)$$

Now from trigonometry, we have the identity

$$A \sin(\alpha + \zeta) + B \sin(\alpha + \eta) = C \sin(\alpha + \beta)$$

where

$$\tan \beta = \frac{A \sin \zeta + B \sin \eta}{A \cos \zeta + B \cos \eta}$$

$$C = \frac{A \cos \zeta + B \cos \eta}{\cos \beta}$$

Hence, using this identity, we may express F'_1, F'_2, F'_3 more concisely as

$$F'_1 = F'_{01} \sin(-\omega t + 1.41)$$
$$F'_2 = F'_{02} \sin(-\omega t - 1.47)$$
$$F'_3 = F'_{03} \sin(-\omega t - 0.14)$$

where

$$F'_{01} = 7.30 \times 10^4 \text{ lb}, \qquad F'_{02} = -4.62 \times 10^4 \text{ lb}, \qquad F'_{03} = -3.16 \times 10^4 \text{ lb}$$

The solutions for the transformed variables Y_1, Y_2, Y_3 are given in terms of the foregoing transformed mass and force components by equations (41)–(43). We find (with $\zeta = 0.08\omega_1 = 0.121$)

$$Y_1 = Y_{01} \sin(-\omega t + 1.52)$$
$$Y_2 = Y_{02} \sin(-\omega t + 1.46)$$
$$Y_3 = Y_{03} \sin(-\omega t - 0.14)$$

where $Y_{01} = 0.459$ ft, $Y_{02} = -0.0168$ ft, $Y_{03} = -4.93 \times 10^{-4}$ ft.

Finally, the displacements U_2, U_3, U_4 of the structural levels are expressible in terms of Y_1, Y_2, Y_3 according to equation (42). The contribution from Y_3 is clearly negligible and we have (using the components of the modal matrix)

$$U_2 = 0.559\,Y_1 - 1.83\,Y_2$$
$$U_3 = 0.720\,Y_1 - 0.819\,Y_2$$
$$U_4 = Y_1 + Y_2$$

Substituting for Y_1 and Y_2 and using the trigonometric identity above, we find (units of feet)

$$U_2 = 0.227 \sin(-\omega t + 1.50)$$
$$U_3 = 0.317 \sin(-\omega t + 1.51)$$
$$U_4 = 0.475 \sin(-\omega t + 1.53)$$

It is of interest to compare the maximum deflection U_4 of 0.475 ft, as determined from the dynamic solution above with that found when inertia and damping forces are neglected. Assuming the same lumped wave forces $\{F^*\}$, the solutions for the displacements U_2, U_3, and U_4 are found, neglecting inertia and damping forces, from equation (26) to be expressible simply as

$$\{U\} = [K]^{-1}\{F^*\}$$

Carrying out the calculation, we find the maximum deck deflection U_4 to equal 0.354 ft. The maximum dynamic deflection is thus seen to be about 35% greater. #

6.4 ITERATIVE METHOD OF MODE CALCULATIONS

In the modal method of solution of the governing dynamic equations of a structure described above, the vibration frequencies and mode shapes were determined directly from equation (29) by setting the determinate of coefficients of the free-vibration displacements equal to zero. For a structure with only two or three identifiable levels, this method is convenient to use since it involves only a limited amount of algebraic manipulation. Most offshore structural designs have, however, many such levels and thus many vibration frequencies and mode shapes. The algebraic method for determining the vibration frequencies and mode shapes accordingly becomes very cumbersome and numerical procedures are more conveniently employed. A commonly used numerical procedure is that of matrix iteration.

To illustrate this method, we consider a structure with n movable levels. The matrix equation governing the vibration frequencies and mode shapes is the same as equation (29), with $\{U_0\}$ now assumed to be a column matrix of order n and $[M^*]$ and $[K]$ assumed to be square matrices of order $n \times n$. For interative purposes, we may write this equation as

$$\{U_0\} = \omega^2[K]^{-1}[M^*]\{U_0\} \tag{46}$$

If we now assume arbitrary values for the elements of matrix $\{U_0\}$, with the last element chosen equal to unity, we may solve for new values (in terms of ω^2) by substituting the assumed values into the right-hand side of equation (46). The new values may next be normalized by factoring out the last element $U_{0n}\omega^2$ and these again used in equation (46) to find another new set of values for the elements of $\{U_0\}$. These are then normalized by factoring out the last element $U_{0n}\omega^2$. If the resulting normalized matrix agrees with that used in its calculation, the mode shape is determined and the associated value of the frequency is determined from the necessary condition $U_{0n}\omega^2 = 1$. If agreement is not obtained, the process is repeated until convergence occurs.

Now, it is a general characteristic of iterative procedures of this kind that the solution will converge to that associated with the smallest value of ω^2, that is to the fundamental mode of vibration. Once this solution is obtained, it is necessary to modify equation (46) so that it will yield the second mode by iteration.

To do this, we use results from the first of equation (32), which may be written in general (assuming level 1 fixed against displacement) for $\lambda_p \neq \lambda_q$ as

$$U_{02}(\lambda_p)U_{02}(\lambda_q)M_2^* + U_{03}(\lambda_p)U_{03}(\lambda_q)M_3^* + \ldots = 0 \tag{47}$$

where $U_{02}(\lambda_p)$ denotes the free-vibration displacement of level 2 associated with mode $\lambda_p = \omega_p^2$, $U_{02}(\lambda_q)$ denotes that associated with mode $\lambda_q = \omega_q^2$, etc. If we now choose $\lambda_p = \lambda_1$ and write $U_{02} = U_{02}(\lambda_q)$, $U_{03} = U_{03}(\lambda_q)$, etc., we have, for example, for a structure with four levels

$$U_{02} = \alpha_3 U_{03} + \alpha_4 U_{04} + \alpha_5 U_{05} \tag{48}$$

where α_3, α_4, and α_5 are defined by

$$\alpha_3 = -\frac{U_{03}(\lambda_1)M_3^*}{U_{02}(\lambda_1)M_2^*}, \qquad \alpha_4 = -\frac{U_{04}(\lambda_1)M_4^*}{U_{02}(\lambda_1)M_1^*}, \qquad \alpha_5 = -\frac{U_{05}(\lambda_1)M_5^*}{U_{02}(\lambda_1)M_2^*}$$

With the identities $U_{03} = U_{03}$, $U_{04} = U_{04}$, and $U_{05} = U_{05}$, we may write the matrix $\{U_0\}$ having these components as

$$\{U_0\} = [S]\{U_0\} \tag{49}$$

where $[S]$ is expressible as

$$[S] = \begin{bmatrix} 0 & \alpha_3 & \alpha_4 & \alpha_5 \\ 0 & 1 & 0 & 0 \\ 0 & 0 & 1 & 0 \\ 0 & 0 & 0 & 1 \end{bmatrix}$$

From equation (47), it can be seen that when the elements of $\{U_0\}$ on the right-hand side of equation (49) are those of any mode other than the first, the equation will simply serve as an identity. When, however, the elements of $\{U_0\}$ are those of the first mode, the right-hand operation will not yield identical values, since equation (47) does not apply for $\lambda_q = \lambda_p$.

Writing the matrix equation (46) as

$$\{U_0\} = \omega^2[K]^{-1}[M^*][S]\{U_0\} \tag{50}$$

the iterative procedure above will thus converge to the lowest mode, which, in this case, is the second mode. The first mode is said to have been swept out of the assumed deflections $\{U_0\}$ on the right-hand side of equation (50) and, for this reason, the matrix $[S]$ is referred to as a *sweeping matrix*.

Once the second mode has been determined, a new sweeping matrix $[S]$ may be constructed to stop both the first and second modes. Using this new sweeping matrix in equation (50), the iterative procedure will then converge to the third mode. The process may be repeated for as many modes as available.

To construct the sweeping matrix to eliminate first and second modes, we use equation (47) and write

$$U_{02} = \beta_3 U_{03} + \beta_4 U_{04} + \beta_5 U_{05} \tag{51}$$

where

$$\beta_3 = -\frac{U_{03}(\lambda_2)M_3^*}{U_{02}(\lambda_2)M_2^*}, \qquad \beta_4 = -\frac{U_{04}(\lambda_2)M_4^*}{U_{02}(\lambda_2)M_2^*}, \qquad \beta_5 = -\frac{U_{05}(\lambda_2)M_5^*}{U_{02}(\lambda_2)M_1^*}$$

Using this equation with equation (48), we may express U_{02} and U_{03} in terms of U_{04} and U_{05} as

$$U_{02} = \gamma_1 U_{04} + \gamma_2 U_{05}$$
$$U_{03} = \gamma_3 U_{04} + \gamma_4 U_{05}$$

where γ_1, γ_2, etc., are constants. The sweeping matrix is thus expressible as

$$[S] = \begin{bmatrix} 0 & 0 & \gamma_1 & \gamma_2 \\ 0 & 0 & \gamma_3 & \gamma_4 \\ 0 & 0 & 1 & 0 \\ 0 & 0 & 0 & 1 \end{bmatrix}$$

It is worth noting that the calculation for the highest mode may also be carried out without using any sweeping matrix. This is because equation (46) can be inverted and written as

$$\{U_0\} = \frac{1}{\omega^2}[M^*]^{-1}[K]\{U_0\} \tag{52}$$

The iterative-solution procedure when applied to this equation will converge to the smallest value of $1/\omega^2$, that is, the highest frequency.

EXAMPLE 6.4-1. Determine by matrix iteration the natural frequencies and mode shapes for the structure of Example 6.3-1.

Using the $[K]$ and $[M^*]$ matrices of Example 6.3-1, we have from equation (46)

$$\begin{Bmatrix} U_{02} \\ U_{03} \\ U_{04} \end{Bmatrix} = \omega^2 \begin{bmatrix} 0.084 & 0.110 & 0.120 \\ 0.066 & 0.140 & 0.180 \\ 0.060 & 0.150 & 0.300 \end{bmatrix} \begin{Bmatrix} U_{02} \\ U_{03} \\ U_{04} \end{Bmatrix}$$

Assuming first unit values of U_{01}, U_{02}, U_{03} on the right-hand side, we find

$$\begin{Bmatrix} U_{02} \\ U_{03} \\ U_{04} \end{Bmatrix} = \omega^2 \begin{bmatrix} 0.084 & 0.110 & 0.120 \\ 0.066 & 0.140 & 0.180 \\ 0.060 & 0.150 & 0.300 \end{bmatrix} \begin{Bmatrix} 1 \\ 1 \\ 1 \end{Bmatrix} = 0.510\omega^2 \begin{Bmatrix} 0.616 \\ 0.757 \\ 1.000 \end{Bmatrix}$$

Repeating the calculation with the new values, we have

$$\begin{Bmatrix} U_{02} \\ U_{03} \\ U_{04} \end{Bmatrix} = \omega^2 \begin{bmatrix} 0.084 & 0.110 & 0.120 \\ 0.066 & 0.140 & 0.180 \\ 0.060 & 0.150 & 0.300 \end{bmatrix} \begin{Bmatrix} 0.616 \\ 0.757 \\ 1.000 \end{Bmatrix} = 0.451\omega^2 \begin{Bmatrix} 0.565 \\ 0.725 \\ 1.000 \end{Bmatrix}$$

Carrying out the procedure several more times, we find

$$\begin{Bmatrix} U_{02} \\ U_{03} \\ U_{04} \end{Bmatrix} = \omega^2 \begin{bmatrix} 0.084 & 0.110 & 0.120 \\ 0.066 & 0.140 & 0.180 \\ 0.060 & 0.150 & 0.300 \end{bmatrix} \begin{Bmatrix} 0.558 \\ 1.719 \\ 1.000 \end{Bmatrix} = 0.441\omega^2 \begin{Bmatrix} 0.558 \\ 0.719 \\ 1.000 \end{Bmatrix}$$

The calculated values on the right-hand side equal those used on left-hand side of the equation. The first-mode frequency, as determined from the necessary relation $0.441\omega^2 = 1$, and the mode shape description are thus

$$\omega_1 = 1.51 \text{ rad/sec}, \qquad U_{02} = 0.558, \qquad U_{03} = 0.719, \qquad U_{04} = 1.0$$

To calculate the second-mode characteristics, we use the sweeping matrix $[S]$, which here takes the form [equation (48)]

$$[S] = \begin{bmatrix} 0 & -2.148 & -3.584 \\ 0 & 1 & 0 \\ 0 & 0 & 1 \end{bmatrix}$$

From equation (50), we have

$$\begin{Bmatrix} U_{02} \\ U_{03} \\ U_{04} \end{Bmatrix} = \omega^2 \begin{bmatrix} 0 & -0.0704 & -0.1811 \\ 0 & -0.0018 & -0.0565 \\ 0 & 0.0211 & 0.0850 \end{bmatrix} \begin{Bmatrix} U_{02} \\ U_{03} \\ U_{04} \end{Bmatrix}$$

Assuming values of unity on the right-hand side, we find

$$\begin{Bmatrix} U_{02} \\ U_{03} \\ U_{04} \end{Bmatrix} = \omega^2 \begin{bmatrix} 0 & -0.0704 & -0.1811 \\ 0 & -0.0018 & -0.0565 \\ 0 & 0.0211 & 0.0850 \end{bmatrix} \begin{Bmatrix} 1 \\ 1 \\ 1 \end{Bmatrix} = 0.1061\omega^2 \begin{Bmatrix} -2.370 \\ -0.549 \\ 1.000 \end{Bmatrix}$$

Carrying out the calculations a few more times, we find

$$\begin{Bmatrix} U_{02} \\ U_{03} \\ U_{04} \end{Bmatrix} = \omega^2 \begin{bmatrix} 0 & -0.0704 & -0.1811 \\ 0 & -0.0018 & -0.0565 \\ 0 & 0.0211 & 0.0850 \end{bmatrix} \begin{Bmatrix} -1.830 \\ -0.809 \\ 1.000 \end{Bmatrix} = 0.0680\omega^2 \begin{Bmatrix} -1.830 \\ -0.809 \\ 1.000 \end{Bmatrix}$$

so that

$$\omega_2 = 3.83 \text{ rad/sec}, \qquad U_{02} = -1.830, \qquad U_{03} = -0.809, \qquad U_{04} = 1.000$$

Finally, for the characteristics of the third mode, we consider equations (48) and (51) in the form

$$U_{02} = -2.148U_{03} - 3.584U_{04}$$
$$U_{02} = -0.737U_{03} + 1.093U_{04}$$

Solving these, we have

$$U_{02} = 3.537U_{04}$$
$$U_{03} = -3.315U_{04}$$

The sweeping matrix is thus expressible as

$$[S] = \begin{bmatrix} 0 & 0 & 3.537 \\ 0 & 0 & -3.315 \\ 0 & 0 & 1 \end{bmatrix}$$

Substituting into equation (50), we have

$$\begin{Bmatrix} U_{02} \\ U_{03} \\ U_{04} \end{Bmatrix} = \omega^2 \begin{bmatrix} 0 & 0 & 0.0525 \\ 0 & 0 & -0.0507 \\ 0 & 0 & 0.0150 \end{bmatrix} \begin{Bmatrix} U_{02} \\ U_{03} \\ U_{04} \end{Bmatrix}$$

Carrying out iterative calculations similar to those above, we find

$$\begin{Bmatrix} U_{02} \\ U_{03} \\ U_{04} \end{Bmatrix} = \omega^2 \begin{bmatrix} 0 & 0 & 0.0525 \\ 0 & 0 & -0.0507 \\ 0 & 0 & 0.0150 \end{bmatrix} \begin{Bmatrix} 3.50 \\ -3.38 \\ 1.00 \end{Bmatrix} = 0.0150\omega^2 \begin{Bmatrix} 3.50 \\ -3.38 \\ 1.00 \end{Bmatrix}$$

so that, for the third mode, we have

$$\omega_3 = 8.16 \text{ rad/sec}, \qquad U_{02} = 3.50, \qquad U_{03} = -3.38, \qquad U_{04} = 1.00$$

It can be seen that these results agree, within round-off error, with those found in Example 6.3-1. #

6.5 STRESS ANALYSIS

Once the horizontal displacements at each level of a structure have been determined, these may be used to calculate the stresses within each member. For this calculation, it is necessary first to determine the vertical displacements and rotations at the ends of the members associated with the horizontal displacements. Consider, for example, member 2–3 of the offshore structure shown earlier in Fig. 6.3. If $\{Q\}$ denotes a column matrix having as components the vertical displacement V_2 and rotation θ_2 at joint 2 and the vertical displacement V_3 and rotation θ_3 at joint 3, we may relate this to the column matrix $\{F^T\}$ of horizontal force components through the equation

$$\{Q\} = [D]\{F^T\} \tag{53}$$

where $[D]$ denotes the appropriate transformation matrix.

The individual components of the matrix $[D]$ can be determined from the static solution associated with horizontal forces applied at the joints. For example, for level 2–6, if we set horizontal forces $F_2 = F_6 = \frac{1}{2}$, so that $F_2^T = 1$, and set all other forces and moments equal to zero, the solution for the displacements V_2, θ_2 and V_3, θ_3 will yield values for the first column $[D]$. Similar treatments will yield the remaining two columns.

The force $\{F^T\}$ is related to the horizontal displacements $\{U\}$ through the matrix $[K]$ developed in connection with equation (20), so that we have the relation connecting the vertical displacements and rotations of the joints with the horizontal displacements expressible as

$$\{Q\} = [D][K]\{U\} \tag{54}$$

With this equation, the vertical displacements and rotations at the ends of the member can be determined at any instant using the known horizontal displacements U_2, U_3, U_4, and the internal forces and moments acting at each end then calculated from the stiffness matrix of the member. From these, the axial and bending stresses in the member may be calculated using the method of Chapter 2.

In addition to being needed for stress analysis, the internal forces and moments acting at the base of the structure, as calculated from the procedures above, are also necessary for determining the stiffness properties of the equivalent free-standing piles beneath the seafloor. The procedure for calculating these properties in a dynamic analysis parallels that used with a static analysis, as described in Chapter 4. In particular, maximum internal horizontal forces and moments acting at the base of the structure (top of the piles) are assumed and these used to calculate equivalent pile stiffnesses according to methods given in Chapter 5. These pile stiffnesses are then used in the analysis above and maximum horizontal forces and moments at the base of the structure calculated. New values of the equivalent pile stiffnesses are next calculated and the analysis repeated. This iterative process is continued until agreement is obtained between the assumed and calculated maximum force and moment at the base of the structure.

It should be emphasized that the member stresses determined by the present analysis are those resulting solely from applied wave loadings. Thus, contributions from wind, current and weight loadings are assumed negligible in comparison with the wave-induced stresses. In assessing whether overstress of a structure from dynamic effects will occur, independent estimates of these stress contributions should therefore be made to ensure that they are negligible or to allow their incorporation into the judgment of possible overstress.

The static contribution from wind forces may be estimated by first determining the associated horizontal displacements of the structural levels from equation (20), with the total-force components all zero except for that associated with the wind loading transmitted to the uppermost level of the structure from the deck. Assuming wave forces dominate the structural response, the equivalent pile stiffnesses needed in this calculation may be assumed the same as those found from the dynamic analysis. With the horizontal displacements of the various levels of the structure so determined, the wind-induced stresses in the various structural members may then be calculated using the same procedures described above for the wave-induced stresses.

A similar estimate may also be made for the static contribution to the member stresses from current forces, provided the current is small in comparison with the average velocity given by equation (11). In this case, the forces acting at each joint of the various structural levels may be estimated using the first term on the right-hand side of equation (3), with u_p replaced by the current u_c. The total-force components of equation (20) are then given by the sum of the joint forces at each structural level.

The contribution to the member stress resulting from weight (and buoyant) loadings may be estimated using the static methods described in Chapter 4, with the simplifying assumption that the water surface is at the still-water level.

Finally, it should be noted that errors exist in the dynamic stresses themselves as a result of the assumption that the wave and inertia loadings are concentrated at the structural levels. To examine these, the structure may be divided into additional (fictitious) structural levels between those already considered and the dynamic analysis repeated. If the member stresses found from this analysis differ significantly from those of the original analysis, further division of the structure into more intermediate levels is recommended until the difference between successive calculations of the member stresses is negligibly small.

EXAMPLE 6.5-1. For the steel structure shown in Fig. 6.6, determine the axial and bending stresses in member 2–3 at joint 2, resulting from horizontal dynamic displacements $U_2 = 0.50$ ft, $U_3 = 0.60$ ft, $U_4 = 1.00$ ft. The stiffness matrix $[K]$ used in the dynamic analysis is as shown:

$$[K] = 10^6 \begin{bmatrix} 4.87 & -5.47 & 1.00 \\ -5.47 & 10.07 & -4.69 \\ 1.00 & -4.69 & 3.70 \end{bmatrix} \quad \text{(ft, lb)}$$

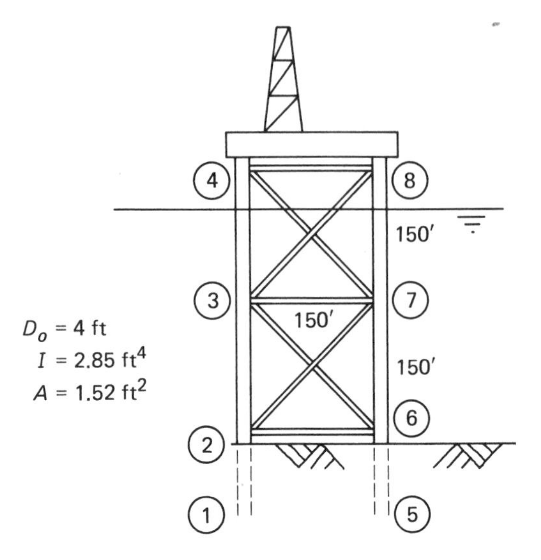

Fig. 6.6

A static analysis with unit horizontal forces F_2^T, F_3^T, F_4^T gives the vertical displacements and rotations indicated:

F_2^T	F_3^T	F_4^T	V_2	θ_2	V_3	θ_3
1	0	0	0.35	4.10	0.46	−1.12
0	1	0	2.32	4.30	3.60	−1.00
0	0	1	4.32	4.30	7.87	−0.88

F's in units of pounds
V's in units of 10^{-8} ft
θ's in units of 10^{-8} rad

The vertical displacements and rotations are related to the horizontal forces through equation (53), which here takes the form

$$\begin{Bmatrix} V_2 \\ \theta_2 \\ V_3 \\ \theta_3 \end{Bmatrix} = 10^{-8} \begin{bmatrix} 0.35 & 2.32 & 4.32 \\ 4.10 & 4.30 & 4.30 \\ 0.46 & 3.60 & 7.87 \\ -1.12 & -1.00 & -0.88 \end{bmatrix} \begin{Bmatrix} F_2^T \\ F_3^T \\ F_4^T \end{Bmatrix}$$

where the elements of the matrix $[D]$ in equation (53) are determined from the given static deflection data. Using the stiffness matrix $[K]$ in accordance with equation (54), we have

$$\begin{Bmatrix} V_2 \\ \theta_2 \\ V_3 \\ \theta_3 \end{Bmatrix} = 10^{-2} \begin{bmatrix} 0.35 & 2.32 & 4.32 \\ 4.10 & 4.30 & 4.30 \\ 0.46 & 3.60 & 7.87 \\ -1.12 & -1.00 & -0.88 \end{bmatrix} \begin{bmatrix} 4.37 & -5.47 & 1.00 \\ -5.47 & 10.07 & -4.69 \\ 1.00 & -4.69 & 3.70 \end{bmatrix} \begin{Bmatrix} U_2 \\ U_3 \\ U_4 \end{Bmatrix}$$

which reduces to

$$\begin{Bmatrix} V_2 \\ \theta_2 \\ V_3 \\ \theta_3 \end{Bmatrix} = 10^{-2} \begin{bmatrix} -6.67 & 1.19 & 5.45 \\ 0.75 & 0.71 & -0.16 \\ -9.58 & -3.17 & 12.70 \\ -0.86 & 0.18 & 0.31 \end{bmatrix} \begin{Bmatrix} U_2 \\ U_3 \\ U_4 \end{Bmatrix}$$

Substituting the dynamic horizontal displacements U_2, U_3, U_4, we find

$$V_2 = 2.82 \times 10^{-2} \text{ ft}, \qquad \theta_2 = 6.42 \times 10^{-3} \text{ rad}$$
$$V_3 = 6.00 \times 10^{-2} \text{ ft}, \qquad \theta_3 = -1.20 \times 10^{-4} \text{ rad}$$

The internal force f_{2y} (positive upward) and moment (positive clockwise) acting at joint 2 is determined from the stiffness matrix of member 2–3 as

$$f_{2y} = \frac{EA}{l}(V_2 - V_3) = -1.39 \times 10^6 \text{ lb}$$

$$m_2 = \frac{6EI}{l^2}(U_2 - U_3) + \frac{2EI}{l}(2\theta_2 - \theta_3) = 1.80 \times 10^6 \text{ ft-lb}$$

Since f_{2y} is acting downward, the member 2–3 is seen to be in tension with axial stress

$$\sigma_a = \frac{f_{2y}}{A} = 9.16 \times 10^5 \text{ lb/ft}^2 \text{ (6360 lb/in}^2\text{)}$$

The bending stress is determined (with outside radius equal to 2 ft) as

$$\sigma_b = \pm\frac{m_2 R}{I} = \pm 1.26 \times 10^6 \text{ lb/ft}^2 \text{ } (\pm 8750 \text{ lb/in}^2) \quad \#$$

6.6 DYNAMIC RESPONSE IN RANDOM WAVES

The discussion above has dealt with the dynamic analysis of structures when subjected to simple sinusoidal design waves. When appreciable dynamic amplification is found from such an analysis, the validity of the design-wave concept itself is open to question since the amplification calculated is very sensitive to the precise wave characteristics chosen and the assumed regular periodic form of the wave. In such circumstances, it is therefore desirable to extend the analysis to include irregular random waves representative more nearly of the actual sea state existing during storm conditions.

For purposes of structural analysis, irregular random waves are customarily regarded as an infinite sum of waves of variable amplitude and random phase, all running in the same (x) direction. The surface deflection η from the still-water level is accordingly expressible as

$$\eta = \sum_n A_n \cos (k_n x - \omega_n t + \epsilon_n) \tag{55}$$

where A_n, k_n, ω_n, and ϵ_n denote, respectively, the amplitude (equal to one-half the wave height), the wavenumber, the frequency, and the phase of the nth wave. The wavenumber and frequency are related through the Airy relation

$$\omega_n^2 = gk_n \tanh k_n h \tag{56}$$

where g denotes the acceleration of gravity, and h denotes water depth.

The phase angle ϵ_n in this expression is assumed to be distributed uniformly over the interval 0 to 2π so that any one value is as equally likely as any other. The amplitude A_n associated with wave frequencies in the small interval $\Delta\omega$ about ω_n is further assumed expressible as

$$\tfrac{1}{2}A_n^2 = S_n \, \Delta\omega \tag{57}$$

where S_n depends on the frequency and is known as the *amplitude* or *energy* spectrum.

The amplitude spectrum can be estimated from wave measurements and numerous expressions for S_η have been proposed. The most widely used relation at the present time is of the form (Bretschneider, 1959; Pierson and Moskowitz, 1963)

$$S_\eta = \frac{A}{\omega^5} e^{-B\omega^{-4}} \tag{58}$$

where A and B are constants. According to Bretschneider,

$$A = 0.169 H_s^2 \left(\frac{2\pi}{T_s}\right)^4, \qquad B = 0.675 \left(\frac{2\pi}{T_s}\right)^4 \tag{59}$$

where H_s (in feet) is the *significant wave height* equal to the average of the highest one-third of the waves in the sea state, and T_s (in seconds) is the corresponding *significant wave period*. The significant wave height and period are related approximately to the mean wave height $\bar{H}$ and mean period $\bar{T}$ by the relations $H_s = 1.6\bar{H}$ and $T_s = 1.1\bar{T}$.

Figure 6.7 shows the Bretschneider spectrum for various significant wave periods and unit significant wave height.

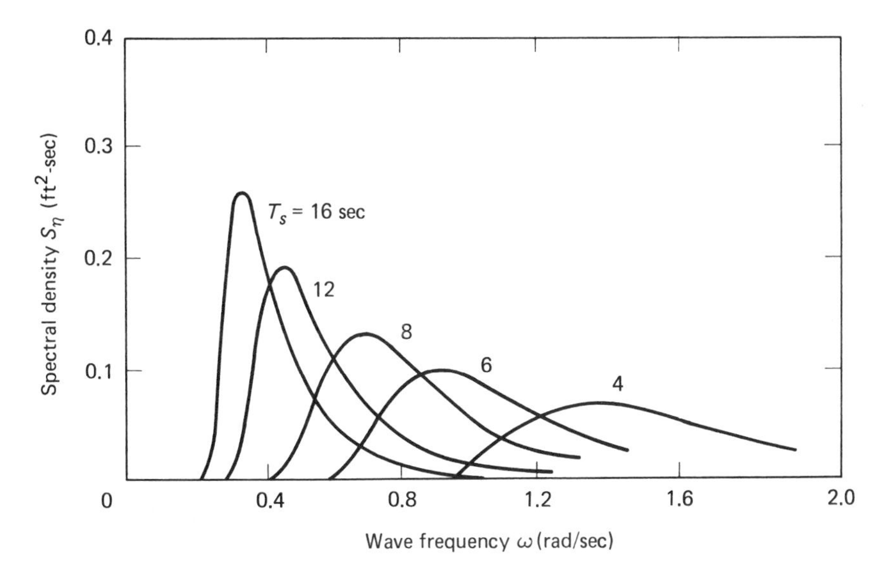

Fig. 6.7. Bretschneider spectrum for unit significant wave height.

For each wave component of equation (55), the associated horizontal velocity and acceleration of the water may be determined from the Airy theory. Assuming a one-dimensional sea where all waves are running the same direction, the horizontal water velocity u and acceleration a in the direction of wave propagation are accordingly given at any vertical position y above the sea floor by the equations

$$u = \sum_n T_u A_n \cos(k_n x - \omega_n t + \epsilon_n) \tag{60}$$

$$a = \sum_n T_a A_n \sin(k_n x - \omega_n t + \epsilon_n) \tag{61}$$

where T_u and T_a are defined by

$$T_u = \omega_n \frac{\cosh k_n y}{\sinh k_n h}, \qquad T_a = -\omega_n^2 \frac{\cosh k_n y}{\sinh k_n h} \tag{62}$$

An important aspect of the representation above of the surface deflection, water velocity, and acceleration is that the time average of their squared values at any place x are expressible in terms of the amplitude spectrum. Consider, for example, the square of the surface deflection η. This may be written from equation (55), with x arbitrarily set equal to zero, as

$$\eta^2 = \sum_m \sum_n A_m A_n \cos(\omega_m t - \epsilon_m) \cos(\omega_n t - \epsilon_n)$$

Taking the time average of both sides and noticing that, because of the random phases, the average of the cosine product will equal zero when $m \neq n$ and will equal $\frac{1}{2}$ when $m = n$, we find, using equation (57),

$$\overline{\eta^2} = \sum_n \tfrac{1}{2} A_n^2 = \sum_n S_\eta \, \Delta\omega$$

where the overhead bar denotes time average. Assuming S_η continuously distributed over the frequency ω, this may finally be written in integral form as

$$\overline{\eta^2} = \int_0^\infty S_\eta \, d\omega \tag{63}$$

In a similar way, we also have

$$\overline{u^2} = \int_0^\infty S_u \, d\omega, \qquad \overline{a^2} = \int_0^\infty S_a \, d\omega \tag{64}$$

where S_u and S_a are defined by

$$S_u = T_u^2 S_\eta, \qquad S_a = T_a^2 S_\eta \tag{65}$$

The above mean-square values are significant, both because their square roots yield characteristic root-mean-square (rms) values and because they can be used to estimate the average of the maximum values that occur over a specified time interval. In particular, if ξ denotes a quantity such as the surface deflection η, the water velocity u, or the acceleration a, and m_0 denotes its mean-square value, the average of its maximum values over time interval T_0 is expressible as (Longuet-Higgins, 1952)

$$\bar{\xi}_{\max} = \left[2m_0 \ln\left(\frac{T_0}{T_s}\right)\right]^{1/2} \tag{66}$$

provided that T_0/T_s is large, say greater than 100.

With these results before us, we may now consider the dynamic response of a typical offshore structure in random waves. We assume, as in the case of the water velocity or acceleration, that any deflection U and stress σ of interest in the structure are linearly related to the wave amplitude. Under random wave loading, they are accordingly expressible as

$$U = \sum_n A_n T_U(\omega) \sin(\omega_n t + \epsilon_n') \tag{67}$$

$$\sigma = \sum_n A_n T_\sigma(\omega) \sin(\omega_n t + \epsilon_n'') \tag{68}$$

where ϵ_n' and ϵ_n'' denote random phase angles. The deflection and stress spectra S_U and S_σ are then expressible in terms of the amplitude spectrum by the equations

$$S_U = T_U^2 S_\eta, \qquad S_\sigma = T_\sigma^2 S_\eta \tag{69}$$

and the mean-square values of the deflection and stress given by

$$\overline{U^2} = \int_0^\infty S_U \, d\omega, \qquad \overline{\sigma^2} = \int_0^\infty S_\sigma \, d\omega \tag{70}$$

It remains to specify the function T_U and T_σ appearing in the above equations. These functions may, in fact, be worked out numerically for any structure using the dynamic anaylsis given earlier for regular waves, provided only that we replace the definition of $\hat{u}_p$ in equation (11) by one that is independent of the wave amplitude. The wave force on the structure will then be linearly dependent on the wave amplitude and this, in turn, will cause the deck deflection and any stress in the structure similarly to be linearly dependent on the wave amplitude, as required in equations (67) and (68).

Using equation (10) with random waves, Borgman (1967) has shown that an appropriate value of $\hat{u}_p$ is expressible in terms of the root-mean-square velocity u_{rms} as

$$\hat{u}_p = \sqrt{\frac{8}{\pi}} \, u_{\text{rms}} = 1.60 u_{\text{rms}} \tag{71}$$

where u_{rms} is evaluated for $y = y_m$ using equation (64).

With this choice, we may select waves of unit amplitude and frequency ω such that the surface deflection is described by

$$\eta = \cos(kx - \omega t) \tag{72}$$

From the methods given earlier for dynamic analysis of structures in regular waves, we may then determine the associated deflection U and stress σ and express the results as

$$\begin{aligned} U &= U_0 \sin(\omega t + \phi_U) \\ \sigma &= \sigma_0 \sin(\omega t + \phi_\sigma) \end{aligned} \tag{73}$$

If we now imagine a random phase angle ϵ added to equation (72), this same phase angle would add to ϕ_U and ϕ_σ, thus making these random angles, say, ϵ' and ϵ'', respectively. Hence, comparing with the individual components of equations (67) and (68), we see that, for any frequency ω, the associated values of T_U and T_σ are given by the amplitudes U_0 and σ_0, respectively, of equation (73). Repeating the calculations for various frequencies within the range where the amplitude-spectrum values differ sensibly from zero, we may thus calculate the functions T_U and T_0 and use these with equation (69) to find the deflection and stress spectra. The mean-square values may then be determined from equation (70) and these finally used with equation (66) to estimate the average of the maximum values occurring over a time interval.

In analyzing offshore pile-supported structures using the foregoing procedure, we must, as in the case of static and regular-wave dynamic analysis, replace the actual embedded piles beneath the seafloor by equivalent free-standing ones whose stiffness properties are determined by methods analogous to those described in Chapter 5 for static analysis. These properties generally depend on the horizontal force and moment at the groundline, which, with irregular random waves, vary from wave to wave. Hence, an approximate representation of the pile stiffnesses is necessary. One method for evaluating these stiffnesses is to assume a regular wave of height equal to the significant wave height and period equal to the significant wave period. By carrying out an iterative dynamic calculation analogous to that described in Chapter 5, the equivalent-pile stiffness properties can then be determined and these used in a further dynamic structural analysis with irregular random waves.

Of course, the average of the maximum values of the member stresses determined from this analysis must be considered together with contributions from wind, current and weight loadings in assessing possible overstress from dynamic effects. These may be estimated using procedures similar to those described above for regular waves, taking into account the pile stiffnesses and the average water velocity, given by equation (71), that are appropriate here.

EXAMPLE 6.6-1. For regular waves of height H (ft) and frequency ω (rad/sec), the maximum deck deflection U (ft) of the structure shown in Fig. 6.8 is described approximately by the indicated equation. Assuming a one-dimensional random sea with $H_s = 20$ ft and $T_s = 9$ sec, determine the root-mean-square value of the

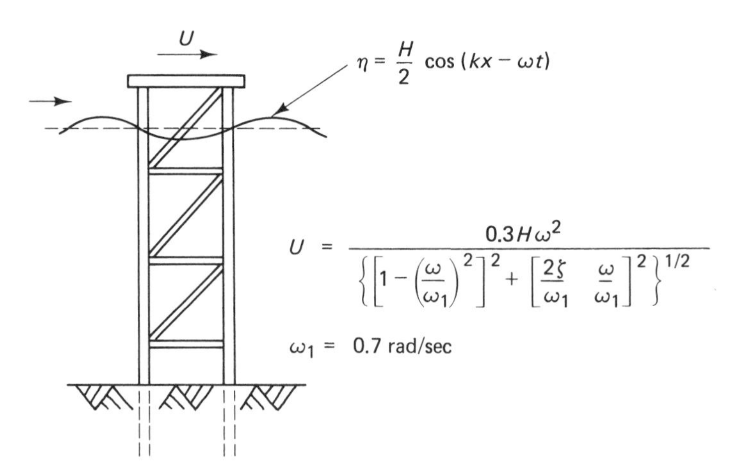

Fig. 6.8

deck deflection and the average of the maximum values over 6 hours. Assume that $\zeta/\omega_1 = 0.05$.

We consider regular waves of unit amplitude ($H/2 = 1$) and various frequencies and calculate the maximum deflections using the given equation. Results are shown in Table 6.4. Values of the amplitude spectrum S_η for the assumed frequencies are also shown, together with values of the associated deflection spectrum S_U as determined from equation (69) using the calculated value of $U = T_U$. Numerical integration of S_U over the interval of significant values gives $\bar{U}^2 = 42.9$ ft², so that $U_{rms} = 6.55$ ft. From equation (66) we then have for $T_0/T_s = 2400$, $m_0 = 42.9$.

Table 6.4

ω (rad/sec)	U (ft)	S_η (ft²-sec)	S_U (ft²-sec)
0.40	0.144	2.9	0.1
0.50	0.308	39.1	3.7
0.55	0.464	54.9	11.8
0.60	0.776	59.6	35.9
0.65	1.528	56.2	131.2
0.70	2.940	48.9	422.7
0.75	1.848	40.7	139.0
0.80	1.176	33.1	45.8
0.90	0.732	21.3	11.4
1.00	0.572	13.7	4.5
1.20	0.444	6.0	1.2
1.40	0.392	2.9	0.4
1.60	0.364	1.4	0.2

$$\bar{U}_{max} = [2(6.58) \ln (2400)]^{1/2} = 25.8 \text{ ft} \quad \#$$

6.7 RESPONSE TO EARTHQUAKE LOADINGS

When a proposed offshore structure is to be located in an area of seismic activity, it is necessary to analyze the structure for its response to earthquake-induced ground motions. Such analysis may be carried out using the basic structural dynamics theory described in Sections 6.2 through 6.5, with wave loadings replaced by those caused by ground acceleration.

Consider, for example, the simple offshore structure shown in Fig. 6.9. As in our previous work, the embedded piles are replaced with equivalent free standing ones. The stiffness properties of these piles may be determined by an iterative process analogous to that considered in Chapter 5 for static analysis.

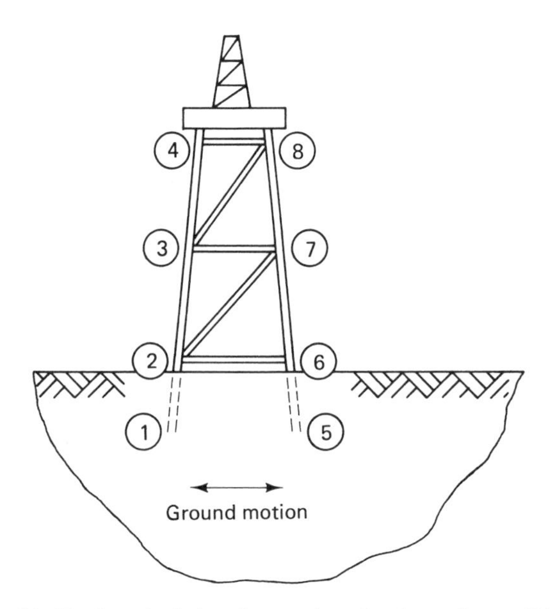

Fig. 6.9. Earthquake-induced ground motion beneath an offshore structure.

The ground motions at the fixed ends of the equivalent piles (level 1–5) are usually expressed in terms of two horizontal components of translational acceleration and a vertical component. The response of the structure to these three components can be calculated by considering each separately and superposing the results. Since the same analytical procedures apply to each component, we restrict attention in the present discussion to the case where the ground motion is in the direction indicated in the figure, that is, parallel to the side faces of the structure.

The structural dynamics is analyzed, as in the case of wave loadings, by considering the three movable levels 2–6, 3–7, and 4–8 of the structure. Resistive damping forces due to the motion of the structure in the water are neglected. To illustrate the main principals involved, we first restrict attention to the idealized case where the lumped masses and forces at levels 2–6 and 3–7 are negligible in comparison with those at level 4–8. The lumped mass at this level is denoted by M_4^*, as described by the appropriate component of equation (22). This includes contributions from the deck and support structure (M_4) and from the added mass (ρB_4) resulting from the acceleration of the structure in water. The horizontal displacements of levels 1–5 and 4–8 are denoted, respectively, by U_1 and U_4.

The net horizontal force F_4 acting at the top of the structure is assumed to consist of the inertia force $-M_4^*\ddot{U}_4$ and a damping force representative of the internal friction in the structure. The latter force is customarily assumed to be proportional to the velocity difference $\dot{U}_4 - \dot{U}_1$ existing between the upper and lower levels of the structure. We thus have the force F_4 expressible as

$$F_4 = -M_4^* \ddot{U}_4 - C(\dot{U}_4 - \dot{U}_1) \tag{74}$$

where C denotes a damping coefficient. From the static methods of Chapter 2, we may also relate the force F_4 to the relative displacement $U_4 - U_1$ existing between the upper and lower levels of the structure through the elastic relation

$$F_4 = K(U_4 - U_1) \tag{75}$$

where K denotes an appropriate stiffness coefficient. Combining equations (74) and (75), we have the equation of motion of the upper level of the structure expressible as

$$M_4^* \ddot{U}_4 + C(\dot{U}_4 - \dot{U}_1) + K(U_4 - U_1) = 0 \tag{76}$$

On writing the relative displacement $U_4 - U_1$ as U and substituting into equation (76), we find finally the governing equation expressible as

$$\ddot{U} + 2\zeta \dot{U} + \Omega^2 U = -\ddot{U}_1 \tag{77}$$

where the damping factor ζ and the natural frequency Ω of the structure are given by

$$\zeta = \frac{C}{2M_4^*}, \qquad \Omega^2 = \frac{K}{M_4^*} \tag{78}$$

and where $\ddot{U}_1$ denotes the earthquake-induced ground acceleration.

Assuming zero initial relative displacement and velocity and small damping such that $\zeta^2 \ll \Omega^2$, the solution of equation (77) may be written as

$$U = -\frac{1}{\Omega} \int_0^t \ddot{U}_1(\tau) e^{-\zeta(t-\tau)} \sin \Omega(t - \tau)\, d\tau \tag{79}$$

If we now assume a time history $\ddot{U}_1$ of the ground acceleration at the site location, we may integrate this equation and determine the time history of the displacement U. The vertical displacement and rotation at the upper joints as well as the horizontal and vertical displacements and the rotations at all other joints in the structure can be related to this horizontal displacement by methods similar to those given in Section 6.5 and the stresses within the individual members then determined. This approach is sometimes referred to as the *time-history method* of determining structural response to earthquake loadings.

One of the simplest ways to define the expected ground acceleration at a site is to use the acceleration record of a past earthquake of given magnitude taken in the vicinity of the site. Figure 6.10 shows, for example, the acceleration record for strong earthquake motion taken in El Centro, California on May 18,

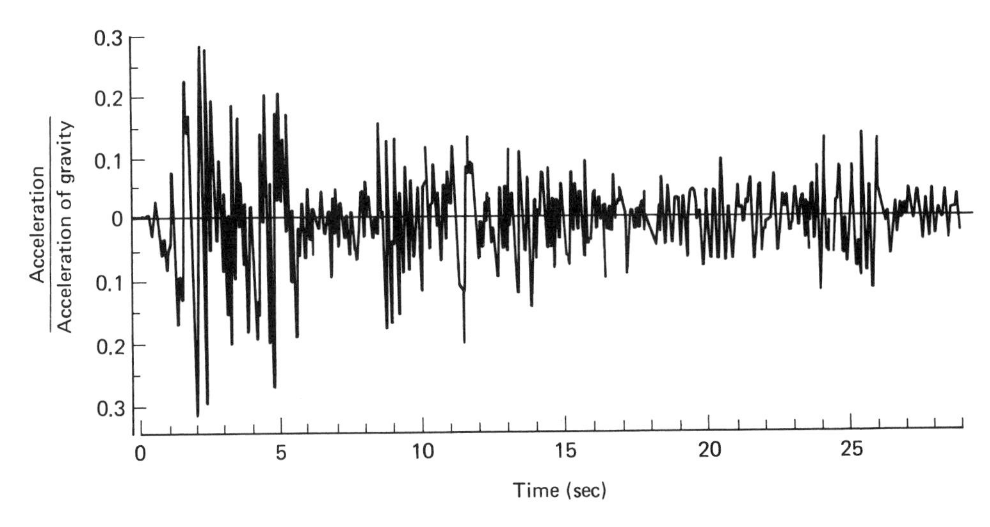

Fig. 6.10. Strong-motion seismogram, El Centro, California, May 18, 1940.

1940. The maximum ground acceleration is seen to be about 30% of the acceleration of gravity.

Unfortunately, however, only limited acceleration records of strong earthquake motion exist and experience has indicated that acceleration records from earthquakes of the same magnitude can vary greatly from one location to another even when the locations are the same distance from the origin of the quake. For this reason, it is thus difficult to estimate design-type acceleration data when actual records of strong-motion earthquakes are not available in the general vicinity of the proposed construction site.

To deal with this problem, attention is generally restricted only to maximum structural response under earthquake loading where it has been found possible to effect some generalization in earthquake description for design purposes. This approach is referred to as the *response-spectrum* method of determining structural response to earthquake loadings.

From equation (79) we have, in particular, the maximum relative displacement U_{max} at the top of the structure given by

$$U_{max} = \frac{1}{\Omega} S_V \tag{80}$$

where S_v is the velocity-response spectrum defined by

$$S_V = \left[\int_0^t \dot{U}_1(\tau) e^{-\zeta(t-\tau)} \sin \Omega (t - \tau) \, d\tau \right]_{max} \tag{81}$$

Notice that this velocity-response spectrum depends, for any given ground acceleration history, only on the natural frequency Ω (or, equivalently, the natural period $T = 2\pi/\Omega$) and the damping parameter ζ. Also notice that we are now dealing with the maximum value of a weighted integral of the ground-acceleration time history rather than the time history itself. This has the effect of reducing the differences that exist between actual acceleration records from earthquakes of the same intensity and, accordingly, makes it possible to define with some confidence appropriate design response spectra associated with earthquake activity in various broad geographic areas.

Usually either the velocity-response spectrum or an acceleration response spectrum is specified as design criterion, where the acceleration-response spectrum S_A is defined by

$$S_A = \Omega S_V \tag{82}$$

Figure 6.11 illustrates a typical design acceleration response spectrum for a damping ratio $\zeta/\Omega = 0.05$. The effective ground acceleration G in this figure is

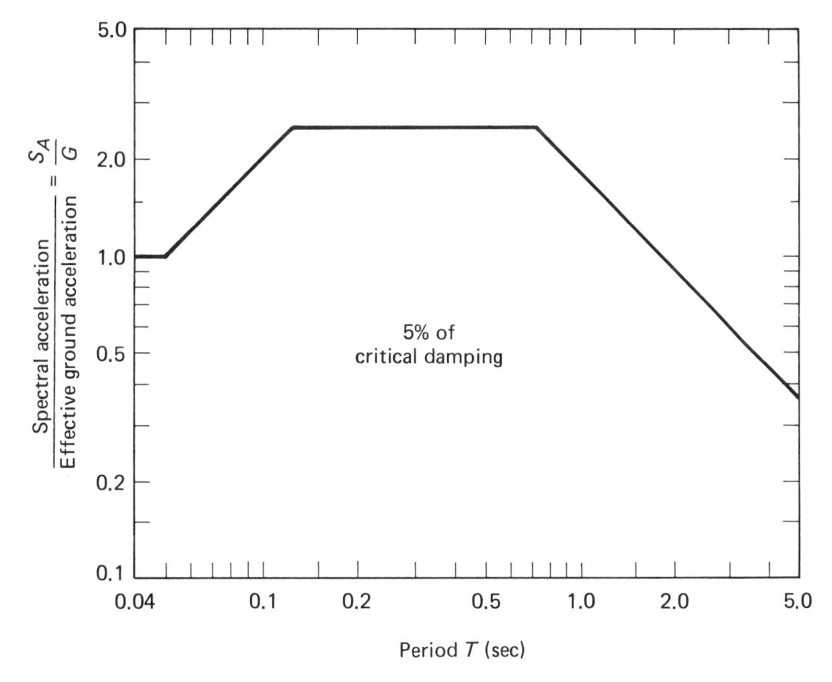

Fig. 6.11. Design-acceleration response spectrum (after API, 1980).

defined as some fraction of the acceleration of gravity depending on the geographic location of interest. Table 6.5 lists typical values of G for selected offshore areas of the United States.

Table 6.5

Location	G (ft/sec²)
Gulf of Mexico	0
Cook Inlet (Alaska)	13
Santa Barbara Channel (California)	8

Source: Data from API (1980).

More detailed acceleration response spectra and associated values of G are given by the American Petroleum Institute (API, 1980).

Knowing the period T of the structure, the ratio S_A/G can be read directly from an acceleration response spectrum such as shown in Fig. 6.11. Using the appropriate value of G, the value of S_A can then be determined and this related to the maximum displacement of the upper level of the structure using equations (80) and (82), that is,

$$U_{\max} = \frac{1}{\Omega^2} S_A = \frac{T^2}{4\pi^2} S_A \tag{83}$$

With the simplified analysis above before us, we may return to the structure of Fig. 6.9 and consider the more realistic case where the lumped masses and forces at levels 2–6 and 3–7 are included along with those at level 4–8. If U_2, U_3, and U_4 denote the horizontal displacements of these levels relative to base level (having absolute displacement U_1), the equations of motion may be written, analogous to equation (76), as

$$\begin{aligned} M_2^*\ddot{U}_2 + C_2\dot{U}_2 + K_{22}U_2 + K_{23}U_3 + K_{24}U_4 &= F_2 \\ M_3^*\ddot{U}_3 + C_3\dot{U}_3 + K_{32}U_2 + K_{33}U_3 + K_{34}U_4 &= F_3 \\ M_4^*\ddot{U}_4 + C_4\dot{U}_4 + K_{42}U_2 + K_{43}U_4 + K_{44}U_4 &= F_4 \end{aligned} \tag{84}$$

where the M^*'s denote the lumped masses at the indicated levels, the C's denote structural damping coefficients, the K's denote appropriate stiffness coefficients, and the F's are defined by

$$F_2 = -M_2^*\ddot{U}_1, \qquad F_3 = -M_3^*\ddot{U}_1, \qquad F_4 = -M_4^*\ddot{U}_1 \tag{85}$$

As in Section 6.2, we assume further that the damping coefficients are proportional to the respective lumped masses and write the equations above in matrix form as

$$[M^*]\{\ddot{U}\} + 2\zeta[M^*]\{\dot{U}\} + [K]\{U\} = \{F\} \tag{86}$$

These equations may be solved using the modal method of analysis described in detail in Section 6.3. In particular, using equation (29) we first solve for the natural frequencies $\omega_1, \omega_2, \omega_3$ of the structure in the form $\lambda_1 = \omega_1^2$, $\lambda_2 = \omega_2^2$, $\lambda_3 = \omega_3^2$, together with the associated mode shapes $U_{01}(\lambda_1)$, $U_{02}(\lambda_1)$, etc. We next form the modal matrix $[P]$ according to equation (33) and introduce new variables Y_1, Y_2, Y_3 such that

$$\{U\} = [P]\{Y\} \tag{87}$$

Equation (86) may then be written as

$$[M']\{\ddot{Y}\} + 2\zeta[M']\{\dot{Y}\} + [K']\{Y\} = \{F'\} \tag{88}$$

where $[M']$ and $[K']$ are diagonal matrices given by

$$[M'] = [P]^T[M^*][P] \tag{89}$$

$$[K'] = [P]^T[K][P] = [M'][\lambda] \tag{90}$$

with $[\lambda]$ being a diagonal matrix having nonzero components λ_1, λ_2, λ_3, and where $\{F'\}$ is defined by

$$\{F'\} = [P]^T\{F\} \tag{91}$$

Equation (88) represents a set of three uncoupled equations for Y_1, Y_2, Y_3 whose solutions may be written, analogous to equation (79), as

$$Y_n = \frac{1}{M'_n\omega_n}\int_0^t F'_n(\tau)e^{-\zeta(t-\tau)} \sin \omega_n(t - \tau)\, d\tau \tag{92}$$

with $n = 1, 2, 3$, respectively. Substituting the components of $\{F\}$ from equation (85) into equation (91), and expanding, we may write this last result in terms of the ground acceleration as

$$Y_n = -\frac{\alpha_n}{M'_n\omega_n}\int_0^t \ddot{U}_1(\tau)e^{-\zeta(t-\tau)} \sin \omega_n(t - \tau)\, d\tau \tag{93}$$

where α_n is defined by

$$\alpha_n = U_{02}(\lambda_n)M_2^* + U_{03}(\lambda_n)M_3^* + U_{04}(\lambda_n)M_4^* \tag{94}$$

If we use a time-history method of analysis where $\ddot{U}_1$ is known, or assumed, the solutions for Y_1, Y_2, Y_3 can be worked out from equation (93) and the horizontal displacements of individual structural levels calculated using equation (87).

Alternatively, we may use the response-spectrum method and determine maximum values of Y_1, Y_2 and Y_3 from a design acceleration response spectrum such as shown in Fig. 6.11. Comparing equation (93) with (79) and using equation (83), we have, in particular,

$$Y_n(\max) = \frac{\alpha_n}{M'_n \omega_n^2} S_{An} \tag{95}$$

where S_{An} denotes the response acceleration for frequency ω_n. The maximum possible values of the horizontal displacements at the various structural levels can then be determined from equation (87) by assuming that maximum values of Y_1, Y_2, Y_3 all occur at the same instant. For the upper level displacement U_4, for example, we have

$$U_4 = U_{04}(\lambda_1) Y_1 + U_{04}(\lambda_2) Y_2 + U_{04}(\lambda_3) Y_3 \tag{96}$$

The fact that the maximum values of Y_1, Y_2, Y_3 cannot, in general, be expected to occur simultaneously is sometimes taken into account by combining the above mode contributions to the displacement as the square root of the sum of the squared values. Equation (96), for example, is then replaced by

$$U_4 = \{[U_{04}(\lambda_1) Y_1]^2 + [U_{04}(\lambda_2) Y_2]^2 + [U_{04}(\lambda_3) Y_3]^2\}^{1/2} \tag{97}$$

Having the maximum value of the horizontal displacements at the various levels of the structure, we may next calculate the associated maximum values of the vertical displacements and rotations at all joints of the structure using the methods of Section 6.5 and, from these, determine the maximum stress in the individual members. These must be superposed on stresses caused by horizontal ground accelerations perpendicular to the direction already considered as well as to stresses caused by vertical accelerations in order to determine the maximum earthquake-induced stresses in the structure. Of course, these stresses must also be superposed on existing stresses caused by weight loadings on the structure to get the stress levels to be compared with allowable values. General practice does not, however, require that earthquake-induced stresses be superposed on those caused by design storm waves as the design of a structure to withstand simultaneously both earthquake forces and storm-induced wave forces would be uneconomical considering the very small chance that both severe conditions would occur together.

EXAMPLE 6.7-1. Consider the steel structure shown in Fig. 6.12 and investigate its response to earthquake-induced ground motion using the acceleration-response spectrum of Fig. 6.11, with $G = 8$ ft/sec².

Using the given mass and stiffness matrices, the natural frequencies and associated mode shapes are determined from equation (29) as

$$\omega_1 = 5.45 \text{ rad/sec}, \quad U_{02} = 0.604, \quad U_{03} = 0.813, \quad U_{04} = 1.00$$
$$\omega_2 = 20.9 \text{ rad/sec}, \quad U_{02} = -2.05, \quad U_{03} = -1.09, \quad U_{04} = 1.00$$
$$\omega_3 = 44.7 \text{ rad/sec}, \quad U_{02} = 5.62, \quad U_{03} = -3.95, \quad U_{04} = 1.00$$

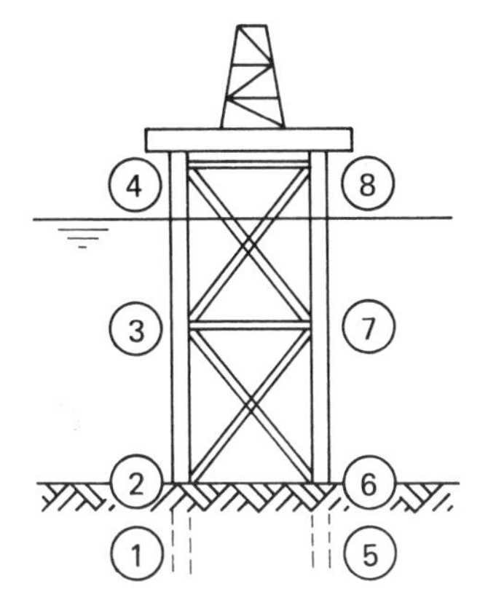

$$[M^*] = 10^4 \begin{bmatrix} 1.44 & 0 & 0 \\ 0 & 2.87 & 0 \\ 0 & 0 & 4.32 \end{bmatrix} \text{(slugs)}$$

$$[K] = 10^7 \begin{bmatrix} 1.67 & -1.64 & 0.35 \\ -1.64 & 3.05 & -1.42 \\ 0.35 & -1.42 & 1.07 \end{bmatrix} \text{(ft-lb)}$$

Fig. 6.12

The modal matrix $[P]$ is thus

$$[P] = \begin{bmatrix} 0.60 & -2.05 & 5.62 \\ 0.81 & -1.09 & -3.95 \\ 1.00 & 1.00 & 1.00 \end{bmatrix}$$

and the masses M'_1, M'_2, M'_3 are determined from equation (89) as

$$M'_1 = 67{,}400 \text{ slugs}, \qquad M'_2 = 138{,}000 \text{ slugs}, \qquad M'_3 = 946{,}000 \text{ slugs}$$

From equation (94), we next determine the factors α_1, α_2, α_3, as

$$\alpha_1 = (0.60)(14{,}440) + (0.81)(28{,}680) + 43{,}200 = 75{,}100$$
$$\alpha_2 = (-2.05)(14{,}440) + (-1.09)(28{,}690) + 43{,}200 = -17{,}700$$
$$\alpha_3 = (5.62)(14{,}440) + (-3.95)(28{,}680) + 43{,}200 = 11{,}100$$

Noting that the periods associated with the frequencies ω_1, ω_2, ω_3 are 1.15, 0.300, and 0.140 sec, respectively, we determine the associated acceleration response from Fig. 6.12 (with $G = 8$ ft/sec²) as

$$S_{A1} = 14.4, \qquad S_{A2} = 20.0, \qquad S_{A3} = 20.0$$

The associated maximum values of Y_1, Y_2, and Y_3 are determined from equation (95) as

$$Y_1 = \frac{(75{,}100)(14.4)}{(67{,}400)(29.7)} = 5.40 \times 10^{-1} \text{ ft}$$

$$Y_2 = \frac{(17{,}700)(20)}{(138{,}000)(436)} = 5.88 \times 10^{-3} \text{ ft}$$

$$Y_3 = \frac{(11{,}100)(20)}{(946{,}000)(1998)} = 1.17 \times 10^{-4} \text{ ft}$$

The maximum horizontal displacements at levels 2, 3, and 4 of the structure are then determined from equation (87), assuming maximum values of Y_1, Y_2, and Y_3 all occur at the same instant. We find

$$U_2 = 0.313 \text{ ft}, \qquad U_3 = 0.431 \text{ ft}, \qquad U_4 = 0.546 \text{ ft} \quad \#$$

REFERENCES

API (1980). *Recommended Practice for Planning, Designing and Constructing Fixed Offshore Platforms*, American Petroleum Institute Publication RP-2A, Dallas, Tex.

Borgman, L. (1967). Spectral Analysis of Ocean Wave Forces on Piling, *Journal of the Waterways and Harbors Division*, ASCE, Vol. 93, pp. 129–156.

Bretschneider, C. L. (1959). Wave Variability and Wave Spectra for Wind Generated Gravity Waves, *U.S. Army Corps of Engineers Beach Erosion Board Memorandum No. 118.*

Longuet-Higgins, M. S. (1952). On the Statistical Distribution of the Heights of Sea Waves, *Journal of Marine Research*, Vol. 11, pp. 245–266.

Pierson, W. S., and L. Moskowitz (1963). A Proposed Spectral Form for Fully Developed Wind Seas Based on the Similarity Theory of S. A. Kitaigorodskii, *U.S. Navy Oceanography Office Report N62306–1042*, 1963.

PROBLEMS

1. Consider the flexible steel structure having side face shown in Fig. P.1. Vertical members have outside diameters of 4 ft and horizontal members

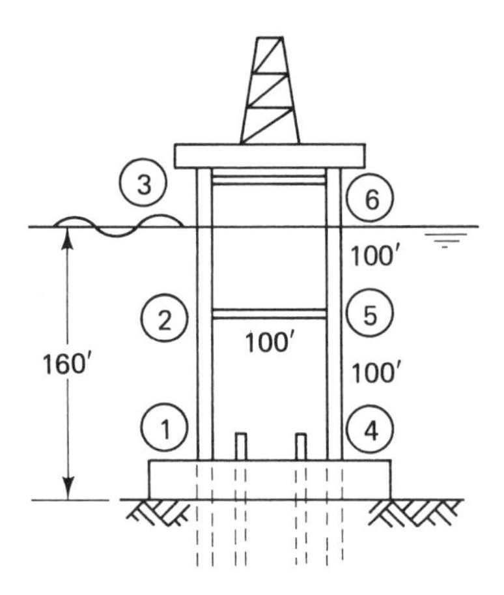

Fig. P.1

have outside diameters of 2 ft. All four faces of the structure are identical. Assuming a wave of height 40 ft and period 9 sec, determine the force matrix $\{F^*\}$ for levels 2–5 and 3–6.

Ans. $\begin{Bmatrix} F_2^* \\ F_3^* \end{Bmatrix} = 10^4 \begin{Bmatrix} 3.26 \sin(-\omega t + 1.34) \\ 3.63 \sin(-\omega t + 1.71) \end{Bmatrix}$ lb

2. For a deck weight of 2000 kips, determine the mass matrix $[M^*]$ for movable levels 2–5 and 3–6 of the structure of Problem 1. The vertical members have wall thicknesses of 1.5 in and the horizontal members have wall thicknesses of 0.5 in. Lower vertical and horizontal members are assumed fully flooded. Upper members are assumed flooded to the still-water level.

Ans. $[M^*] = 10^4 \begin{bmatrix} 1.46 & 0 \\ 0 & 3.69 \end{bmatrix}$ slugs

3. A static analysis of the frame of Problem 1 gives, for $F_{2x}^T = 1.0$ lb and all other forces and moments equal to zero, average displacements U_2, U_3 at levels 2–5 and 3–6, respectively, of

$$U_2 = 8.80 \times 10^{-6} \text{ ft}, \qquad U_3 = 1.80 \times 10^{-5} \text{ ft}$$

Similarly, for $F_{3x}^T = 1.0$ lb and all other forces and moments equal to zero, we find

$$U_2 = 1.80 \times 10^{-5} \text{ ft}, \qquad U_3 = 5.32 \times 10^{-5} \text{ ft}$$

Using these values, determine the stiffness matrix of the frame connecting horizontal forces at levels 2–5 and 3–6 with horizontal displacements at these levels.

Ans. $[K] = 10^5 \begin{bmatrix} 3.64 & -1.23 \\ -1.23 & 0.60 \end{bmatrix}$ lb/ft

4. Determine from algebraic solution of equation (29) the free-vibration frequencies and associated mode shapes of the frame of Problem 1, using results from Problems 2 and 3.

Ans. $\omega_1 = 0.693$ rad/sec, $U_{02}/U_{03} = 0.345$
$\omega_2 = 5.11$ rad/sec, $U_{02}/U_{03} = -7.60$

5. Determine expressions for the horizontal displacements U_2, U_3 of the levels 2–5 and 3–6 of the structure of Problem 1, using the results from Problems 1 to 4. Assume that $\zeta/\omega_1 = 0.08$.

Ans. $U_2 = 5.38 \sin(-\omega t - 3.0)$ ft, $U_3 = 15.6 \sin(-\omega t - 3.0)$ ft.

6. For the frame of Problem 1, a static analysis gives vertical displacement and rotation at joint 2 for $F_{2x}^T = 1.0$ lb, and all other forces and moments equal to zero, of

$$V_2 = 1.77 \times 10^{-9} \text{ ft}, \qquad \theta_2 = 1.09 \times 10^{-7} \text{ rad}$$

Similarly, for $F^T_{3x} = 1.0$ lb, we have

$$V_2 = 6.17 \times 10^{-9} \text{ ft}, \qquad \theta_2 = 2.92 \times 10^{-7} \text{ rad}$$

Use the results from Problem 3 to find the matrix connecting the vertical displacement and rotation at joint 2 with the horizontal displacements at levels 2–5 and 3–6.

Ans. $\begin{Bmatrix} V_2 \\ \theta_2 \end{Bmatrix} = 10^{-4} \begin{bmatrix} -1.15 & 1.52 \\ 37.6 & 41.1 \end{bmatrix} \begin{Bmatrix} U_2 \\ U_3 \end{Bmatrix}$ (ft, rad)

7. Use the results from Problems 5 and 6 to find the maximum bending stress in member 1–2 at joint 1 of the structure of Problem 1.
Ans. $\sigma = 92.3$ kips/in².

8. Calculate the maximum static deflections U_{2s}, U_{3s} associated with the structure and forces of Problem 1, neglecting inertia and damping forces and using the equation

$$[K]\{U_s\} = \{F^*\}$$

9. Use the results from Problems 5 and 8 to determine the dynamic amplification factor for the deck displacement of the structure of Problem 1.
Ans. $U_3/U_{3s} = 6.2$.

10. Compare the results from Problem 9 with those found from the simplified analysis of Section 4.10. Use data of Problem 3.
Ans. $U_3/U_{3s} = 5.8$ (Chapter 4).

11. Use the results from Problems 6 and 8 to find the maximum static bending stress in member 1–2 at joint 1 of the structure of Problem 1.
Ans. $\sigma = 16{,}900$ lb/in².

12. A single-pile structure has a mass matrix $[M^*]$ and a stiffness matrix $[K]$ connecting horizontal forces and displacements, as indicated in Fig. P.12. Determine the free-vibration frequencies and mode shapes.

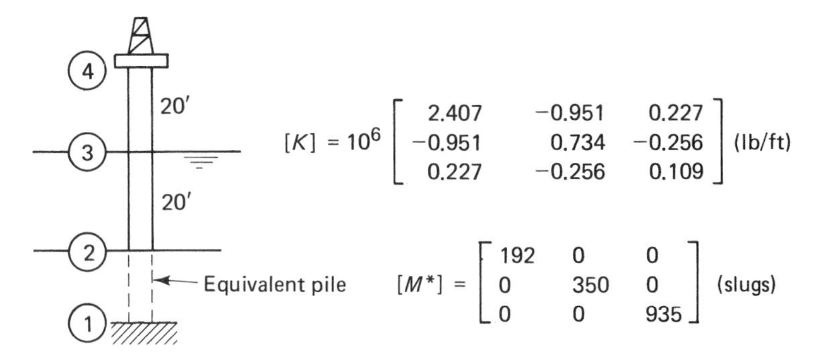

Fig. P.12

Ans. $\omega_1 = 3.14$ rad/sec, $U_{02}/U_{04} = 0.0918$, $U_{03}/U_{04} = 0.470$
$\omega_2 = 30.7$ rad/sec, $U_{02}/U_{04} = -2.38$, $U_{03}/U_{04} = -4.96$
$\omega_3 = 117$ rad/sec, $U_{02}/U_{04} = 32.9$, $U_{03}/U_{04} = -7.73$

13. Recalculate the free-vibration frequencies and mode shapes in Problem 4 using matrix iteration.

14. Determine the free-vibration frequencies and mode shapes of the structure of Example 6.7-1 using matrix iteration.

15. Determine the root-mean-square surface deflection of a random one-dimensional sea governed by the Bretschneider spectrum with $H_s = 20$ ft and $T_s = 8$ sec.
Ans. $\eta_{\rm rms} = 5.0$ ft.

16. For the sea state of Problem 15, with water depth equal to 700 ft, determine the root-mean-square horizontal velocity at the still-water level. [*Hint*: Use tanh $k_n h \approx 1$ and integrate equation (64) directly. Justify this approximation.]
Ans. $u_{\rm rms} = 4.74$ ft/sec.

17. For the sea state of Problem 15, determine the average of the maximum values of the surface deflection over 12 hours.
Ans. $\eta_{\rm max} = 20.7$ ft.

18. The maximum bending stress σ in the member 1–2 at joint 1 of the structure shown in Fig. P.18 is described approximately in terms of regular

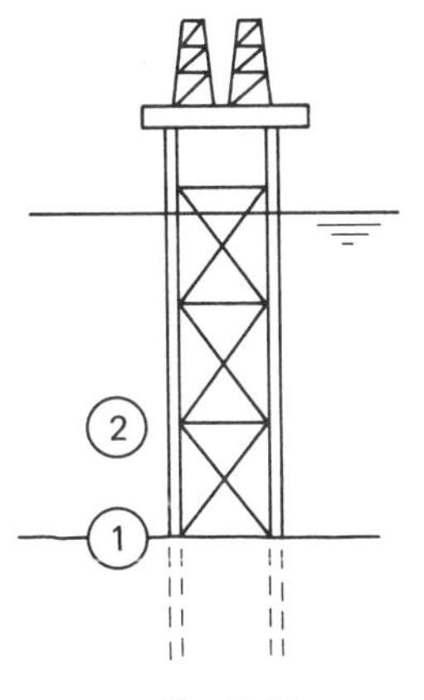

Fig. P.18

waves of height H (ft) and frequency ω (rad/sec) by

$$\sigma = \frac{700H\omega^2}{\left\{\left[1 - \left(\frac{\omega}{\omega_1}\right)^2\right]^2 + \left(\frac{2\zeta}{\omega_1}\frac{\omega}{\omega_1}\right)^2\right\}^{1/2}} \qquad \text{lb/in}^2$$

where $\omega_1 = 1.25$ rad/sec. Assuming that $2\zeta/\omega_1 = 0.1$, determine the root-mean-square value of σ for an irregular sea having Bretschneider spectrum with $H_s = 35$ ft and $T_s = 10$ sec.

19. Use the root-mean-square value of stress found in Problem 18 to calculate the average of the maximum values of the stress over 12 hours.

20. For the structure of Problem 1, determine the root-mean-square value of the bending stress in member 1–2 at joint 1 for a one-dimensional random sea having $H_s = 20$ ft and $T_s = 9$ sec. Values of u_{rms} at levels 2-5 and the still-water level are 1.41 ft/sec and 4.22 ft/sec, respectively.
Ans. $\sigma_{\text{rms}} = 9.7$ kips/in².

21. Use the result from Problem 20 to determine the average of the maximum values of stress over 6 hours.
Ans. $\bar{\sigma}_{\text{rms}} = 38.3$ kips/in².

22. The height and period of a regular design wave is typically assumed to equal $2H_s$ and T_s, where H_s is the significant wave height and T_s is the significant wave period of an irregular sea during storm conditions. Using this, compare the results from Problem 21 with those from Problem 7 and give an explanation for the much larger stress found in the regular wave analysis.

23. An idealized earthquake-induced ground motion is described by

$$\ddot{U}_1 = 10 \sin 12t$$

By direct or numerical integration of equation (79), determine the maximum deck deflection U of a structure having natural frequency $\Omega = 3.0$ rad/sec.

24. For the structure of Problem 23, determine the maximum deck deflection using the response spectrum of Fig. 6.11, with $G = 13$ ft/sec².
Ans. $U_{\max} \approx 4.6$ ft.

25. The concrete gravity structure shown in Fig. P.25 has a deck mass of 15,000 slugs. The mass of the column, including added mass, is 20,000 slugs.

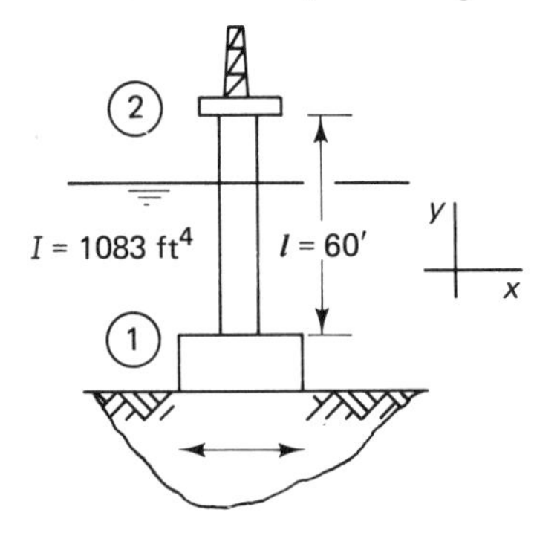

Fig. P.25

Assuming the natural frequency of the structure described by

$$\Omega^2 = \frac{3EI}{M_2^* l^3}$$

with $E = 4 \times 10^6$ lb/in², determine:

(a) The maximum earthquake-induced deck deflection in the x-direction using the response spectrum of Fig. 6.11, with $G = 13$ ft/sec².

(b) The maximum associated bending stress at the base of the column.

26. For the structure of Problem 1, determine the earthquake-induced horizontal displacements at levels 2–5 and 3–6 using the response spectrum of Fig. 6.11, with $G = 13$ ft/sec².

27. Use the displacements found in Problem 26 to determine the bending stress in member 1–2 at joint 1. See Problem 6.

APPENDIX

MATRIX ALGEBRA

A.1 DETERMINANTS

An nth-order determinant is defined as an array of numbers, called *elements*, having n rows and n columns and is expressed as

$$D = \begin{vmatrix} a_{11} & a_{12} & \cdots & a_{1n} \\ a_{21} & a_{22} & \cdots & a_{2n} \\ \cdot & & & \\ \cdot & & & \\ \cdot & & & \\ a_{n1} & a_{n2} & \cdots & a_{nn} \end{vmatrix}$$

A determinant has a unique value defined as the algebraic sum of all possible products of n factors such that:

1. Each product has as its factors one and only one element from each row and each column.
2. Each product is preceded by a plus or minus sign according as the number of inversions (larger numbers preceeding smaller ones) in the row subscripts is even or odd, after the factors have been written so that column subscripts appear from left to right in order of magnitude.

It follows from condition 1 that there will be $n \times (n-1) \times (n-2) \times \ldots \times 1$ products in all, with the sign of each determined by condition 2.

As an example, consider the second-order determinant

$$D = \begin{vmatrix} a_{11} & a_{12} \\ a_{21} & a_{22} \end{vmatrix}$$

There are $2 \times 1 = 2$ products to consider in evaluating the determinant. Written so that column (second) subscripts appear in order of magnitude, these are

$$a_{11}a_{22}, \qquad a_{21}a_{12}$$

The row (first) subscripts for the first product are in numerical order so that no inversion exists. The row subscripts for the second product are in order 2, 1 so that one inversion exists. Thus, the value of the determinate is

$$D = a_{11}a_{22} - a_{21}a_{12}$$

Next consider the third-order determinant

$$D = \begin{vmatrix} a_{11} & a_{12} & a_{13} \\ a_{21} & a_{22} & a_{23} \\ a_{31} & a_{32} & a_{33} \end{vmatrix}$$

There are $3 \times 2 \times 1 = 6$ products to consider in evaluating this determinant. Again, writing these so that column subscripts appear in order of magnitude, these are

$$a_{11}a_{22}a_{33}, \quad a_{31}a_{12}a_{23}, \quad a_{21}a_{32}a_{13}$$

$$a_{31}a_{22}a_{13}, \quad a_{21}a_{12}a_{33}, \quad a_{11}a_{32}a_{23}$$

The number of inversions in the row subscripts of these products are 0, 2, 2, 3, 1, 1, respectively. Thus, the value of the determinant is

$$D = a_{11}a_{22}a_{33} + a_{31}a_{12}a_{23} + a_{21}a_{32}a_{13} - a_{31}a_{22}a_{13} - a_{21}a_{13}a_{33} - a_{11}a_{32}a_{23}$$

The *cofactor* c_{ij} of element a_{ij} of a determinant is defined by the expression

$$c_{ij} = (-1)^{i+j}m_{ij}$$

where m_{ij} denotes the minor of element a_{ij}, defined as the determinant formed by deleting the ith row and jth column from the original determinant. It follows from the definition above that the value of an nth-order determinant is expressible as

$$D = \sum_{j=1}^{n} a_{ij}c_{ij} = \sum_{j=1}^{n} a_{ij}c_{ij}$$

where i in the first sum and j in the second sum can have any value from 1 to n.

Example: Given the determinant

$$D = \begin{vmatrix} 2 & 1 & 3 \\ 1 & 3 & 2 \\ 0 & 1 & 4 \end{vmatrix}$$

its value is calculated using the first of the sums above, with $i = 1$, as

$$D = a_{11}c_{11} + a_{12}c_{12} + a_{13}c_{13}$$

where, using our earlier expression for the value of a second-order determinant,

$$c_{11} = (-1)^2 \begin{vmatrix} 3 & 2 \\ 1 & 4 \end{vmatrix} = +(12 - 2) = +10$$

$$c_{12} = (-1)^3 \begin{vmatrix} 1 & 2 \\ 0 & 4 \end{vmatrix} = -(4 - 0) = -4$$

$$c_{13} = (-1)^4 \begin{vmatrix} 1 & 3 \\ 0 & 1 \end{vmatrix} = +(1 - 0) = 1$$

Thus, the value of the determinant is

$$D = 2(10) + 1(-4) + 3(1) = 19$$

A.2 MATRICES

A rectangular array of terms arranged in m rows and n columns is called a matrix of order $m \times n$. For example

$$[A] = \begin{bmatrix} a_{11} & a_{12} & a_{13} & a_{14} \\ a_{21} & a_{22} & a_{23} & a_{24} \\ a_{31} & a_{32} & a_{33} & a_{34} \end{bmatrix}$$

is a 3×4 matrix. Unlike a determinant, a matrix has no defined numerical value. It is simply an array of numbers or terms.

A square matrix is one in which the number of rows is equal to the number of columns. It is referred to as an $n \times n$ matrix or a matrix of order n. A square matrix is said to be *symmetric* if the elements on the upper right half can be obtained by flipping the matrix about the diagonal; e.g.,

$$[A] = \begin{bmatrix} 2 & 1 & 3 \\ 1 & 5 & 0 \\ 3 & 0 & 1 \end{bmatrix}$$

A row matrix has $m = 1$; e.g.,

$$[A] = [a_1 a_2 a_3]$$

and a column matrix has $n = 1$; e.g.,

$$A = \begin{Bmatrix} a_1 \\ a_2 \\ a_3 \end{Bmatrix}$$

Note that curly brackets are used to denote a column matrix and distinguish it from others.

A diagonal matrix is a square matrix having all nondiagonal elements equal to zero; e.g.,

$$[A] = \begin{bmatrix} 2 & 0 & 0 \\ 0 & 4 & 0 \\ 0 & 0 & 3 \end{bmatrix}$$

The unit matrix $[I]$ is a diagonal matrix in which the diagonal elements are all equal to unity.

The transpose $[A]^T$ of a matrix $[A]$ is one in which the rows and columns are interchanged. For example,

$$[A] = \begin{bmatrix} a_{11} & a_{12} & a_{13} \\ a_{21} & a_{22} & a_{23} \end{bmatrix}, \qquad [A]^T = \begin{bmatrix} a_{11} & a_{21} \\ a_{12} & a_{22} \\ a_{13} & a_{23} \end{bmatrix}$$

The minor m_{ij} and cofactor c_{ij} of a square matrix are defined as the corresponding minor and cofactor of the determinant formed from the matrix.

An adjoint matrix of a square matrix $[A]$ is a transpose of the matrix of cofactors of $[A]$. Thus, if the cofactor matrix of $[A]$ is

$$[C] = \begin{bmatrix} c_{11} & c_{12} & c_{13} \\ c_{21} & c_{22} & c_{23} \\ c_{31} & c_{32} & c_{33} \end{bmatrix}$$

the adjoint is

$$\text{adj } A = [C]^T = \begin{bmatrix} c_{11} & c_{21} & c_{31} \\ c_{12} & c_{22} & c_{32} \\ c_{13} & c_{23} & c_{33} \end{bmatrix}$$

Matrix Operations

Two matrices having the same number of rows and columns may be added by summing the corresponding elements.

Example:

$$\begin{bmatrix} 1 & 3 & 2 \\ 4 & 1 & 1 \end{bmatrix} + \begin{bmatrix} 2 & 0 & 4 \\ 1 & -2 & -3 \end{bmatrix} = \begin{bmatrix} 3 & 3 & 6 \\ 5 & -1 & -2 \end{bmatrix}$$

The product of two matrices $[A]$ and $[B]$ is another matrix $[C]$.

$$[C] = [A][B]$$

The element c_{ij} of $[C]$ is determined by multiplying the elements of the ith row in $[A]$ by the elements of the jth column in $[B]$ according to the rule

$$c_{ij} = \sum_{k=1}^{n} a_{ik} b_{kj}$$

Example:

$$[A] = \begin{bmatrix} 1 & 1 & 1 \\ 1 & 2 & 2 \\ 1 & 2 & 3 \end{bmatrix}, \qquad [B] = \begin{bmatrix} 2 & 0 \\ 0 & 1 \\ 3 & -1 \end{bmatrix}$$

$$[C] = [A][B] = \begin{bmatrix} 1 & 1 & 1 \\ 1 & 2 & 2 \\ 1 & 2 & 3 \end{bmatrix} \begin{bmatrix} 2 & 0 \\ 0 & 1 \\ 3 & -1 \end{bmatrix} = \begin{bmatrix} 5 & 0 \\ 8 & 0 \\ 11 & -1 \end{bmatrix}$$

It is evident that, for multiplication, the number of columns in $[A]$ must equal the number of rows in $[B]$. We also note that $[A][B] \neq [B][A]$.

The multiplication of a matrix $[A]$ by a scalar c is defined as a matrix having as elements the corresponding elements of $[A]$ multiplied by c. For example,

$$c[A] = \begin{bmatrix} ca_{11} & ca_{12} \\ ca_{21} & ca_{22} \end{bmatrix}$$

The postmultiplication of a matrix by a column matrix results in a column matrix. Thus,

$$\begin{bmatrix} 1 & 1 & 1 \\ 1 & 5 & 2 \\ 2 & 1 & 3 \end{bmatrix} \begin{Bmatrix} 1 \\ 3 \\ 2 \end{Bmatrix} = \begin{Bmatrix} 6 \\ 20 \\ 11 \end{Bmatrix}$$

Premultiplication of a matrix by a row matrix (or transpose of a column matrix) results in a row matrix; e.g.

$$[1 \quad 3 \quad 2] \begin{bmatrix} 1 & 1 & 1 \\ 1 & 5 & 2 \\ 2 & 1 & 3 \end{bmatrix} = [8 \quad 18 \quad 13]$$

The inverse $[A]^{-1}$ of a square matrix $[A]$ satisfies the relationship

$$[A]^{-1}[A] = [A][A]^{-1} = [I]$$

and an orthogonal (square) matrix $[A]$ satisfies the relationship

$$[A]^T[A] = [A][A]^T = [I]$$

From the definition of an inverse matrix it is evident that for an orthogonal matrix $[A]^T = [A]^{-1}$.

The inverse of a matrix is determined by the relation

$$[A]^{-1} = \frac{1}{A} \text{ adj } [A]$$

where A denotes the value of the determinant of $[A]$.

Example: Find the inverse of the matrix

$$[A] = \begin{bmatrix} 1 & 0 & 1 \\ 2 & 3 & 2 \\ 1 & 4 & 3 \end{bmatrix}$$

The determinant of the matrix is evaluated as $A = 6$. The cofactors are determined as

$$c_{11} = (-1)^2\begin{bmatrix} 3 & 2 \\ 4 & 3 \end{bmatrix} = 1$$

$$c_{12} = (-1)^3\begin{bmatrix} 2 & 2 \\ 1 & 3 \end{bmatrix} = -4$$

etc.

and the matrix of cofactors is

$$[C] = \begin{bmatrix} 1 & -4 & 5 \\ 4 & 2 & -4 \\ -3 & 0 & 3 \end{bmatrix}$$

The adjoint matrix is the transpose of $[C]$, so that the inverse is determined from the expression above as

$$[A]^{-1} = \frac{1}{6}\begin{bmatrix} 1 & 4 & -3 \\ -4 & 2 & 0 \\ 5 & -4 & 3 \end{bmatrix} = \begin{bmatrix} \frac{1}{6} & \frac{2}{3} & -\frac{1}{2} \\ -\frac{2}{3} & \frac{1}{3} & 0 \\ \frac{5}{6} & -\frac{2}{3} & \frac{1}{2} \end{bmatrix}$$

Partitioned Matrix

A matrix may be partitioned into submatrices by horizontal and vertical lines as shown by the example below.

$$\left[\begin{array}{cc:c} 2 & 4 & -1 \\ 0 & -3 & 4 \\ 1 & 2 & 2 \\ \hdashline 3 & -1 & -5 \end{array}\right] = \begin{bmatrix} [A] & \{B\} \\ [C] & [D] \end{bmatrix}$$

where the submatrices are

$$[A] = \begin{bmatrix} 2 & 4 \\ 0 & -3 \\ 1 & 2 \end{bmatrix}, \qquad \{B\} = \begin{Bmatrix} -1 \\ 4 \\ 2 \end{Bmatrix}, \qquad [C] = [3 \quad -1], \qquad [D] = [-5]$$

Partioned matrices obey the normal rules of matrix algebra and can be added, subtracted, and multiplied as though the submatrices were ordinary matrix elements.

A.3 SOLUTION OF LINEAR EQUATIONS

Consider the set of linear equations

$$a_{11}x_1 + a_{12}x_2 + a_{13}x_3 = y_1$$
$$a_{21}x_1 + a_{22}x_2 + a_{23}x_3 = y_2$$
$$a_{31}x_1 + a_{32}x_2 + a_{33}x_3 = y_3$$

with the a's and y's assumed known and the x's unknown. These equations can be expressed in matrix notation as

$$[A]\{X\} = \{Y\}$$

where $\{X\}$, $\{Y\}$, and $[A]$ are defined by

$$\{X\} = \begin{Bmatrix} x_1 \\ x_2 \\ x_3 \end{Bmatrix}, \qquad \{Y\} = \begin{Bmatrix} y_1 \\ y_2 \\ y_3 \end{Bmatrix}, \qquad [A] = \begin{bmatrix} a_{11} & a_{12} & a_{13} \\ a_{21} & a_{22} & a_{23} \\ a_{31} & a_{32} & a_{33} \end{bmatrix}$$

Multiplying through by the inverse $[A]^{-1}$, we have

$$[A]^{-1}[A]\{X\} = [A]^{-1}\{Y\}$$

But, from the definition of $[A]^{-1}$, we have

$$[A]^{-1}[A]\{X\} = [I]\{X\} = \{X\}$$

Hence, the solution for $\{X\}$ is

$$\{X\} = [A]^{-1}\{Y\}$$

which shows that the solution for x_1, x_2, x_3 can be determined by finding the inverse of the matrix $[A]$ and multiplying the column matrix $\{Y\}$ by it.

We have illustrated this matrix-solution technique for three equations and three unknowns. The procedure is, of course, equally valid for n equations and n unknowns. The matrix operations described above, including the determination of the inverse, are conveniently carried out with a digital computer.

INDEX

A

T